STUDENT SOLUTIONS MANUAL
to accompany

CALCULUS

PRELIMINARY EDITION

Deborah Hughes-Hallett
Harvard University

Andrew Gleason
Harvard University

et al.

Prepared By: Kenny Ching
Eric Connally
Stephen A. Mallozzi
Michael Mitzenmacher

John Wiley & Sons, Inc.
New York • Chichester • Brisbane • Toronto • Singapore

ISBN 0-471-58359-6

Printed in the United States of America

Printed and bound by the Courier Companies, Inc.

10 9 8 7 6 5 4 3

Chapter 1

1.1 Solutions

1. **(i)** The first graph does not match any of the given stories. In this picture, the person keeps going away from home, but the speed decreases as time passes. So a story for this might be: *I started walking to school at a good pace, but since I stayed up all night studying calculus, I got more and more tired the farther I walked.*

   (ii) This graph matches (b), the flat tire story. Note the long period of time during which the distance from home did not change (the horizontal part).

   (iii) This one matches (c), in which the person started calmly but sped up.

   (iv) This one is (a), in which the person forgot his or her books and had to return home.

3.

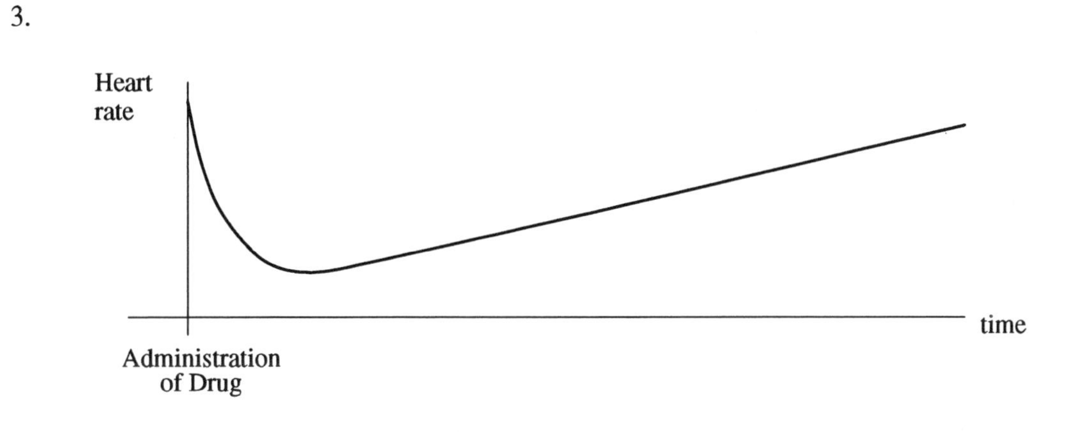

5. At the beginning, electric potential is zero. After a short time, the potential increases rapidly (in a linear fashion) to a peak; after that, it declines sharply, but as it approaches zero the rate of decrease slows, so that it approaches zero very slowly.

7. At first, as the number of workers increases, the productivity also increases. As a result, the graph of the curve goes up initially. After a certain point the curve goes downward; in other words, as the number of workers increases beyond that point, the productivity decreases. This might be due either to the inefficiency inherent in large organizations or simply due to workers getting in each other's way as too many are crammed on the same line.

9.

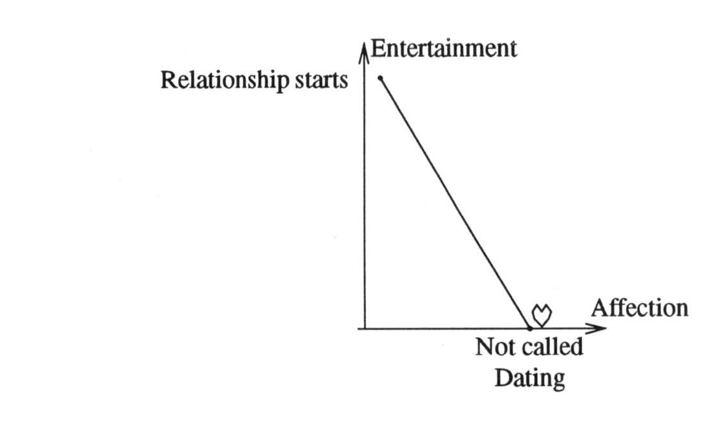

11. The price p_1 represents the maximum price any consumer would pay for the good. The quantity q_1 is the quantity of the good that could be given away if the item were free.

13. The graph is

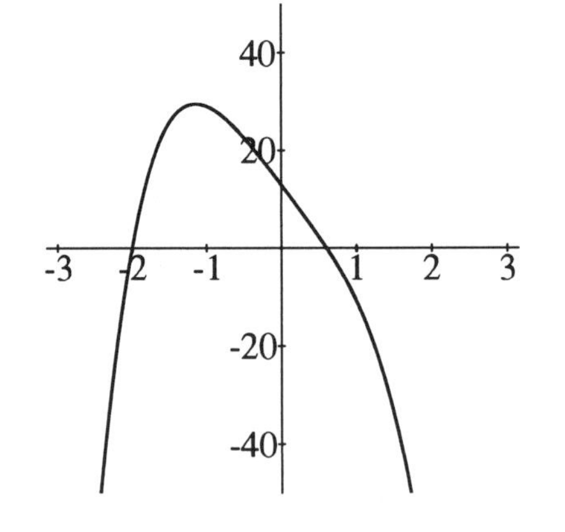

(a) The range appears to be $y \leq 30$.

(b) The function has two zeros.

15. The values $x \geq 2$ and $x \leq -2$ do not determine real values for f, because at those points either the denominator is zero or the square root is of a negative number.

If $f(t) = 5$ then $\frac{1}{\sqrt{4-t^2}} = 5$, or $\sqrt{4 - t^2} = \frac{1}{5}$. Solving for t, we have

$$t = \pm\sqrt{\frac{99}{25}} = \pm\frac{3}{5}\sqrt{11}.$$

[1]Originally, the economists thought of price as the dependent variable and put it on the vertical axis. Unfortunately, when the point of view changed, the axes did not.

1.2 Solutions

1. Rewriting the equation as $y = -\frac{5}{2}x + 4$ shows that the slope is $-\frac{5}{2}$ and the vertical intercept is 4.

3. $y - c = m(x - a)$

5. The line $y + 4x = 7$ has slope -4. Therefore the parallel line has slope -4 and equation $y - 5 = -4(x - 1)$ or $y = -4x + 9$. The perpendicular line has slope $\frac{-1}{(-4)} = \frac{1}{4}$ and equation $y - 5 = \frac{1}{4}(x - 1)$ or $y = 0.25x + 4.75$.

7. (a) E (b) D (c) A (d) F (e) B (f) C

9. (a) Finding slope (-50) and intercept gives $q = 1000 - 50p$
   (b) Solving for p gives $p = 20 - 0.02q$

11. For the line $3x + 4y = -12$, the x-intercept is $(-4, 0)$ and the y-intercept is $(0, -3)$. The distance between these two points is

$$d = \sqrt{(-4-0)^2 + (0-(-3))^2} = \sqrt{16+9} = \sqrt{25} = 5$$

13. (a) Given the two points (0,32) and (100,212), and assuming the graph is a line,

$$\text{Slope} = \frac{212-32}{100} = \frac{180}{100} = 1.8.$$

   (b) The F-intercept is (0,32), so

$$°\text{Fahrenheit} = 1.8°\text{Centigrade} + 32.$$

   (c) If the temperature is 20°Centigrade, then

$$°\text{Fahrenheit} = 1.8(20) + 32 = 68°\text{Fahrenheit}.$$

   (d) If °Fahrenheit = °Centigrade then

$$\begin{aligned} °\text{Centigrade} &= 1.8°\text{Centigrade} + 32 \\ -32 &= 0.8°\text{Centigrade} \\ °\text{Centigrade} &= -40° = °\text{Fahrenheit} \end{aligned}$$

15.

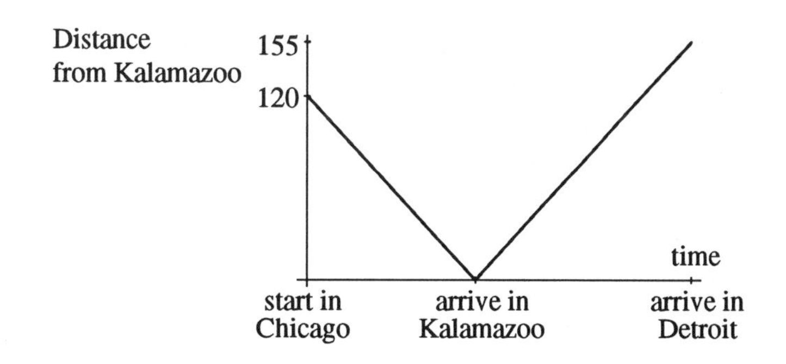

17. (a) $k = p_1 s + p_2 l$ where s = # of liters of soda and l = # of liters of oil.

(b) If $s = 0$, then $l = \frac{k}{p_2}$. Similarly, if $l = 0$, then $s = \frac{k}{p_1}$. These two points give you enough information to draw a line containing the points which satisfy the equation.

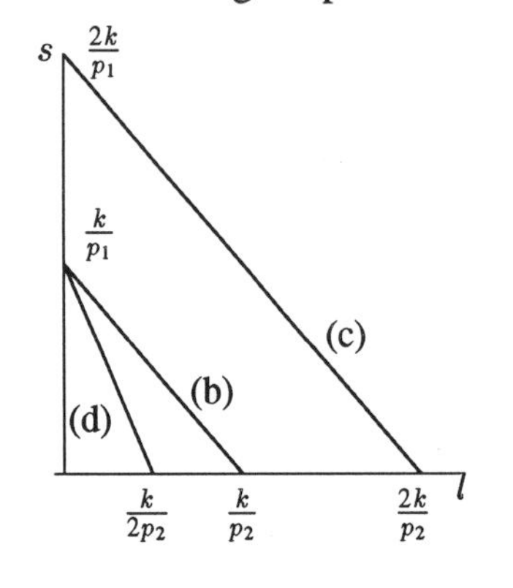

(c) If the budget is doubled, we have the constraint: $2k = p_1 s + p_2 l$. We find the intercepts as before. If $s = 0$, then $l = \frac{2k}{p_2}$; if $l = 0$, then $s = \frac{2k}{p_1}$. The intercepts are both twice what they were before.

(d) If the price of oil doubles, our constraint would be $k = p_1 s + 2p_2 l$. Then, calculating the intercepts gives that the s intercept remains the same, but the l intercept gets cut in half. $s = 0 \Rightarrow l = \frac{k}{2p_2} = \frac{1}{2}\frac{k}{p_2}$. Therefore the maximum amount of oil you can buy is half of what it was previously.

19. Given $l - l_0 = al_0(t - t_0)$ with l_0, t_0 and a all constant,

(a) We have $l = al_0(t - t_0) + l_0 = al_0 t - al_0 t_0 + l_0$, which is a linear function of t with slope al_0 and y-intercept at $(0, -al_0 t_0 + l_0)$.

(b) If $l_0 = 100$, $t_0 = 60°$F and $a = 10^{-5}$, then

$$\begin{aligned} l = 10^{-5}(100)t - 10^{-5}(100)(60) + 100 &= 10^{-3}t + 99.94 \\ &= 0.001t + 99.94 \end{aligned}$$

(c) If the slope is positive, (as in (b)), then as the temperature rises, the length of the metal increases and it expands. If the slope were negative, then the metal would contract as the temperature rises.

21. (a) $R = k(20 - H)$, where k is a positive constant
For $H > 20$, R is negative, indicating that the coffee is cooling.

(b)

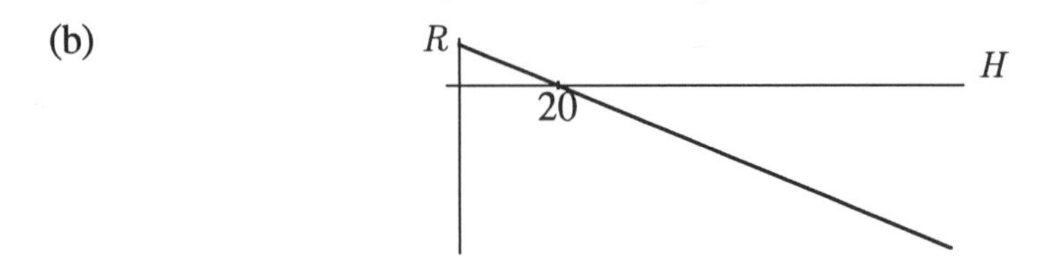

1.3 Solutions

1.

3.

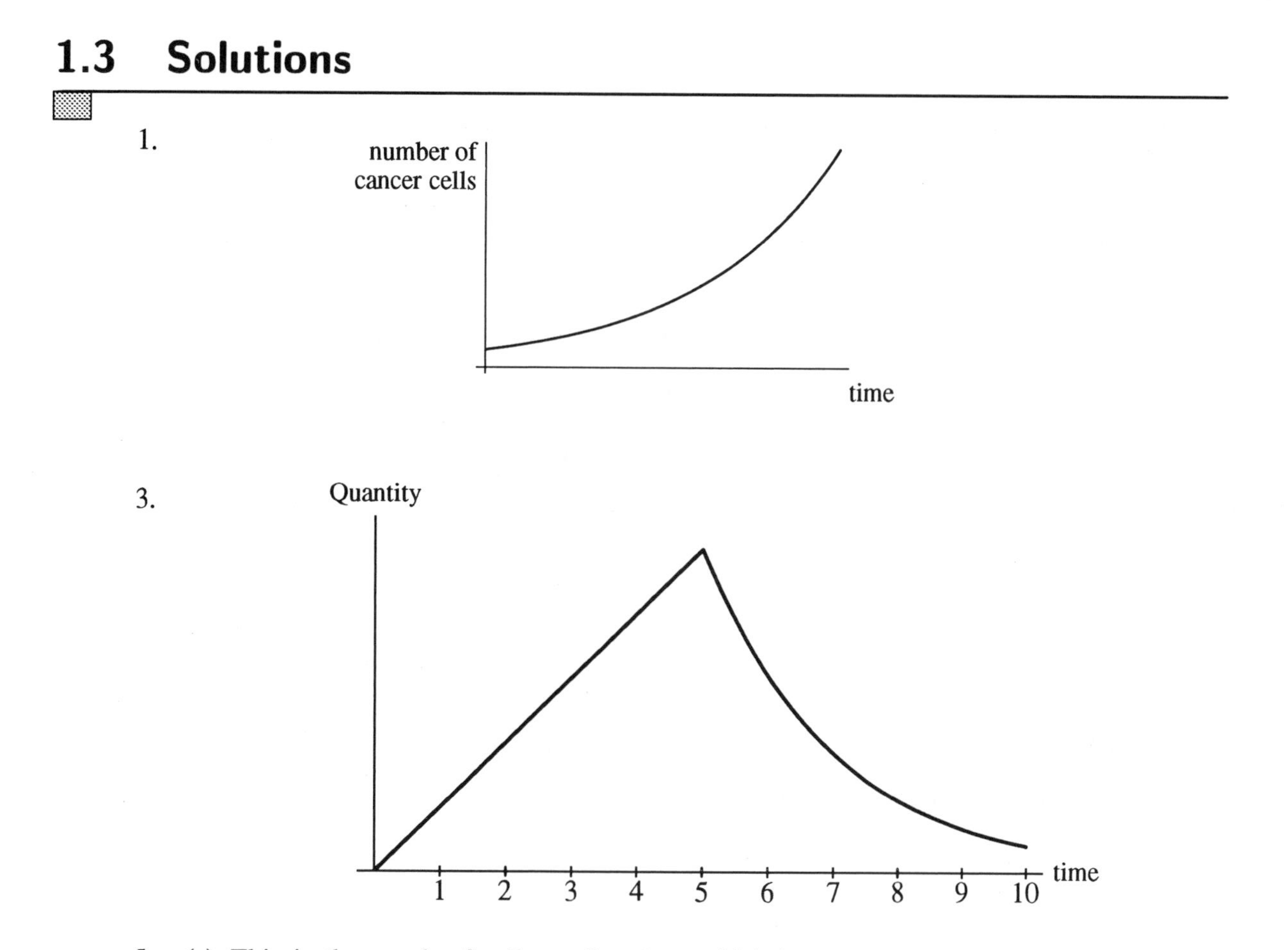

5. (a) This is the graph of a linear function, which increases at a constant rate, and thus corresponds to $k(t)$, which increases by 0.3 over each interval of 1.

   (b) This graph is concave down, so it corresponds to a function in which the increases are getting smaller, as is the case with $h(t)$, whose increases are 10, 9, 8, 7, and 6.

   (c) This graph is concave up, so it corresponds to a function in which the increases are getting bigger, as is the case with $g(t)$, whose increases are 1, 2, 3, 4, and 5.

7. $f(s) = 2(1.1)^s$, $g(s) = 3(1.05)^s$, $h(s) = (1.03)^s$.

9. Each increase of 1 in t seems to cause $g(t)$ to decrease be a factor of 0.8, so we expect an exponential function with base 0.8. To make our solution agree with the data at $t = 0$, we need a coefficient of 5.50, so our completed equation is

$$g(t) = 5.50(0.8)^t.$$

11. $y = 2(3^x)$ is possible.

13. $y = 4(1 - 2^{-x})$ is possible.

15. The doubling time is approximately 2.3. For example, the population is 20,000 at time 3.7, 40,000 at time 6, and 80,000 at time 8.3.

17. The quantity, Q, of the substance at time t, can be represented by an equation of the form

$$Q = Q_0 a^t$$

We are given $Q = 0.70Q_0$ when $t = 10$, so we have

$$\begin{aligned} 0.70Q_0 &= Q_0 a^{10} \\ 0.70 &= a^{10} \\ a &= (0.70)^{\frac{1}{10}} \\ a &\approx 0.965 \end{aligned}$$

Thus $Q = Q_0(0.70)^{\frac{1}{10}t}$, or $Q \approx Q_0(0.965)^t$

In 50 years, $Q \approx Q_0(0.965)^{50} \approx 0.168Q - 0$, so about 17% of the original quantity is left.

To find the half life, we want to find t such that

$$\begin{aligned} 0.5Q_0 &= Q_0(0.965)^t \\ 0.5 &= (0.965)^t \end{aligned}$$

Trying different values for t, we find

$$\begin{aligned} (0.965)^{19} &\approx 0.51 \\ (0.965)^{20} &\approx 0.49, \end{aligned}$$

so the half life is about 19.5 years.

20% will be left if

$$\begin{aligned} 0.20Q_0 &= Q_0(0.965)^t \\ 0.20 &= (0.965)^t \end{aligned}$$

Again, trying different values for t we find

$$(0.965)^{45.2} \approx 0.20,$$

so 20% will be left after 45.2 years.

We could solve for when there is 10% remaining as above. Instead, however, we could note that since 10% is half of 20%, it should take about 19.5 years for 20% to decay to 10%. Thus the time to decay to 10% of the original amount is about $45.2 + 19.5 = 64.7$ years.

19. (a) Compounding 33% interest 12 times should be the same as compounding the yearly rate R once, so we get

$$\left(1+\frac{R}{100}\right)^1 = \left(1+\frac{33}{100}\right)^{12}$$

Solving for R, we obtain R=2963.51. The yearly rate, R, is 2963.51%.

(b) The monthly rate r satisfies

$$\begin{aligned}\left(1+\frac{4.6}{100}\right)^1 &= \left(1+\frac{r}{100}\right)^{12}\\ 1.046^{\frac{1}{12}} &= 1+\frac{r}{100}\\ r &= 100(1.046^{\frac{1}{12}}-1) = 0.3755\end{aligned}$$

The monthly rate is 0.3755%.

21. (a) We compound the daily inflation rate 30 times to get a monthly rate r:

$$\left(1+\frac{r}{100}\right)^1 = \left(1+\frac{1.3}{100}\right)^{30},$$

solving for r, we get r=47.3, so the monthly rate was 47.3%.

(b) We compound the daily inflation rate 365 times to get a yearly rate R:

$$\left(1+\frac{R}{100}\right)^1 = \left(1+\frac{1.3}{100}\right)^{365},$$

solving for R, we get R=11054.4, so the yearly rate was 11054.4% during 1988. We could have obtained the same (large) result by compounding the monthly rate 12 times.

1.4 Solutions

1. (a) We have $P_0 = 1$ million, and $k = 0.02$, so $P(t) = (1{,}000{,}000)(e^{0.02t})$.

(b)

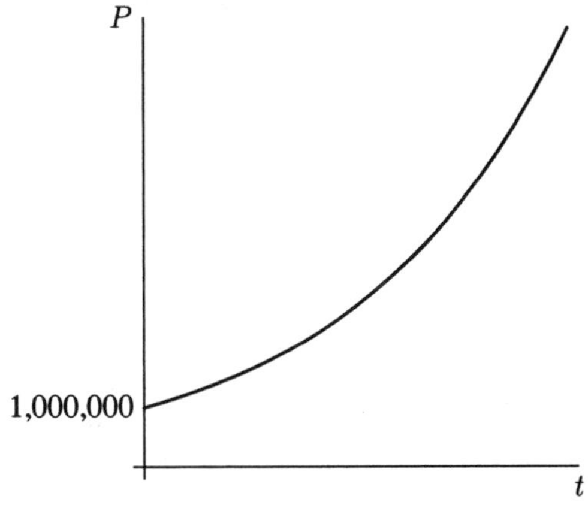

3. 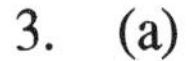 (a)

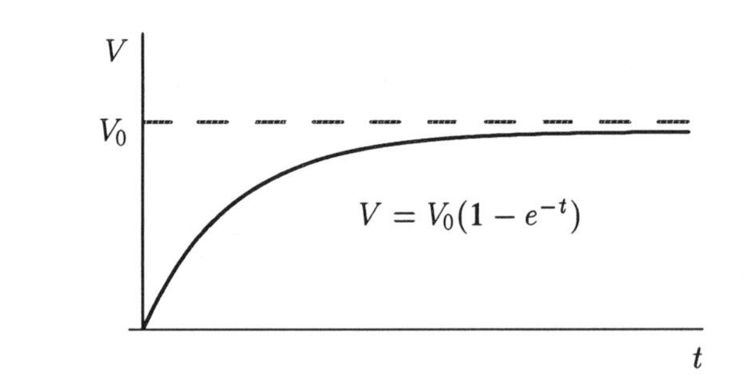

(b) V_0 represents the terminal velocity of the raindrop, or the maximum speed it can attain as it falls (although a raindrop starting at rest will never quite reach V_0 exactly).

5. For $20 \le x \le 100$, $0 \le y \le 1.2$, this function looks like a horizontal line at $y = 1.0725\ldots$ (In fact, the graph approaches this line from below.) Now, $e^{0.07} = 1.0725\ldots$, so this strongly suggests that, as we already know,

$$\text{As } x \to \infty,\ \left(1 + \frac{0.07}{x}\right)^x \to e^{0.07}.$$

y; 1.1; 1; .9; $y = 1.07250$; $y = (1 + \frac{0.07}{x})^x$

7. We need to compute

$$\left(1 + \frac{0.05}{n}\right)^n$$

for larger and larger values of n to observe an upper bound.

$$\begin{aligned}
\left(1 + \frac{0.05}{1000}\right)^{1000} &= 1.05126978... \\
\left(1 + \frac{0.05}{10000}\right)^{10000} &= 1.05127096... \\
\left(1 + \frac{0.05}{100000}\right)^{100000} &= 1.05127108... \\
\left(1 + \frac{0.05}{1000000}\right)^{1000000} &= 1.05127109...
\end{aligned}$$

So it appears that the effective interest rate with continuous compounding is approximately 5.127%.

9. $e^{0.06} = 1.0618365$, so the effective annual rate $\approx 6.18365\%$.

11. We know that for a given annual rate, the higher the frequency of compounding, the higher the effective annual yield. So the effective yield of (a) will be greater than that of (c) which

is greater than that of (b). Also, the effective annual yield of (e) will be greater than that of (d). Now the effective annual yield of (e) will be less than the effective annual yield of 5.5% annual rate, compounded twice a year, and the latter will be less than the yield from (b). Thus d<e<b<c<a. Matching these up with our choices, we get
(d) A, (e) B, (b) C, (c) D, (a) E.

13. We use the formula $A = A_0(1 + \frac{r}{n})^{nt}$, where t is in years, r is annual rate of interest and n is number of times interest is compounded per year.

    (a) Compounding daily,

$$\begin{aligned} A &= 450,000\left(1 + \frac{6}{365}\right)^{(213)(365)} \\ &= 450,000\,(1.00016438)^{77745} \\ &\approx \$1.59602561 \times 10^{11} \end{aligned}$$

    This amounts to approximately \$160 billion.

    (b) Compounding yearly,

$$\begin{aligned} A &= 450,000\,(1+0.06)^{213} \\ &= 450,000(1.06)^{213} = 450,000(245555.29) \\ &= \$1.10499882 \times 10^{11} \end{aligned}$$

    This is only \$110.5 billion.

    (c) We first wish to find the interest that will accrue during 1990. For 1990, the principal is $\$1.105 \times 10^{11}$. At 6% annual interest, during 1990 the money will earn

$$0.06 \times \$1.105 \times 10^{11} = \$6.63 \times 10^{9}.$$

    The number of seconds in a year is

$$\left(365\frac{\text{day}}{\text{year}}\right)\left(24\frac{\text{hr}}{\text{day}}\right)\left(60\frac{\text{min}}{\text{hr}}\right)\left(60\frac{\text{sec}}{\text{min}}\right) = 31536000 \text{ sec.}$$

    Thus, over 1990, interest is accumulating at the rate of

$$\frac{\$6.63 \times 10^{9}}{31536000\text{sec}} \approx \$210.24\text{ /sec.}$$

1.5 Solutions

1. (a) $8^{\frac{2}{3}} = (8^{\frac{1}{3}})^2 = 2^2 = 4.$

   (b) $9^{(-\frac{3}{2})} = (9^{\frac{1}{2}})^{(-3)} = 3^{(-3)} = \dfrac{1}{3^3} = \dfrac{1}{27}.$

3.

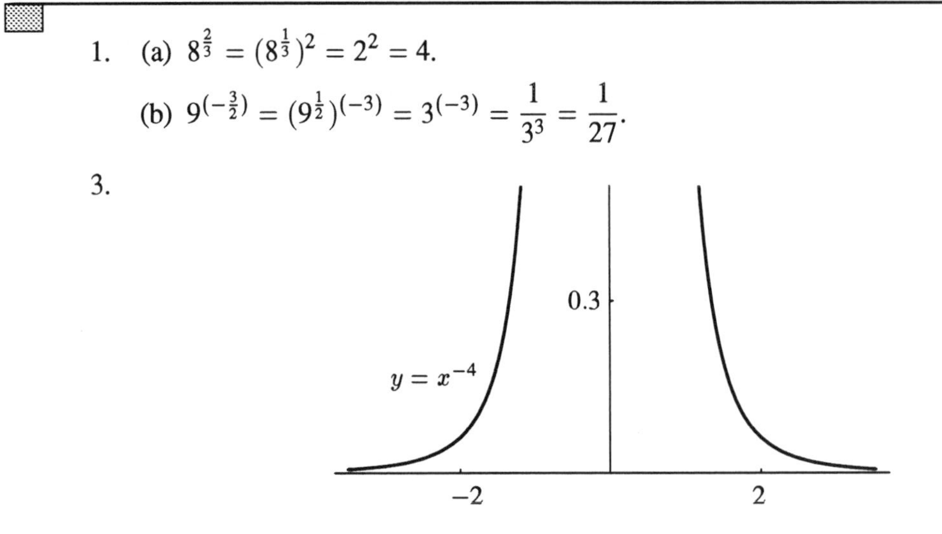

5. (a) $y = x^4$ goes to positive infinity in both cases.

   (b) $y = -x^7$ goes to negative infinity as $x \to \infty$, and goes to positive infinity as $x \to -\infty$.

7. As $x \to \infty$, $f(x) = x^5$ has the largest positive values. As $x \to -\infty$, $g(x) = -x^3$ has the largest positive values.

9.

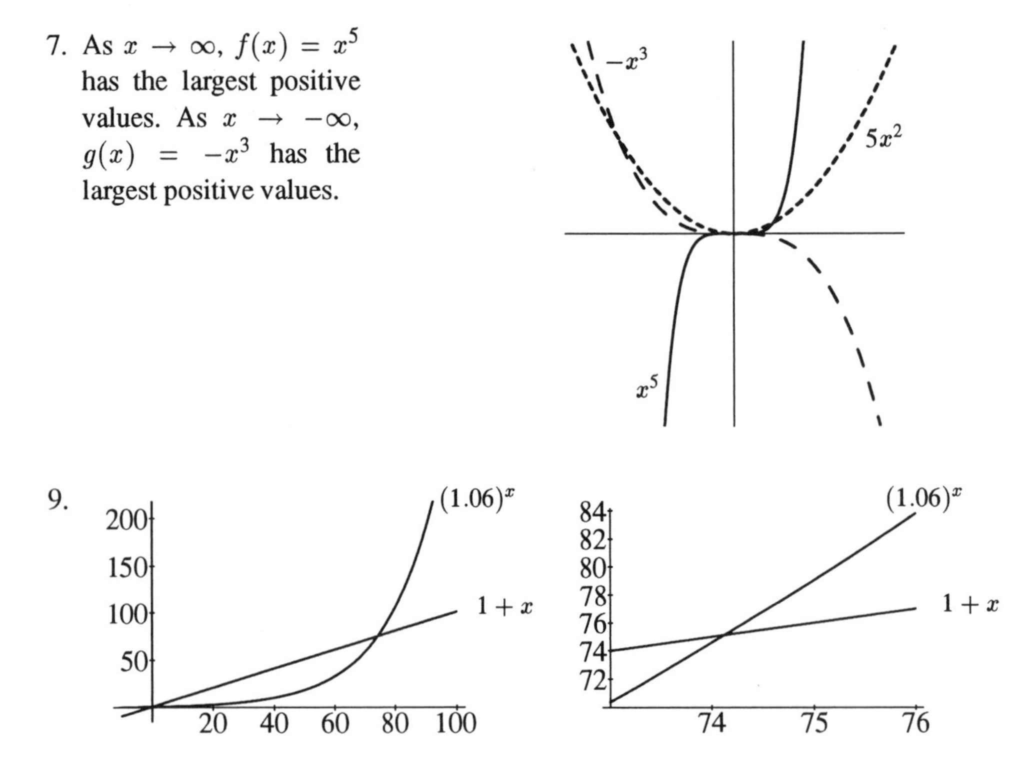

Looking at the large-scale graph on the left, we see that the functions intersect at $x = 0$ and near $x = 75$, at about 74.1, as we see by zooming in with the graph on the right.

11. 3^x is always positive while x^3 is negative for $x < 0$, so we know that $3^x > x^3$ for $x < 0$. Looking at the large-scale graph, we see that for $x > 4$, 3^x is also clearly bigger than x^3. We zoom in on the interval $(0, 4)$ to see the behavior there, shown by the graph on the right. We see that the graphs approach very close to each other near $x = 3$ (where the values are equal), but elsewhere, $3^x > x^3$. To figure out what is going on near $x = 3$, we zoom in again, and notice that on the interval from about 2.5 to 3, $x^3 > 3^x$. Thus $3^x > x^3$ if $x < 2.5$ (approximately) or $x > 3$.

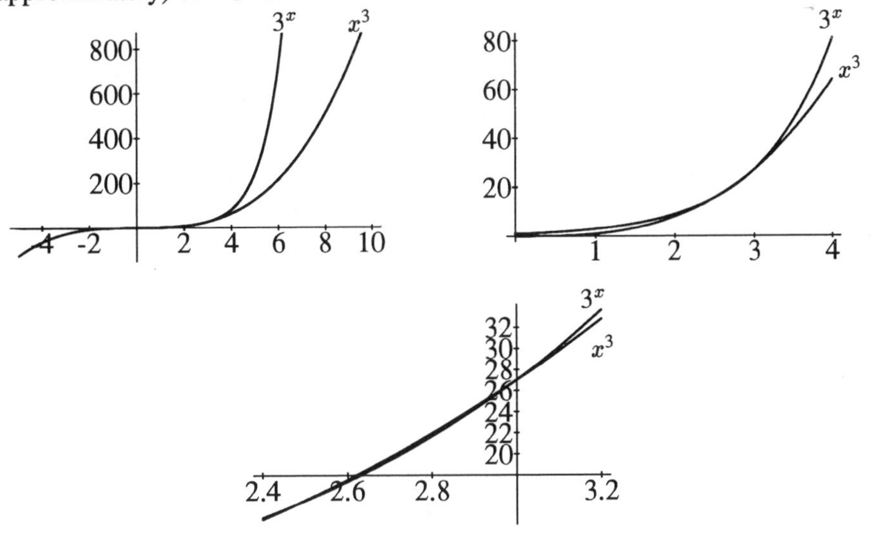

Alternatively, graph $f(x) = 3^x - x^3$ for $0 \le x \le 3.5$ and $-2 \le y \le 4$. This shows $3^x > x^3$ for $x > 3$ and $x < 2.48$.

13. (a) Since the rate R varies directly with the fourth power of the radius r, we have the formula

$$R = kr^4$$

where k is a constant.

(b) Given $R = 400$ for $r = 3$, we can determine the constant k.

$$\begin{aligned} 400 &= k(3)^4 \\ 400 &= k(81) \\ k &= \frac{400}{81} \approx 4.938. \end{aligned}$$

So the formula is

$$R = 4.938r^4$$

(c) We plug in $r = 5$, yielding:

$$R = 4.928(5)^4 = 3086.42\frac{\text{cm}^3}{\text{sec}}.$$

15. Looking at g, we see that the ratio of the values is:

$$\frac{3.12}{3.74} = \frac{3.74}{4.49} = \frac{4.49}{5.39} = \frac{5.39}{6.47} = \frac{6.47}{7.76} \approx 0.83.$$

Thus g is an exponential function, and so f and k are the power functions. Each is of the form ax^2 or ax^3, and since $k(1.0) = 9.01$ we see that for k, the constant coefficient is 9.01. Trial and error gives

$$k(x) = 9.01x^2,$$

since $k(2.2) = 43.61 \approx 9.01(4.84) = 9.01(2.2)^2$. Thus $f(x) = ax^3$ and we find a by noting that $f(9) = 7.29 = a(9^3)$ so

$$a = \frac{7.29}{9^3} = 0.01.$$

and $f(x) = 0.01x^3$.

17.

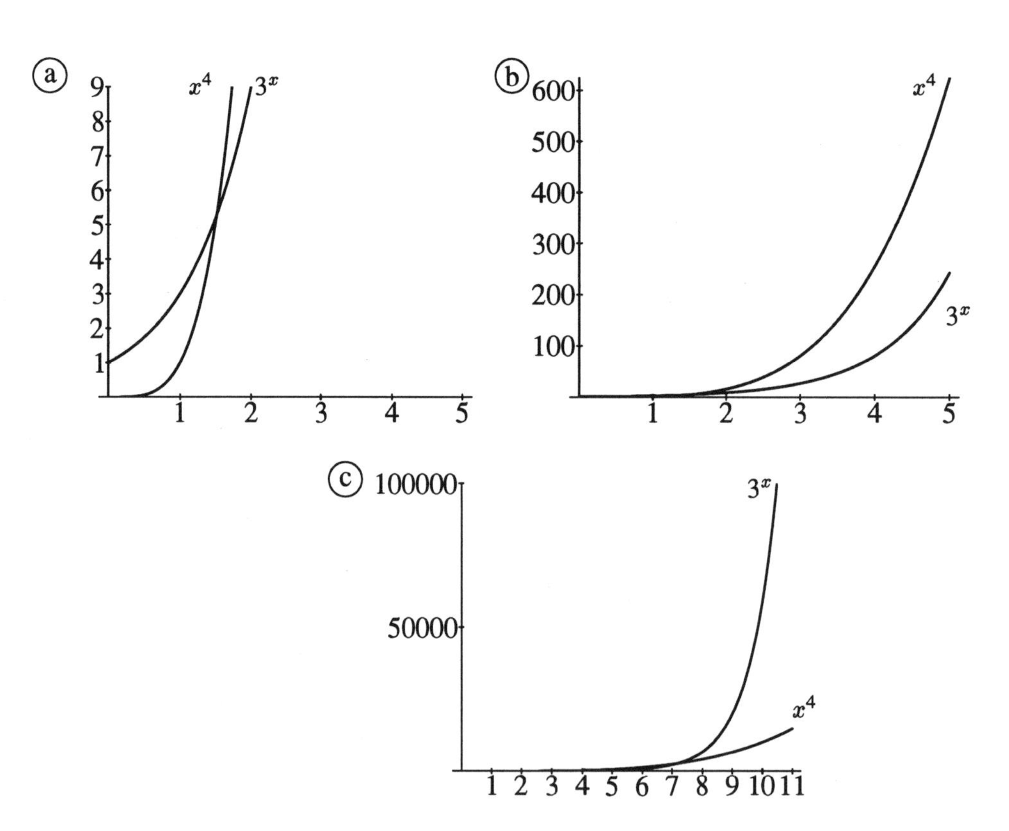

19. (a) The domain is $(0, 4000)$, the range is $(0, 10^8)$.
(b) The domain is $(0, 3000)$, the range is $(0, 10^7)$.
(c) The domain is $(0, 0.2)$, the range is $(0, 0.04)$.

1.6 Solutions

1. $f^{-1}(75)$ is the length of the column of mercury in the thermometer when the temperature is 75°F.

3. Not invertible. This is because that given a certain number of customers, say, $f(t) = 1500$, there could be many times instants that corresponds to this number. So we just don't know which time instant is the right one.

5. Not invertible, since it costs the same to mail a 50-gram letter as it does to mail a 51-gram letter.

7.

x	3	−7	19	4	178	2	1
$f^{-1}(x)$	1	2	3	4	5	6	7

The domain of f^{-1} is the set consisting of the integers $\{3, -7, 19, 4, 178, 2, 1\}$.

9.

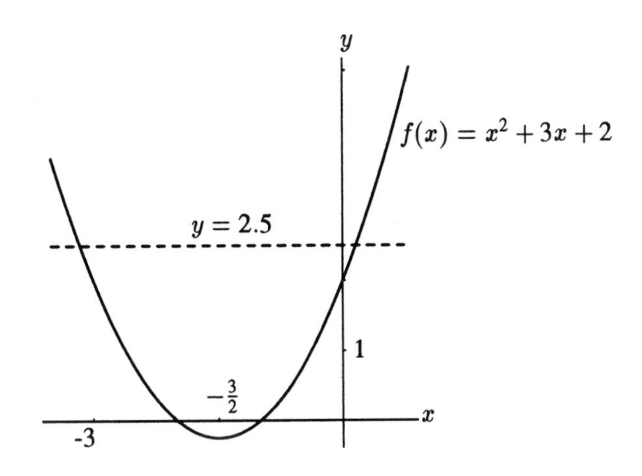

Since a horizontal line cuts the graph of $f(x) = x^2 + 3x + 2$ two times, f is not invertible.

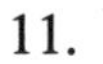

11.

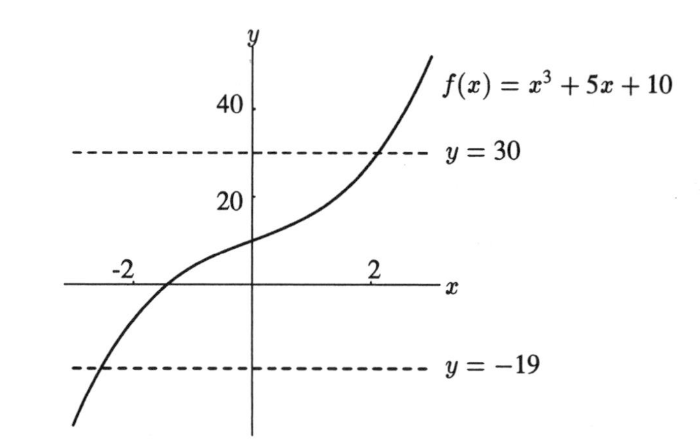

Since any horizontal line cuts the graph once, f is invertible.

13. (a) The function f tells us C in terms of q. To get the inverse, we want q in terms of C, which we find by solving for q:

$$\begin{aligned} C &= 100 + 2q, \\ C - 100 &= 2q, \\ q &= \frac{C - 100}{2} = f^{-1}(C). \end{aligned}$$

(b) The inverse function tells us the number of articles that can be produced for a given cost.

15. The definition of increasing says that for f increasing, as x gets larger, $f(x)$ gets larger for all x. But since, in this case, we know x has an inverse, we can let $f(x)$ increase and see what happens to the corresponding values of x. We know they must increase; otherwise, increasing x would cause $f(x)$ to decrease (since $f(x)$ would increase when x decreases). So we know that the inverse function increases.

17. Yes, f is invertible, since f is increasing.

(b) $f^{-1}(400)$ is the year in which 400 million motor vehicles were registered in the world. From the picture, we see that $f^{-1}(400)$ is around 1979.

(c) Since the graph of f^{-1} is the reflection of the graph of f in the line $y = x$, we get the following picture:

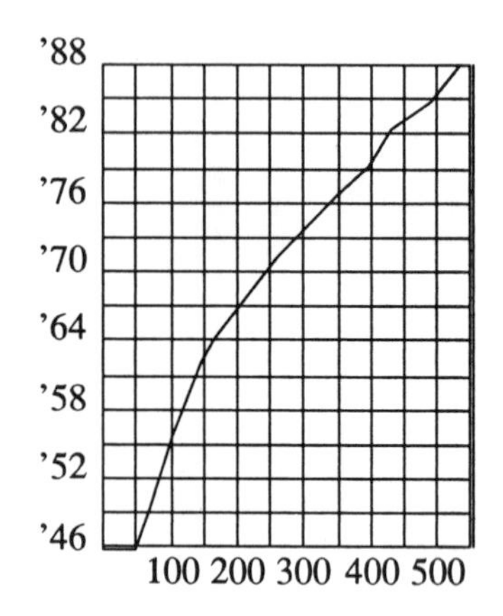

Figure 1.1: The Graph of f^{-1}

1.7 Solutions

1.

x	1	2	3	4	5	6	7	8	9	10
$f(x)$	0	0.30	0.48	0.60	0.70	0.78	0.85	0.90	0.95	1.00
$g(x)$	1.00	1.41	1.73	2.00	2.24	2.45	2.65	2.83	3.00	3.16

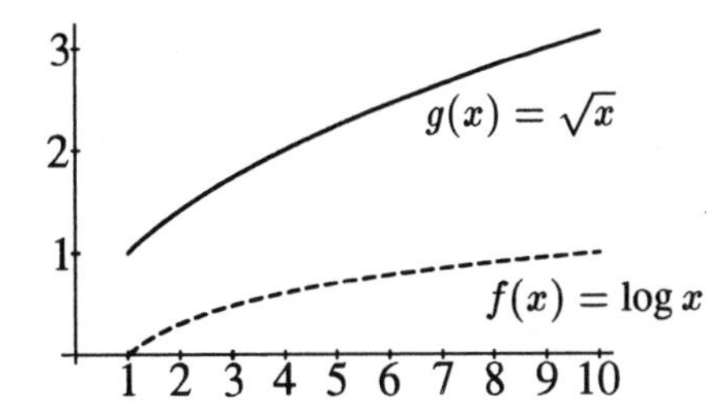

3. Since $10^{\log x} = x$ for $x > 0$, this equation is $y = x$ for $x > 0$. Its graph is therefore a straight line , with slope 1, to the right the origin.

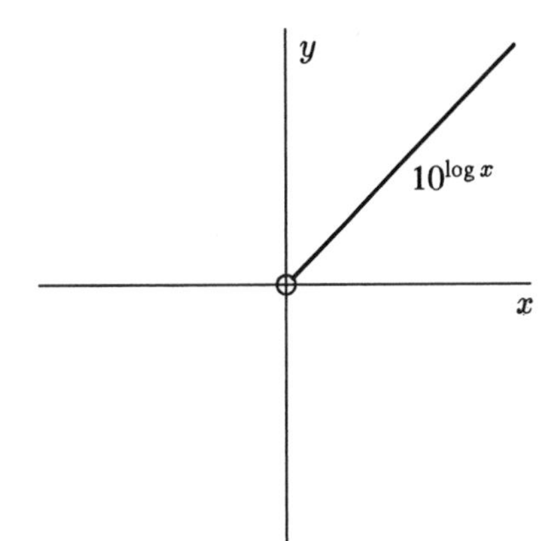

5. $t = \dfrac{\log 2}{\log 1.02} \approx 35.003.$

7. Collecting similar factors yields $\left(\frac{1.04}{1.03}\right)^t = \frac{12.01}{5.02}$. Solving for t yields

$$t = \frac{\log\left(\frac{12.01}{5.02}\right)}{\log\left(\frac{1.04}{1.03}\right)} \approx 90.283.$$

9.

$$t = \frac{\log\left(\frac{P}{P_0}\right)}{\log a} = \frac{\log P - \log P_0}{\log a}.$$

11. Collecting similar terms yields

$$\left(\frac{a}{b}\right)^t = \frac{Q_0}{P_0}.$$

Hence

$$t = \frac{\log\left(\frac{Q_0}{P_0}\right)}{\log\left(\frac{a}{b}\right)}.$$

13.

$$\begin{aligned} 2\log\alpha - 3\log B - \frac{\log\alpha}{2} &= \frac{3}{2}\log\alpha - 3\log B \\ &= 3\log\frac{\sqrt{\alpha}}{B} = \frac{3}{2}\log\frac{\alpha}{B^2}. \end{aligned}$$

15.

$$\log(10^{x+7}) = x + 7.$$

17.

$$10^{2\log Q} = 10^{\log Q^2} = Q^2.$$

19.

$$10^{-(\log B)/2} = \left[10^{\log B}\right]^{-\frac{1}{2}} = \frac{1}{\sqrt{B}}.$$

21. Since the factor by which the prices have increased after time t is given by $(1.05)^t$, the time after which the prices have doubled solves

$$\begin{aligned} 2 &= (1.05)^t \\ \log 2 &= \log(1.05^t) = t\log(1.05) \\ t &= \frac{\log 2}{\log 1.05} \approx 14.21 \text{ years.} \end{aligned}$$

23. The population has increased by a factor of $\frac{56{,}000{,}000}{40{,}000{,}000} = 1.4$ in 10 years. Thus we have the formula

$$P = 40{,}000{,}000(1.4)^{\frac{t}{10}},$$

thus $\frac{t}{10}$ gives the number of 10-year periods that have passed since 1980.

In 1980, $\frac{t}{10} = 0$, so we have $P = 40{,}000{,}000$.

In 1990, $\frac{t}{10} = 1$, so $P = 40{,}000{,}000(1.4) = 56{,}000{,}000$.

In 2000, $\frac{t}{10} = 2$, so $P = 40{,}000{,}000(1.4)^2 = 78{,}400{,}000$.

To find the doubling time, solve $80{,}000{,}000 = 40{,}000{,}000(1.4)^{\frac{t}{10}}$, to get $t \approx 20.6$ years.

25. (a) Using the formula for exponential decay, $A = A_0 e^{-kt}$, we plug in $A = 10.32$ when $t = 0$. Since $e^0 = 1$, $A_0 = 10.32$

Since the half-life is 12 days, when $t = 12$, $A = \frac{10.32}{2} = 5.16$. Plugging this into the formula, we have

$$\begin{aligned} 5.16 &= 10.32e^{-12k}, \\ 0.5 &= e^{-12k}, \quad \text{and, taking ln of both sides,} \\ \ln 0.5 &= -12k, \\ k &= -\frac{\ln(0.5)}{12} \approx 0.057762. \end{aligned}$$

The full equation is

$$A = 10.32e^{-0.057762t}.$$

(b) We want to solve for t when $A = 1$. Plugging into the equation from (a) yields

$$\begin{aligned} 1 &= 10.32e^{-0.057762t}, \\ \frac{1}{10.32} = 0.096899 &= e^{-0.057762t}, \\ \ln 0.096899 &= -0.057762t, \\ t &= \frac{-2.33408}{-0.057762} = 40.41 \text{ days.} \end{aligned}$$

27. Assuming a rate of inflation of 4.6% a year, prices increase from one year to the next by a factor of 1.046. Letting t be time in years from 1990, the price P of a stamp in dollars is given, in our model, by the equation

$$P(t) = 0.29(1.046)^t,$$

where the 0.29 comes from plugging in the condition that at $t = 0$, $P(t) = 29¢$. To find the time when it costs a dollar to mail a letter, we solve

$$\begin{aligned} 0.29(1.046)^t &= 1, \\ \log 0.29(1.046)^t &= \log 1, \\ \log 0.29 + t \log 1.046 &= 0, \\ t &= -\frac{\log 0.29}{\log 1.046} \approx 27.52. \end{aligned}$$

So a stamp should cost \$1 by the middle of the year $1990 + 27.52 = 2017$.

29. We want an x so that 2^x is the distance from the earth to the Moon in *inches*.

$$2^x = 239{,}000\,\text{miles}\,\frac{5{,}280\,\text{feet}}{\text{mile}}\,\frac{12\,\text{inches}}{\text{foot}} = 15{,}143{,}040{,}000\,(\approx 1.5 \times 10^{10})$$

Applying the logarithm function to both sides, we have:

$$x = \frac{\log 15{,}143{,}040{,}000}{\log 2} \approx 33.82 \text{ inches.}$$

(The outer radius of the moon, by a similar calculation, is at $x = 33.85$ inches. The graph doesn't take long to pass through the moon!)

31. We want an x so that 2^x is the distance from the earth to Proxima Centauri in *inches*.

$$2^x = 1.3\,\text{parsecs}\,\frac{1.917 \times 10^{13}\,\text{miles}}{\text{parsec}}\,\frac{5{,}280\,\text{feet}}{\text{mile}}\,\frac{12\,\text{inches}}{\text{foot}} \approx 1.57899 \times 10^{18}.$$

Applying the logarithmic function to both sides, we have:

$$x \log 2 \approx \log(1.57899 \times 10^{18}), \quad \text{so} \quad x \approx \frac{\log(1.57899 \times 10^{18})}{\log 2} \approx 60.45 \text{ inches.}$$

33. We want an x so that 2^x is the distance from the earth to the Andromeda Galaxy in *inches*.

$$2^x = 6.7 \times 10^5\,\text{parsecs}\,\frac{1.917 \times 10^{13}\,\text{miles}}{\text{parsec}}\,\frac{5{,}280\,\text{feet}}{\text{mile}}\,\frac{12\,\text{inches}}{\text{foot}} \approx 8.13790 \times 10^{23}.$$

Applying the logarithmic function to both sides, we have:

$$x \log 2 \approx \log(8.13790 \times 10^{23}); \quad \text{so,} \quad x \approx \frac{\log(8.13790 \times 10^{23})}{\log 2} \approx 79.43 \text{ inches.}$$

1.8 Solutions

1.

x	1	2	3	4	5	6	7	8	9	10
$f(x)$	0	0.30	0.48	0.60	0.70	0.78	0.85	0.90	0.95	1.00
$g(x)$	0	0.69	1.10	1.39	1.61	1.79	1.95	2.08	2.20	2.30

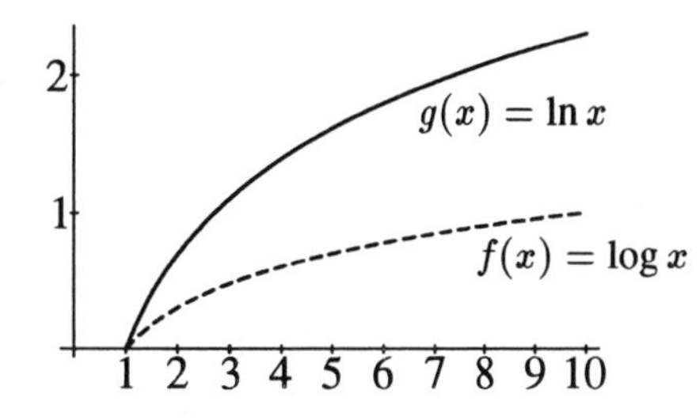

3. Since $e^{\ln x} = x$ for $x > 0$, this equation is $y = x$ for $x > 0$. Its graph is therefore a straight line , with slope 1, to the right of the origin.

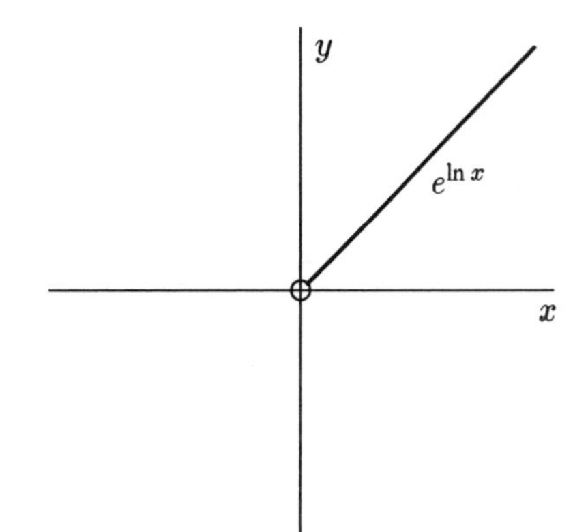

5. $t = \ln \frac{a}{b}$

7. $\ln a + kt = bt$, so $t = \dfrac{\ln a}{b-k}$, since $k \neq b$.

9. (a) $P = 2.5e^{0.028t}$ million, where t is the number of years since 1986.
   (b) If $P = 2.5e^{0.028t}$, then $t = \frac{1}{0.028} \ln \frac{P}{2.5}$.
   For $P = 5$, $t \approx 24.8$, so the population reaches 5 million in 2010.
   For $P = 10$, $t \approx 49.5$, so the population reaches 10 million in 2035.
   For $P = 20$, $t \approx 74.3$, so the population reaches 20 million in 2060.

11. Say the bacterial population has initial size P_0. Since the population doubles every 5 hours, we have exponential growth. This gives us the equation

$$2P_0 = P_0 e^{k \cdot 5}, \qquad \text{where } k \text{ is the rate constant.}$$

This implies that

$$2 = e^{5k},$$

and so,

$$k = \frac{1}{5} \ln 2.$$

In order to find the time t at which the population triples, or equals $3P_0$, we set

$$\begin{aligned} 3P_0 &= P_0\, e^{\frac{1}{5} t \ln 2} \\ 3 &= e^{\frac{1}{5} t \ln 2} \\ \ln 3 &= \frac{1}{5} t \ln 2 \\ t &= 5 \cdot \frac{\ln 3}{\ln 2} \\ &\approx 7.92 \quad \text{hours.} \end{aligned}$$

13. (a) We want to find t such that

$$0.15 Q_0 = Q_0 e^{-0.000121t},$$

so $0.15 = e^{-0.000121t} \Rightarrow \ln 0.15 = -0.000121t \Rightarrow t = \frac{\ln 0.15}{-0.000121} \approx 15{,}678.7$ years.

(b) Let T be the half-life of Carbon-14. Then

$$0.5 Q_0 = Q_0 e^{-0.000121T},$$

so $0.5 = e^{-0.000121T} \Rightarrow T = \frac{\ln 0.5}{-0.000121} \approx 5{,}728.5$ years.

15. Since f is increasing, f has an inverse. To find the inverse of $f(t) = 50e^{0.1t}$, we replace t with $f^{-1}(t)$, and, since $f(f^{-1}(t)) = t$, we have

$$t = 50e^{0.1 f^{-1}(t)}$$

We then solve for $f^{-1}(t)$:

$$\begin{aligned} t &= 50e^{0.1 f^{-1}(t)} \\ \frac{t}{50} &= e^{0.1 f^{-1}(t)} \\ \ln\left(\frac{t}{50}\right) &= 0.1 f^{-1}(t) \\ f^{-1}(t) &= \frac{1}{0.1} \ln\left(\frac{t}{50}\right) = 10 \ln\left(\frac{t}{50}\right) \end{aligned}$$

1.9 Solutions

1. (a) The equation is $y = 2x^2 + 1$. Note that its graph is narrower than the graph of $y = x^2$ which appears in grey.

   (b) $y = 2(x^2 + 1)$ moves the graph up one unit and *then* stretches it by a factor of two.

   (c) No, the graphs are not the same. Note that stretching it vertically leaves any point whose y-value is zero in the same place but moves any other point. This is the source of the difference because if you stretch it first, its lowest point stays at the origin. Then you shift it up by one and its lowest point is $(0, 1)$. Alternatively, if you shift it first, its lowest point is $(0, 1)$ which, when stretched by 2, becomes $(0, 2)$.

3. $\ln(\ln(x))$ means take the ln of the value of the function $\ln x$. On the other hand, $\ln^2(x)$ means take the function $\ln x$ and square it. For example, consider each of these functions evaluated at e. Since $\ln e = 1$, $\ln^2 e = 1^2 = 1$, but $\ln(\ln(e) = \ln(e) = 0$. See the graphs below.

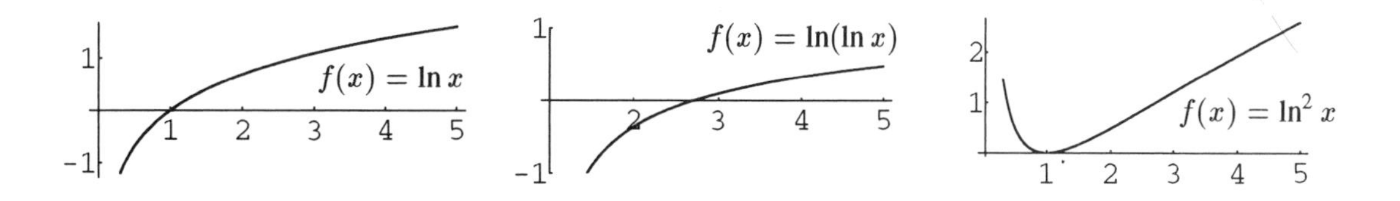

5. (a) $f(g(t)) = f(\frac{1}{t+1}) = \left(\frac{1}{t+1}+7\right)^2$

(b) $g(f(t)) = g((t+7)^2) = \frac{1}{(t+7)^2+1}$

(c) $f(t^2) = (t^2+7)^2$

(d) $g(t-1) = \frac{1}{(t-1)+1} = \frac{1}{t}$

7. (a) $f(g(100)) = f(\log 100) = f(2) = 10^2 = 100$

(b) $g(f(3)) = g(10^3) = g(1000) = \log(1000) = 3$

(c) $f(g(x)) = 10^{\log x} = x$

(d) $g(f(x)) = \log(10^x) = x$

9. $m(z+h) - m(z) = (z+h)^2 - z^2 = 2zh + h^2$.

11. $m(z+h) - m(z-h) = (z+h)^2 - (z-h)^2 = z^2 + 2hz + h^2 - (z^2 - 2hz + h^2) = 4hz$.

13. $f(x) = x^3$, $g(x) = x + 1$.

15. $f(x) = \ln x$, $g(x) = x^3$. (Another possibility: $f(x) = 3x$, $g(x) = \ln x$.)

17. Here are the graphs.

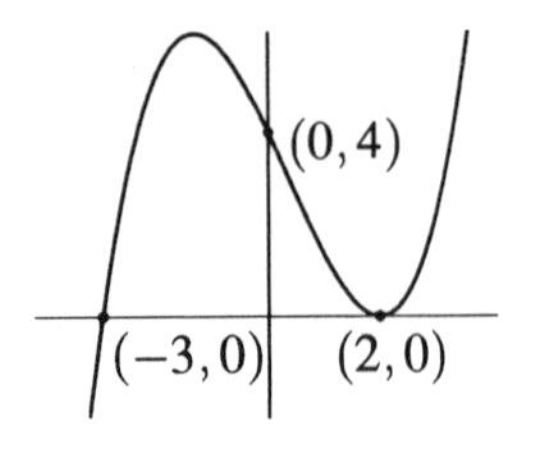

Figure 1.2: (a) $y = 2f(x)$

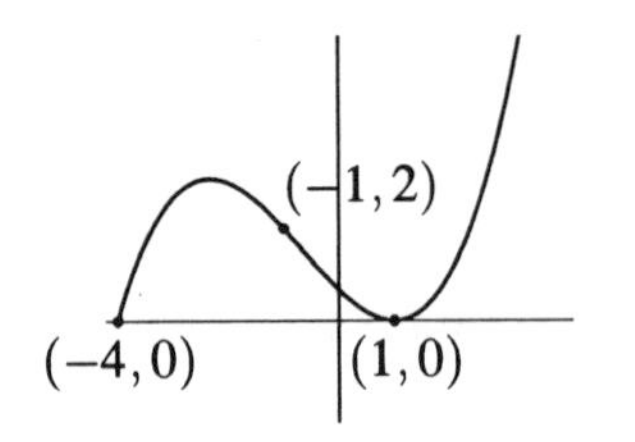

Figure 1.3: (b) $y = f(x+1)$

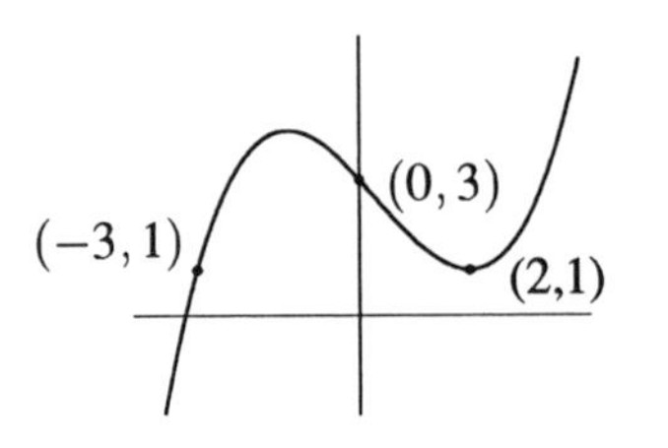

Figure 1.4: (c) $y = f(x) + 1$

19.

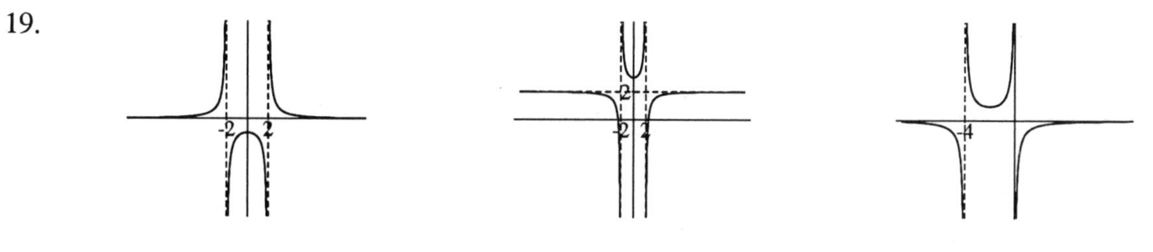

Figure 1.5: (a) $y = -f(x)$ **Figure 1.6**: (b) $y = f(x) + 2$

Figure 1.7: (c) $y = f(x + 2)$

21. Given that $f(x)$ and $g(x)$ are odd, if $h(x) = f(x) + g(x)$,

$$h(-x) = f(-x) + g(-x) = -f(x) - g(x) = -h(x)$$

so $h(x)$ is odd. If $h(x) = f(x)g(x)$,

$$h(-x) = f(-x)g(-x) = (-1)f(x)(-1)g(x) = f(x)g(x) = h(x)$$

so $h(x)$ is even.

23. $f(x) = x^n$ is even for n an even integer, and odd for n an odd integer.

25. $\dfrac{1}{x}$ is odd; $\ln|x|$ is even, e^x is neither.

27.

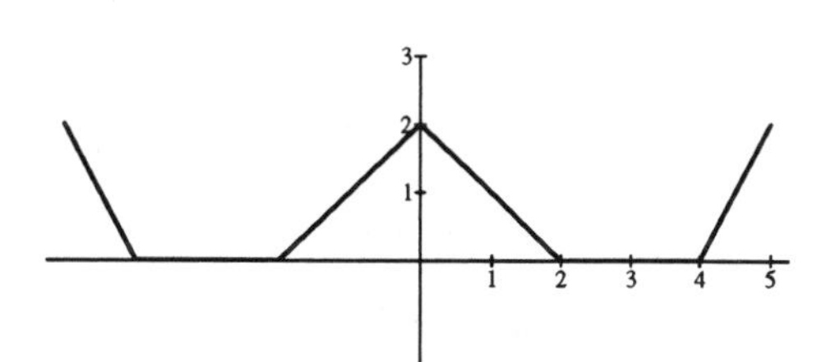

29. $f(g(1)) = f(2) = 2$.

30. $g(f(2)) = g(2) = 2$.

31. $f(f(1)) = f(0) \approx -0.7$.

33. Using the same way to compute $f(g(x))$ as in Problem 30, we get the following table. Then we can plot the graph of $f(g(x))$.

x	$f(x)$	$g(f(x))$
-3	3	0
-2.5	1	2
-2	0	-0.3
-1.5	$-.6$	-1.6
-1	-1	-2
-0.5	-0.9	-1.9
0	-0.7	-1.8
0.5	-0.5	-1.5
1	0	-0.3
1.5	1	2
2	2	2
2.5	2.3	1.8
3	2.6	1.1

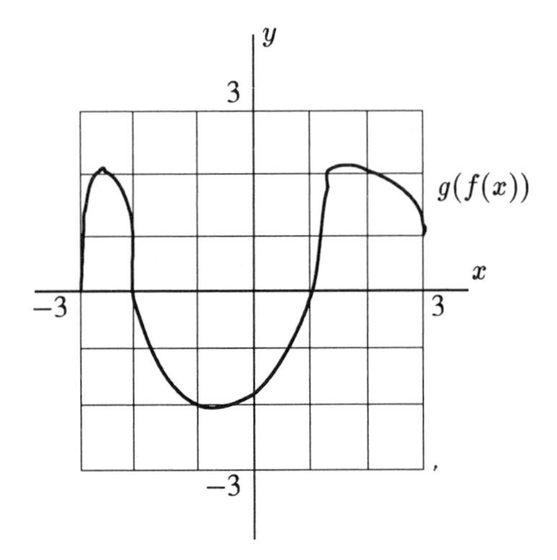

35.

x	$f(x)$	$g(x)$	$h(x)$
-3	0	0	0
-2	2	2	-2
-1	2	2	-2
0	0	0	0
1	2	-2	-2
2	2	-2	-2
3	0	0	0

Table 1.1: Completed chart.
(for Problem 35)

37.

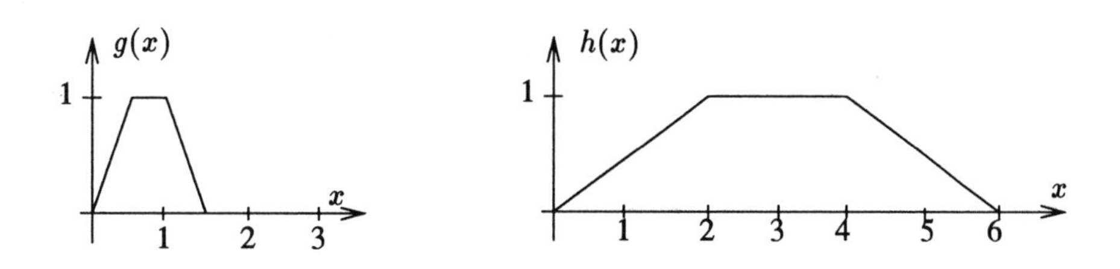

The graph of $f(ax)$ is what you get when you squeeze the graph of $f(x)$ by a factor of a in the horizontal direction. If $a < 0$, then the graph is flipped about the y-axis in addition to being squeezed.

1.10 Solutions

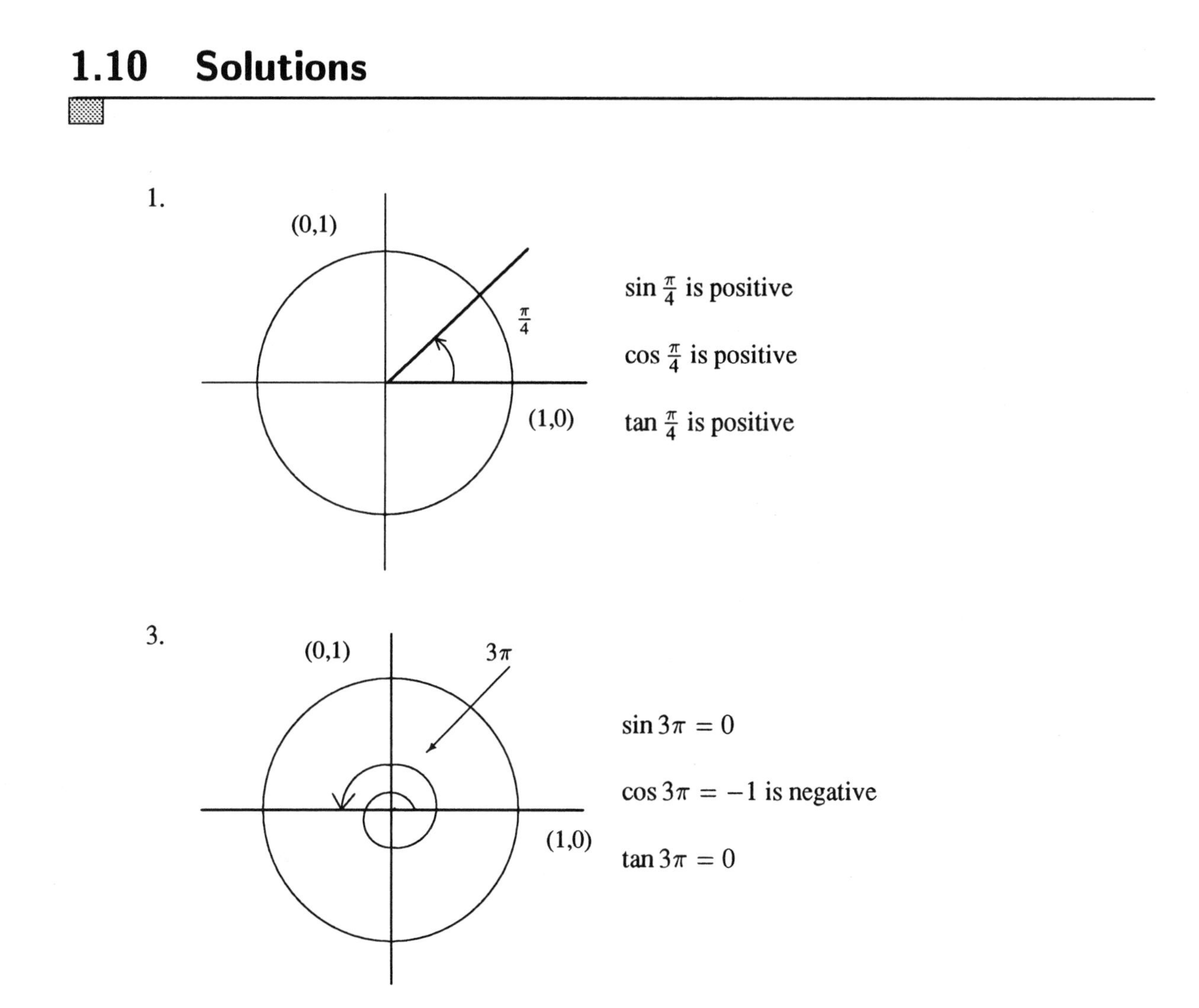

5.

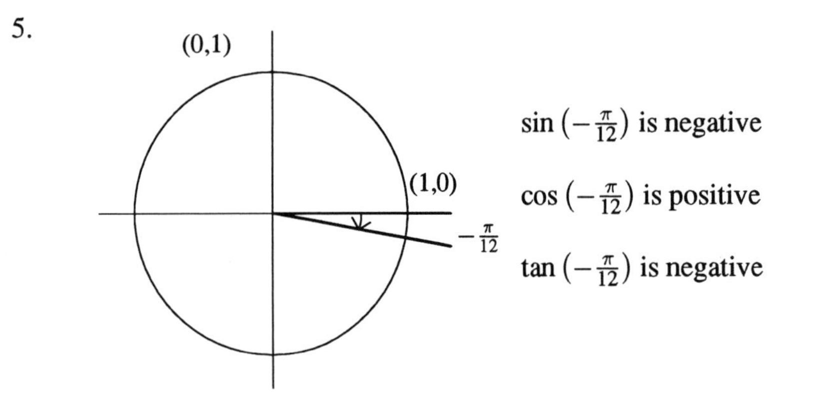

7. $4 \text{ radians} \cdot \frac{180°}{\pi \text{ radians}} = \left(\frac{720}{\pi}\right)^{\circ} \approx 240°$

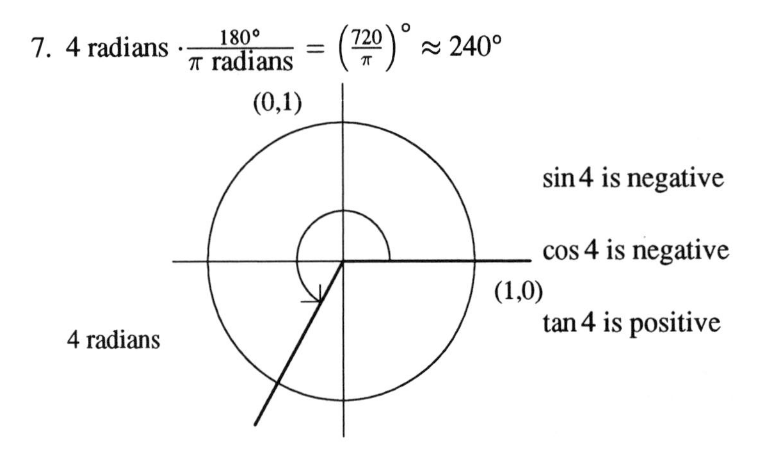

9.

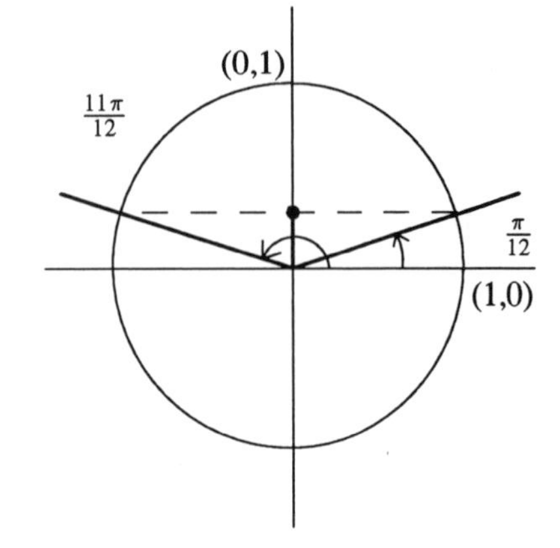

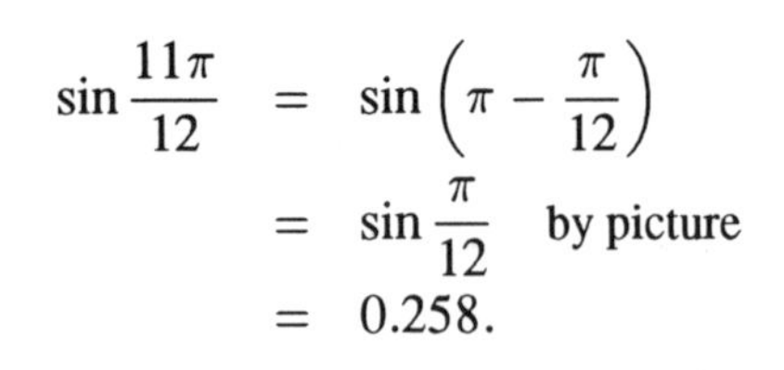

$$\begin{aligned} \sin\frac{11\pi}{12} &= \sin\left(\pi - \frac{\pi}{12}\right) \\ &= \sin\frac{\pi}{12} \quad \text{by picture} \\ &= 0.258. \end{aligned}$$

11.

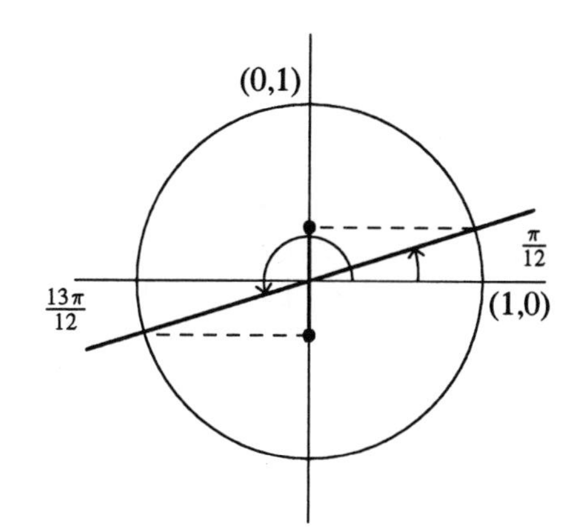

$$\begin{aligned}\sin\frac{13\pi}{12} &= \sin\left(\pi+\frac{\pi}{12}\right)\\ &= -\sin\frac{\pi}{12} \quad \text{by picture}\\ &= -0.258.\end{aligned}$$

13.

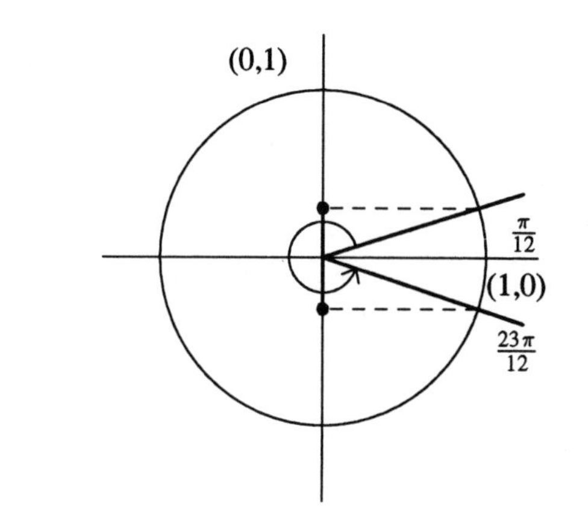

$$\begin{aligned}\sin\frac{23\pi}{12} &= \sin\left(2\pi-\frac{\pi}{12}\right)\\ &= -\sin\frac{\pi}{12} \quad \text{by picture}\\ &= -0.258.\end{aligned}$$

15.

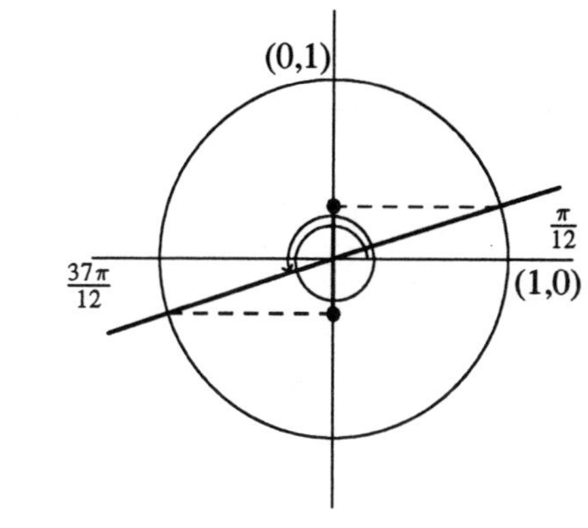

$$\begin{aligned}\sin\frac{37\pi}{12} &= \sin\left(3\pi+\frac{\pi}{12}\right)\\ &= -\sin\frac{\pi}{12} \quad \text{by picture}\\ &= -0.258.\end{aligned}$$

17.

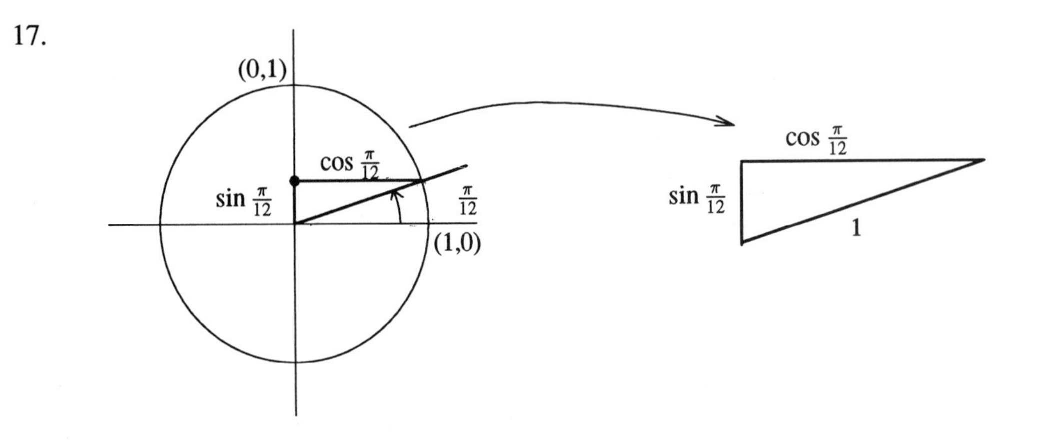

By the Pythagorean Theorem, $(\cos\frac{\pi}{12})^2 + (\sin\frac{\pi}{12})^2 = 1^2$; so $(\cos\frac{\pi}{12})^2 = 1 - (\sin\frac{\pi}{12})^2$ and $\cos\frac{\pi}{12} = \sqrt{1-(\sin\frac{\pi}{12})^2} = \sqrt{1-(0.258)^2} \approx 0.966$.
We take the positive square root since by the picture we know that $\cos\frac{\pi}{12}$ is positive.

19. From the example, we have $h = 5 + 4.9\cos(\frac{\pi}{6}t)$, where t represents hours after midnight and h represents the height of the water.

$$\text{At 3:00 am, } t = 3 \text{ so } h = 5 + 4.9\cos(\frac{\pi}{6}\cdot 3) = 5 + 4.9(0) = 5 \text{ feet}$$

$$\text{At 4:00 am, } t = 4 \text{ so } h = 5 + 4.9\cos(\frac{\pi}{6}\cdot 4) = 5 + 4.9(-0.5) = 2.55 \text{ feet}$$

$$\text{At 5:00 pm, } t = 17 \text{ so } h = 5 + 4.9\cos(\frac{\pi}{6}\cdot 17) \approx 5 + 4.9(-0.866) \approx 0.76 \text{ feet}$$

21. The moon makes one revolution around the Earth in about 27.3 days, so its period is 27.3 days $\approx$ one month.

23. One hour.

25. Using the fact that 1 revolution $= 2\pi$ radians and 1 minute $= 60$ seconds, we have

$$\begin{aligned} 33\frac{1}{3}\frac{\text{rev}}{\text{min}} &= (33\frac{1}{3})\cdot 2\pi\frac{\text{rad}}{\text{min}} = 33\frac{1}{3}\cdot 2\pi\frac{1 \text{ rad}}{60 \text{ sec}} \\ &\approx \frac{(33.333)(6.283)}{60} \\ &\approx 3.491 \text{ radians per second} \end{aligned}$$

27. $\sin x^2$ is by convention $\sin(x^2)$, which means you square the x first and then take the sine. $\sin^2 x = (\sin x)^2$ means find $\sin x$ and then square it.

$\sin(\sin x)$ means find $\sin x$ and then take the sine of that.

Expressing each as a composition: If $f(x) = \sin x$ and $g(x) = x^2$, then
$\sin x^2 = f(g(x))$
$\sin^2 x = g(f(x))$
$\sin(\sin x) = f(f(x))$.

29.

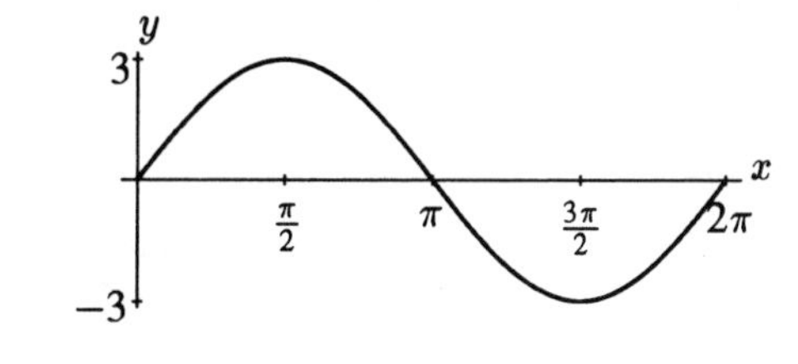

The amplitude is 3 and the period is 2π.

31.

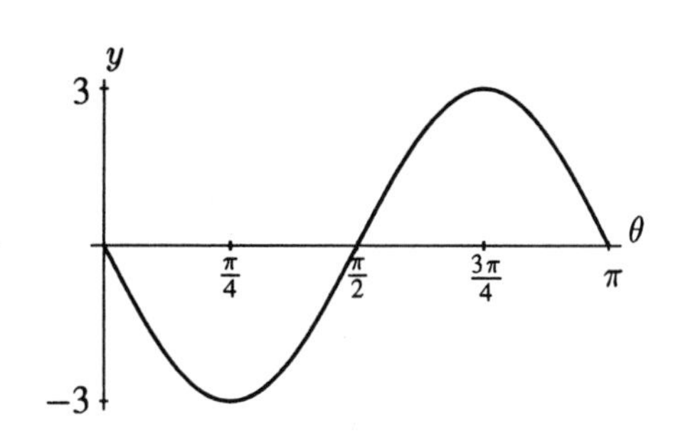

The amplitude is 3 and the period is π.

33.

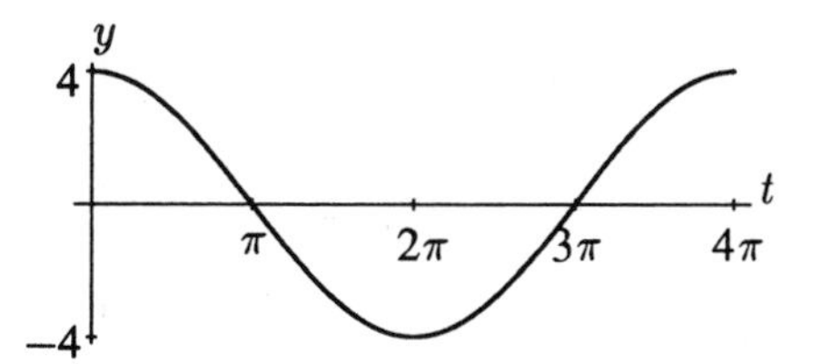

The amplitude is 4 and the period is 4π.

35. (a) This graph is a sine curve with period 8π and amplitude 2, so it is given by $f(x) = 2\sin(\frac{x}{4})$.

(b) This graph is the same as (a) but shifted up by 2, so it is given by $f(x) = 2\sin(\frac{x}{4}) + 2$.

(c) This graph is a cosine curve with period 6π and amplitude 5, so it is given by $f(x) = 5\cos(\frac{x}{3})$.

(d) This graph is an inverted sine curve with amplitude 4 and period π, so it is given by $f(x) = -4\sin(2x)$.

(e) This graph is an inverted cosine curve with amplitude 8 and period 20π, so it is given by $f(x) = -8\cos(\frac{x}{10})$.

37. From a plot of the two functions, we can see that there are three obvious points of intersection in the domain $|x| \leq 2$. There is one at $(0, 0)$, another at about $(1.31, 2.29)$ and another at about $(-1.31, -2.29)$. Now we show that there are no points of intersection for these graphs, provided that $|x| > 2$.

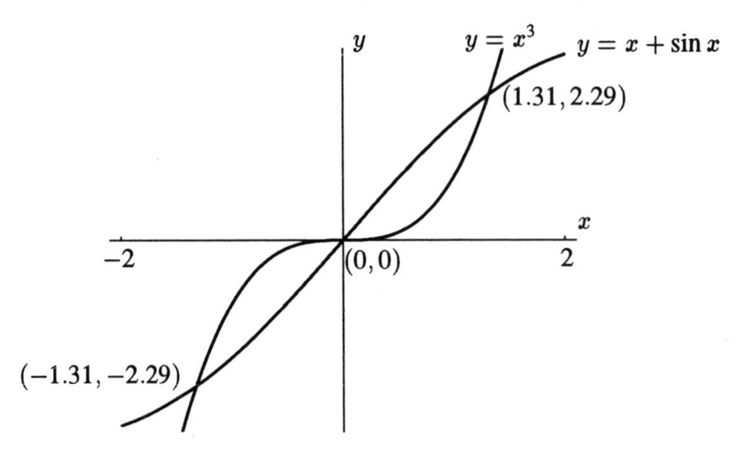

We know that $x + \sin x$ will never exceed $x + 1$, since $\sin x$ is never greater than 1. Now, for $x > 2$,

$$x + 1 < 2x < x^2 < x^3.$$

So,

$$x + \sin x \leq x + 1 < x^3.$$

Therefore, the curves cannot intersect when $x > 2$. A similar argument applies when $x < -2$.

39. (a) Beginning at time $t = 0$, the voltage will have oscillated through a complete cycle when $\cos(120\pi t) = \cos(2\pi)$, hence when $t = \frac{1}{60}$ second. The period is $\frac{1}{60}$ second.

(b) V_0 represents the amplitude of the oscillation.

(c)

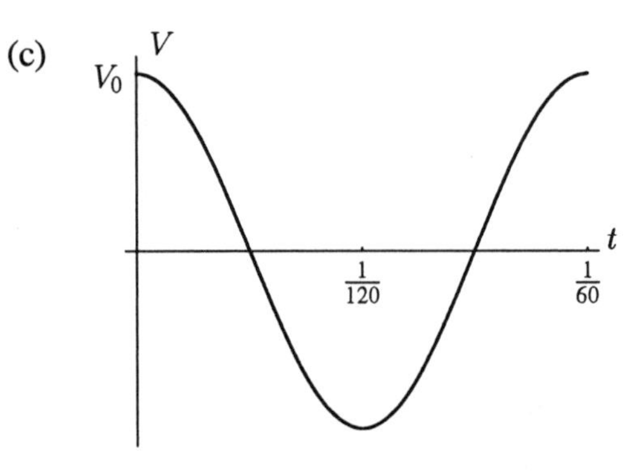

41. (a) Table of values of $g(x) = \arccos x$:

x	-1	-0.8	-0.6	-0.4	-0.2	0	0.2	0.4	0.6	0.8	1
$\arccos x$	3.14	2.50	2.21	1.98	1.77	1.57	1.37	1.16	0.93	0.64	0

(b)

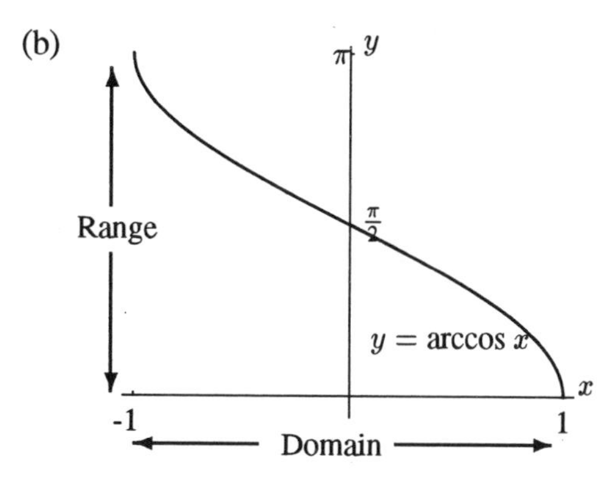

(c) Domain is $-1 \leq x \leq 1$. Range is $0 \leq y \leq \pi$.

(d) The domain of arccos and arcsin are the same, $-1 \leq x \leq 1$, since their inverses (sine and cosine) only take on values in this range.

(e) The domain of the original sine function was restricted to the the interval $[-\frac{\pi}{2}, \frac{\pi}{2}]$ to construct the arcsine function. Hence, the range of arcsine is also $[-\frac{\pi}{2}, \frac{\pi}{2}]$.

Now, if we restrict the domain of cosine in the same way, we obtain an arccosine curve which is not a function:
For example, for $x = 0$, $y = \arccos x$ will have two values, $-\frac{\pi}{2}$, and $\frac{\pi}{2}$. Also, it gives no values for $x < 0$, so it is not very useful.
The domain of cosine should instead be restricted to $[0, \pi]$, so that $y = \arccos x$ gives a unique y for each value of x.

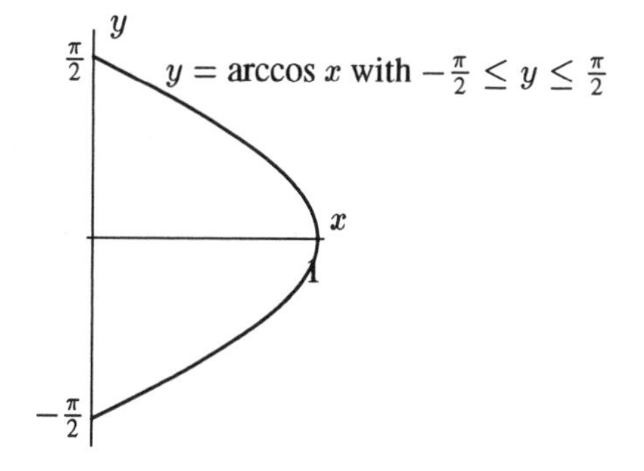

43. (a) Yes, they must intersect at $3.64 - \pi \approx 0.5$.
    (b) They also intersect at $3.64 + \pi \approx 6.78$.
    (c) $3.64 - 2\pi \approx -2.64$.

45. (a) The period is 2π.
    (b) The period of $\sin 3\alpha$ is $\frac{2}{3}\pi$. Period of $\cos\alpha$ is 2π.
    (c) The combined function repeats when each part repeats separately—although $3\sin\alpha$ repeats every $\frac{2}{3}\pi$, the combined function must 'wait' until $\cos\alpha$ repeats for it to return to its original value.

1.11 Solutions

1. (a) Degree ≥ 3, leading coefficient negative.
   (b) Degree ≥ 4, leading coefficient positive.
   (c) Degree ≥ 4, leading coefficient negative.
   (d) Degree ≥ 5, leading coefficient negative.
   (e) Degree ≥ 5, leading coefficient positive.

3.

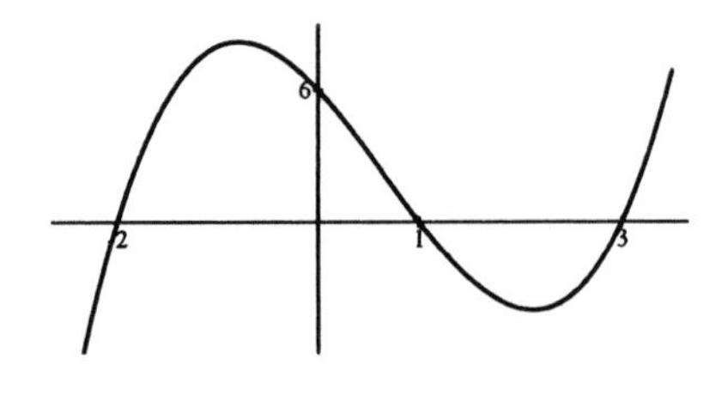

5.

7. To find vertical asymptote(s), look at the behavior of y as x approaches a value for which the denominator is 0.

$$2x + 2 = 0 \quad \text{when} \quad x = -1.$$

If we plug in values for x near -1, we will see that y goes to $+\infty$ as x approaches -1 from the right, but y goes to $-\infty$ as x approaches -1 from the left. Clearly, $x = -1$ is a vertical asymptote.

To find horizontal asymptote(s), look at the behavior of y as x goes to $+\infty$ and as x goes to $-\infty$. Note that as $x \to \pm\infty$, only the highest power of x matters, so that the 1 and the 2 become insignificant compared to x for large values of x. Thus,

$$y = \frac{1 - 4x}{2x + 2} \approx \frac{-4x}{2x} = -2$$

Clearly, $y = -2$ is a horizontal asymptote.

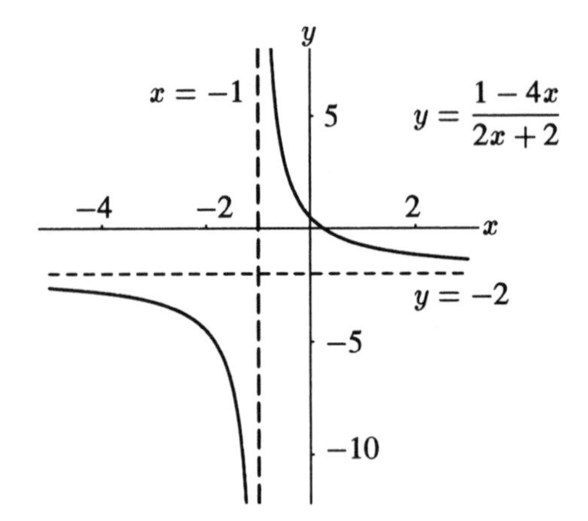

9. To find vertical asymptote(s), look at the behavior of y as x approaches a value for which the denominator is 0.

$$x^2 - 4 = 0 \quad \text{when} \quad x = \pm 2.$$

If we plug in values for x near -2 and near $+2$, we will see that

$y \longrightarrow +\infty$ as $x \longrightarrow +2$ from the right, but
$y \longrightarrow -\infty$ as $x \longrightarrow +2$ from the left, and
$y \longrightarrow -\infty$ as $x \longrightarrow -2$ from the right, but
$y \longrightarrow +\infty$ as $x \longrightarrow -2$ from the left.

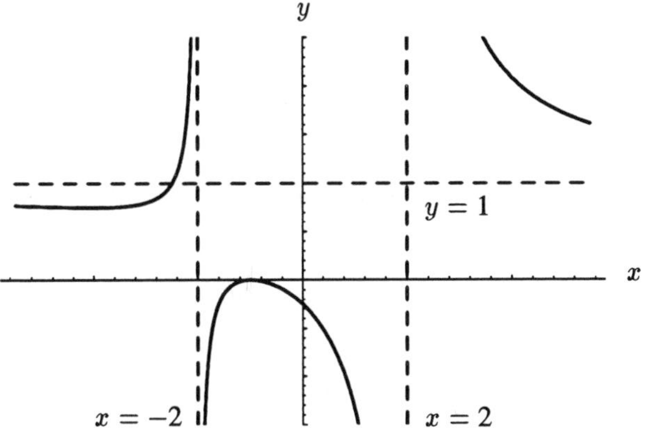

Clearly, $x = -2$ and $x = 2$ are vertical asymptotes.

To find horizontal asymptote(s), look at the behavior of y as x goes to $+\infty$ and as x goes to $-\infty$. Note that as $x \to \pm\infty$, only the highest power of x matters, so that the $2x$, the 1, and the -4 become insignificant compared to the x^2 terms for large values of x. Thus,

$$y = \frac{x^2 + 2x + 1}{x^2 - 4} \approx \frac{x^2}{x^2} = 1$$

Clearly, $y = 1$ is a horizontal asymptote.

11. (a) If (1,1) is on the graph, we know that

$$\begin{aligned} 1 &= a(1)^2 + b(1) + c \\ \text{and } 1 &= a + b + c. \end{aligned}$$

(b) If (1,1) is the vertex, then the axis of symmetry is $x = 1$, so

$$\begin{aligned} -\frac{b}{2a} &= 1 \\ \text{and so } a = -\frac{b}{2} & \quad \text{or} \quad b = -2a. \end{aligned}$$

But to be the vertex, (1,1) must also be on the graph, so we know that the result of (a) holds:

$$a + b + c = 1.$$

Substituting $b = -2a$, we get $-a + c = 1$, which we can rewrite as

$$a = c - 1 \text{ or } c = 1 + a.$$

(c) For (0,6) to be on the graph, we must have f(0)=6. But

$$f(0) = a(0^2) + b(0) + c = c$$

so $c = 6$.

(d) To satisfy all the conditions, we must first, from (c), have $c = 6$. From (b), $a = c - 1$ so $a = 5$. Also from (b), $b = -2a$, so $b = -10$. Thus the completed equation is

$$y = f(x) = 5x^2 - 10x + 6,$$

which satisfies all the given conditions.

13. (a) The object starts at $t = 0$, when $s = v_0(0) - \frac{1}{2}g(0)^2 = 0$. Thus it starts on the ground, with zero height.

(b) The object hits the ground when $s = 0$. This is satisfied at $t = 0$, before it has left the ground, and at some later time t that we must solve for.

$$\begin{aligned} s = 0 &= v_0 t - \frac{1}{2}gt^2 \\ 0 &= t(v_0 - \frac{1}{2}gt) \end{aligned}$$

Thus $s = 0$ when $t = 0$ and when $v_0 - \frac{1}{2}gt = 0$, i.e. when $t = \frac{2v_0}{g}$. $t = 0$ is the starting time, so it must hit the ground at time $t = \frac{2v_0}{g}$.

(c) The object reaches its maximum height halfway between when it is released and when it hits the ground, or at

$$t = \frac{1}{2}\left(\frac{2v_0}{g}\right) = \frac{v_0}{g}.$$

(d) Since we know the time at which the object reaches its maximum height, to find the height it actually reaches we just use the given formula, which tells us s at any given t. Plugging in $t = \frac{v_0}{g}$,

$$\begin{aligned} s &= v_0\left(\frac{v_0}{g}\right) - \frac{1}{2}g\left(\frac{v_0^2}{g^2}\right) = \frac{v_0^2}{g} - \frac{v_0^2}{2g} \\ &= \frac{2v_0^2 - v_0^2}{2g} = \frac{v_0^2}{2g}. \end{aligned}$$

15. (a) i. $f(x) = k(x+3)(x-1)(x-4) = k(x^3 - 2x^2 - 11x + 12)$, where $k < 0$. ($k \approx -\frac{1}{6}$ if the horizontal and vertical scales are equal; otherwise one can't tell how large k is.)

ii. $f(x) = kx(x+3)(x-4) = k(x^3 - x^2 - 12x)$, where $k < 0$. ($k \approx -\frac{2}{9}$ if the horizontal and vertical scales are equal; otherwise one can't tell how large k is.)

iii. $f(x) = k(x+2)(x-1)(x-3)(x-5) = k(x^4 - 7x^3 + 5x^2 + 31x - 30)$, where $k > 0$. ($k \approx \frac{1}{15}$ if the horizontal and vertical scales are equal; otherwise one can't tell how large k is.)

iv. $f(x) = k(x+2)(x-2)^2(x-5) = k(x^4 - 7x^3 + 6x^2 + 28x - 40)$, where $k < 0$. ($k \approx -\frac{1}{15}$ if the scales are equal; otherwise one can't tell how large k is.)

(b) i. This graph appears to be increasing for $-1.5 < x < 3$, decreasing for $x < -1.5$ and for $x > 3$.

ii. This graph appears to be increasing for $-1.5 < x < 2.5$, decreasing for $x < -1.5$ and for $x > 2.5$.

iii. This graph appears to be increasing for $-1 < x < 2$ and for $x > 4.3$, decreasing for $x < -1$ and for $2 < x < 4.3$.

iv. This graph is increasing for $x < -1$ and for $2 < x < 4$, decreasing for $-1 < x < 2$ and for $4 < x$.

17. $p(x) = 1$.

19. $p(x) = ax^2 - ax + 1$ for any a.

21. (a) $a(v) = \frac{1}{m}(\text{ENGINE} - \text{WIND}) = \frac{1}{m}(F_E - kv^2)$, where k is a positive constant.

    (b)

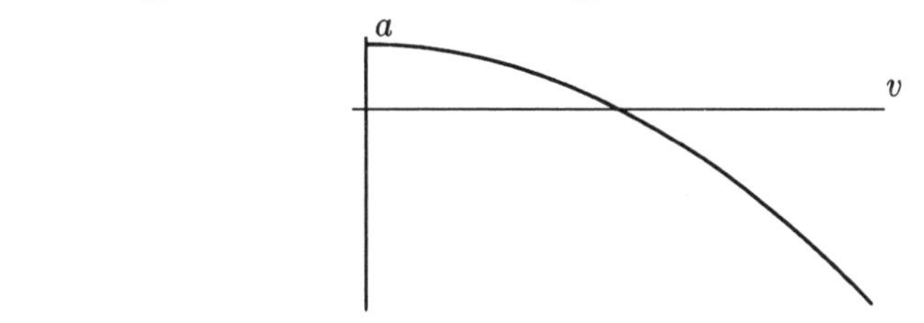

1.12 Solutions

1. (a) The root is between 0.3 and 0.4, at about 0.35.

   (b) The root is between 1.5 and 1.6, at about 1.55.

   (c) The root is between -1.8 and -1.9, at about -1.85.

3. The root is between -1.7 and -1.8, at about -1.75.

5. There is one root at $x = -1$ and another at about $x = 1.35$.

7. The root occurs at about 0.9, since the function changes sign between 0.8 and 1.

9. The root occurs between 0.6 and 0.7, at about 0.65.

11. Zoom in on graph: $t = \pm 0.824$. [Note: t must be in radians; one must zoom in two or three times.]

13. (a) Let $F(x) = \sin x - 2^{-x}$. Then $F(x) = 0$ will have a root where $f(x)$ and $g(x)$ cross. The first positive value of x for which the functions intersect is $x \approx 0.7$.

    (b) The functions intersect for $x \approx 0.4$.

15. (a) Since f is continuous, there must be one zero between $\theta = 1.4$ and $\theta = 1.6$, and another between $\theta = 1.6$ and $\theta = 1.8$. These are the only clear cases. We might also want to investigate the interval $0.6 \leq \theta \leq 0.8$ since $f(\theta)$ takes on values close to zero on at least part of this interval. Now, $\theta = 0.7$ is in this interval, and $f(0.7) = -0.01 < 0$, so f changes sign twice between $\theta = 0.6$ and $\theta = 0.8$ and hence has two zeros on this interval (assuming f is not *really* wiggly here, which it's not). There are a total of 4 zeros.

(b) As an example, we find the zero of f between $\theta = 0.6$ and $\theta = 0.7$. $f(0.65)$ is positive; $f(0.66)$ is negative. So this zero is contained in $[0.65, 0.66]$. The other zeros are contained in the intervals $[0.72, 0.73]$, $[1.43, 1.44]$, and $[1.7, 1.71]$.

(c) You've found all the zeros. A picture will confirm this.

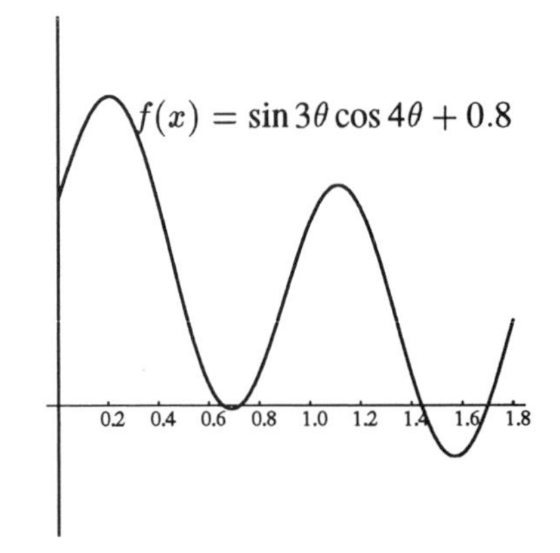

17.

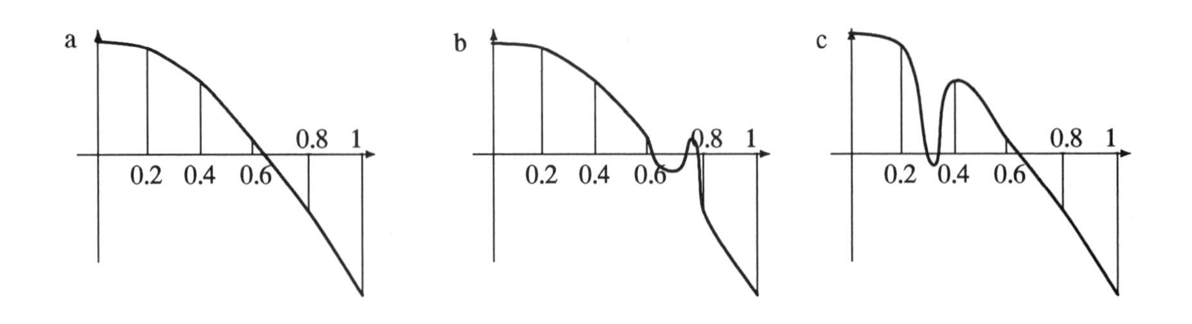

19. Starting with $x = 0$, and repeatedly taking the cosine, we get the numbers below. Continuing until the first three decimal places remain fixed under iteration, we have this list:

0	0.735069
1	0.7401473
0.5403023	0.7356047
0.8575532	0.7414251
0.6542898	0.7375069
0.7934804	0.7401473
0.7013688	0.7383692
0.7639597	0.7395672
0.7221024	0.7387603
0.7504178	0.7393039
0.7314043	0.7389378
0.7442374	etc.

and this diagram:

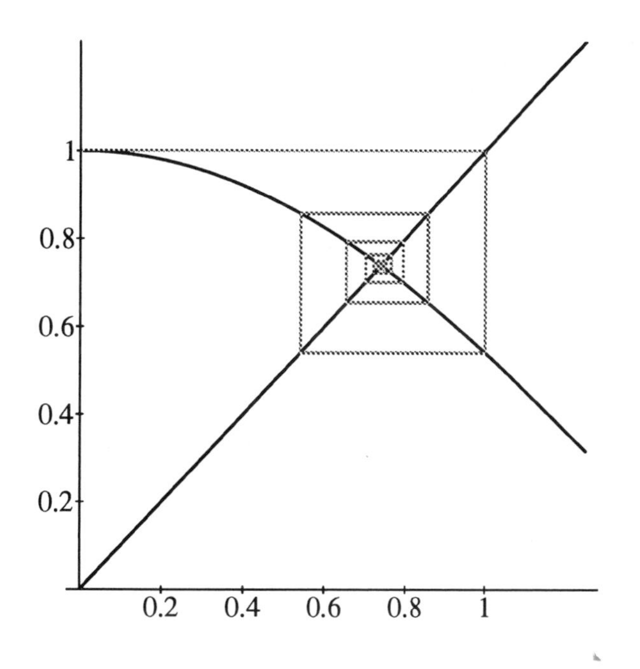

1.13 Answers to Miscellaneous Exercises for Chapter 1

1.

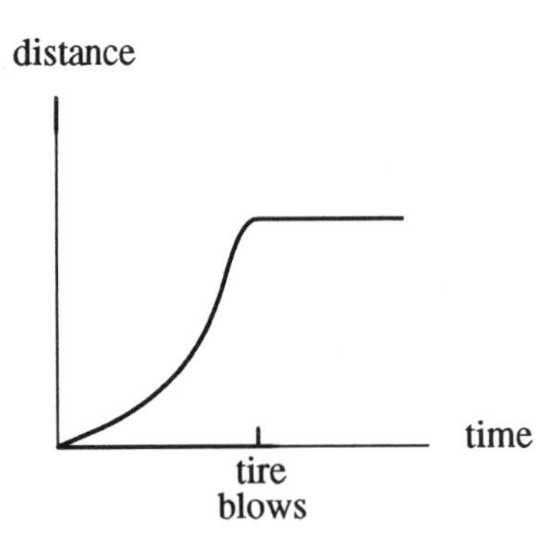

3. There are three peaks in the curve, each one of which occurs around one of the three meal times. The first peak is about 7 o'clock in the morning, the next one right before noon and the third one around 6 o'clock in the evening. From the relative magnitudes of the peaks, more gas is used to cook dinner than lunch, and more gas is used to cook lunch than breakfast.

5. (a) It was decreasing from March 2 to March 5 and increasing from March 5 to March 9.

   (b) From March 5 to 8, the average temperature increased, but the rate of increase went down, from 12° between March 5 and 6 to 4° between March 6 and 7 to 2° between March 7 and 8.

   From March 7 to 9, the average temperature increased, and the rate of increase went up, from 2° between March 7 and 8 to 9° between March 8 and 9.

7. We will let

$$\begin{aligned} T &= \text{amount of fuel for take-off} \\ L &= \text{amount of fuel for landing} \\ P &= \text{amount of fuel per mile in the air} \\ m &= \text{the length of the trip in miles} \end{aligned}$$

   Then Q, the total amount of fuel needed, is given by

$$Q(m) = T + L + Pm$$

9. (a) $f(t+1) = (t+1)^2 + 1 = t^2 + 2t + 1 + 1 = t^2 + 2t + 2$

   (b) $f(t^2+1) = (t^2+1)^2 + 1 = t^4 + 2t^2 + 1 + 1 = t^4 + 2t^2 + 2$

   (c) $f(2) = 2^2 + 1 = 5$

   (d) $2f(t) = 2(t^2+1) = 2t^2 + 2$

   (e) $(f(t))^2 + 1 = (t^2+1)^2 + 1 = t^4 + 2t^2 + 1 + 1 = t^4 + 2t^2 + 2$

11. (a) $f(n) + g(n) = (3n^2 - 2) + (n + 1) = 3n^2 + n - 1$

(b) $f(n)g(n) = (3n^2 - 2)(n + 1) = 3n^3 + 3n^2 - 2n - 2$

(c) the domain of $\frac{f(n)}{g(n)}$ is defined everywhere where $g(n) \neq 0$, i.e. for all $n \neq -1$.

(d) $f(g(n)) = 3(n + 1)^2 - 2 = 3n^2 + 6n + 1$

(e) $g(f(n)) = (3n^2 - 2) + 1 = 3n^2 - 1$

13. One possible graph is given below.

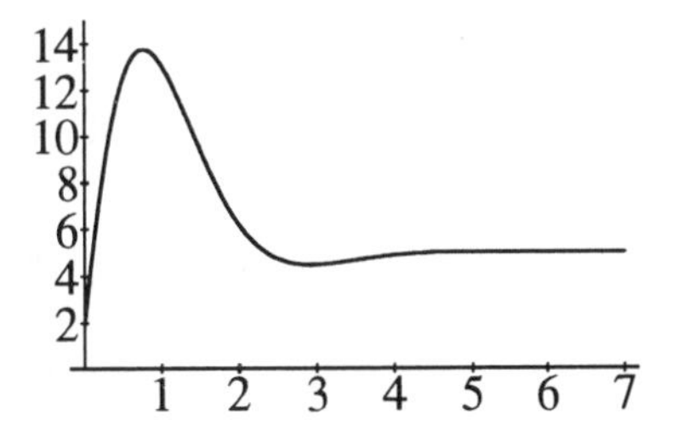

15. (a) $f(x)$ assumes its maximum value, 87.04, at $x = -0.5$.

    (b) $f(x)$ increases fastest between -1 and -0.5, where it increases by 78.9. $f(x)$ decreases fastest between -2.5 and -2, where it decreases by 61.92.

    (c) Since $f(x)$ changes sign between $x = 1$ and $x = 1.5$, it appears likely that $f(x) = 0$ somewhere between these two values. For the same reason, it is zero somewhere between $x = 2.5$ and $x = 3$.

    (d) You don't have to look far: The only place where the function is increasing and the rate of increase is increasing (in the data we're given) is the increase from the rate between -1.5 and -1 to the rate between -1 and -0.5. The same thing happens between 2 and 3. There are also places where the function is decreasing but the rate of decrease is becoming smaller (so the rate of increase is becoming a larger but still negative number), but that's not what the question asked for.

17. We use the equation $B = Pe^{rt}$. We want to have a balance of $B = \$20{,}000$ in $t = 6$ years, with an annual interest rate of 10%.

$$\begin{aligned} 20{,}000 &= Pe^{(0.1)6} \\ P &= 20{,}000e^{-0.6} \\ &\approx \$10{,}976.23. \end{aligned}$$

19. $e^{0.08} = 1.0833$ so the effective annual yield$= 8.33\%$.

21. If C is the amount of Carbon-14 originally present in the skull, then after 5730 years, $\frac{1}{2}C$ is left. Hence,

$$\frac{1}{2}C = Ce^{-k \cdot 5730}$$

$$\begin{aligned}
\frac{1}{2} &= e^{-k\cdot 5730} \\
\ln\frac{1}{2} &= -k\cdot 5730 \\
k &= \frac{-\ln\frac{1}{2}}{5730} = \frac{\ln 2}{5730}.
\end{aligned}$$

The question asks how long it took for the skull to lose 80% of its Carbon-14. In other words, when will the remaining Carbon-14= $\frac{1}{5}C$?

$$\begin{aligned}
\text{Remaining Carbon-14} &= \frac{1}{5}C = Ce^{-\frac{\ln 2}{5730}\cdot t} \\
\frac{1}{5} &= e^{-\frac{\ln 2}{5730}\cdot t} \\
\ln\frac{1}{5} &= -\frac{\ln 2}{5730}\cdot t.
\end{aligned}$$

Since $\ln\frac{1}{5} = -\ln 5$, we have

$$t = \frac{\ln 5}{\ln 2}\cdot 5730t \approx 13{,}300 \text{ years},$$

the approximate age of the skull.

23. (a) After one year, prices in Argentina were 450% higher, so they were multiplied by $1+4.5 = 5.5$. If d is the daily rate of inflation, then after one day prices will be multiplied by $(1+d)$ and after one year they will be multiplied by $(1+d)^{365}$. Thus to find d, we solve $(1+d)^{365} = 5.5$ to get $1+d = (5.5)^{\frac{1}{365}} \approx 1.00468$ so $d = 0.00468 = 0.468\%$ daily inflation.

(b) Since $(1.00468)^{150} \approx 2$, the doubling time was about five months.

25. (a)

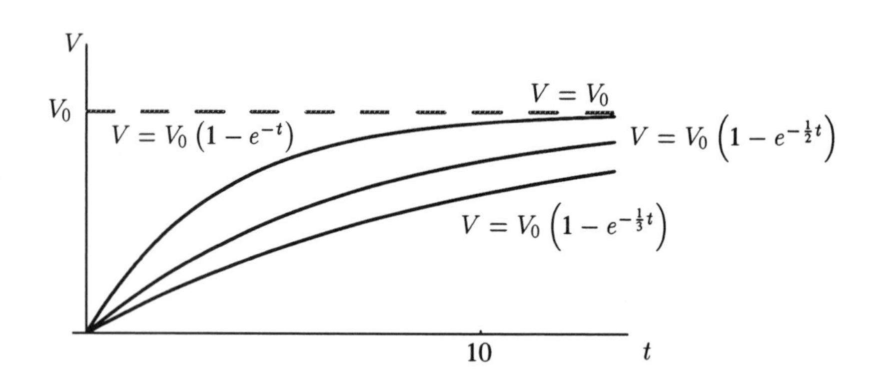

(b) As is plain from (a), increasing the value of t_0 stretches the curve to the right. In other words, a raindrop falling for t_0 large does not speed up as quickly as a raindrop falling for t_0 small.

27. (a) $y = -kx(x+5) = -k(x^2+5x)$, where $k > 0$ is any constant.
    (b) $y = k(x+2)(x+1)(x-1) = k(x^3+2x^2-x-2)$, where $k > 0$ is any constant.
    (c) $x = ky(y-4) = k(y^2-4y)$, where $k > 0$ is any constant.

29. (a) There are at least 3 roots in the interval: one between $x = 1$ and $x = 2$, one between $x = 2$ and $x = 3$, and one between $x = 3$ and $x = 4$.
    (b) The picture of the graph to the right shows that there are a total of five roots: there are *three*, not just one, between $x = 3$ and $x = 4$.

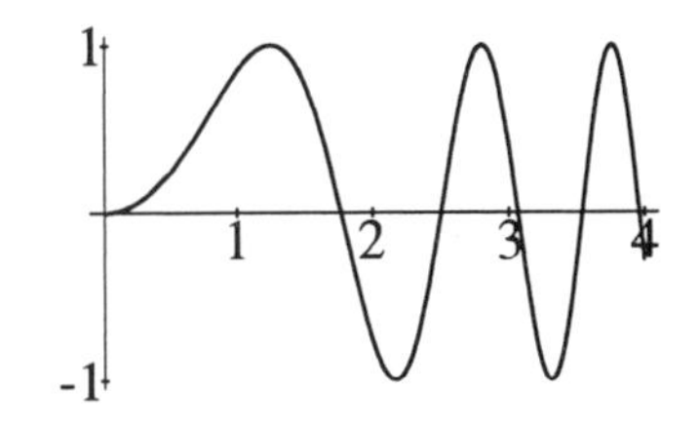

    (c) In order, the roots are approximately 1.8, 2.5, 3.1, 3.5, and 4.0.
    (d) $\sin x = 0$ only when x is an integer multiple of π, so $\sin(x^2) = 0$ only when t^2 is a multiple of π, say $k\pi$ where k is an integer. Thus the positive roots of $\sin(t^2)$ are $\sqrt{\pi}$, $\sqrt{2\pi}$, $\sqrt{3\pi}$, etc. The smallest positive root is therefore $\sqrt{\pi}$.
    (e) $\sqrt{\pi} \approx 1.8$, $\sqrt{2\pi} \approx 2.5$, $\sqrt{3\pi} \approx 3.1$, $\sqrt{4\pi} \approx 3.5$, $\sqrt{5\pi} \approx 4.0$.

31. (a)

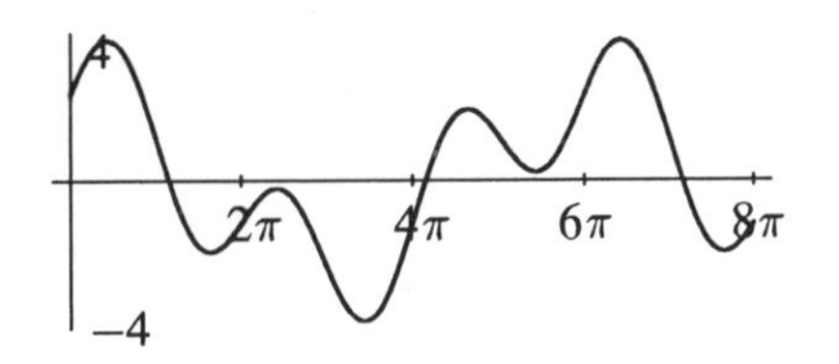

    From the graph, the period appears to be 6π.
    (b) The period of $\sin\alpha$ is 2π. The period of $\cos\frac{\alpha}{3}$ is 6π.
    (c) Since $\sin\alpha$ repeats itself after every 2π, it also repeats itself after every 6π, since it goes through three cycles and comes back to where it started. The function we graphed in (a) then was the sum of two functions with period 6π and thus is periodic with period 6π, as we observed. (Multiplying by constant factors doesn't affect the periodicity.)

33. (a) goes with H, (b) goes with E, (c) goes with G, (d) goes with A, (e) goes with C, (f) goes with F, (g) goes with B, (h) goes with D, and (i) goes with I.

35. (a) i. The inverse function is

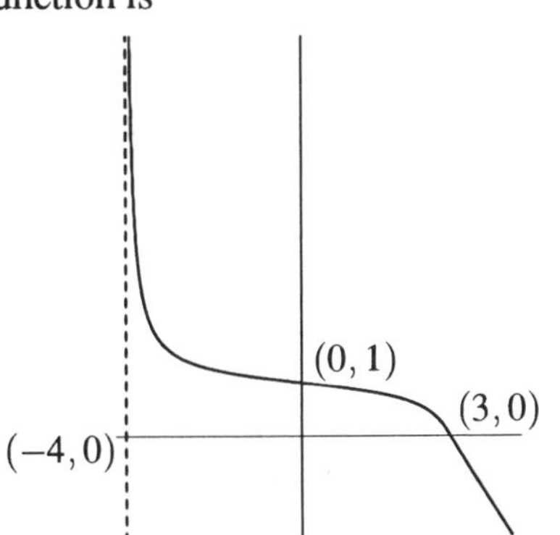

ii. The reciprocal function is

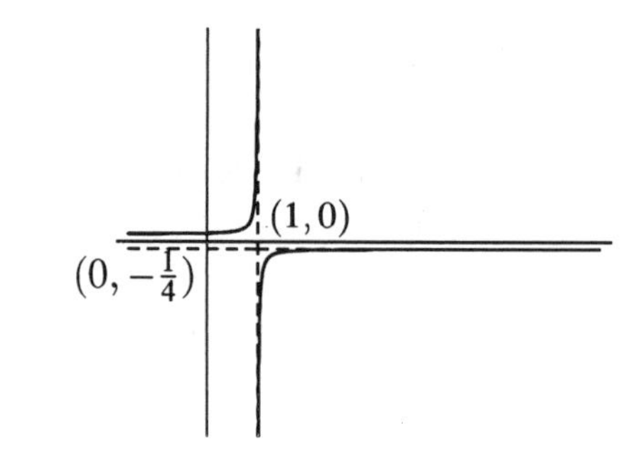

(b) The horizontal asymptote at $y = -4$ in the original function becomes a vertical asymptote at $x = -4$ in the inverse function.

37. (a) This moves the graph one unit to the left.

(b) A non-constant polynomial tends toward $+\infty$ or $-\infty$ as $x \to \infty$. This polynomial p does not. Therefore p must be a constant function, i.e. its graph is a horizontal line.

39. (a) $r(p) = kp(A - p)$, where $k > 0$ is a constant.

(b) $p = A/2$.

41. (a) The rate of change of the fish population is the difference between the rate they are reproducing, and the rate they are being caught. We are given that they are reproducing at a rate of 5% of P where P is their population, and being caught at a constant rate Y. Thus the net rate of change R (in fish per year) is given by

$$R = (0.05)P - Y$$

(b)

43. (a) $S(0) = 12$ since the days are always 12 hours long at the equator.

(b) Since $S(0) = 12$ from part (a) and the formula gives $S(0) = a$, we have $a = 12$. Since $S(x)$ must be continuous at $x = x_0$, and the formula gives $S(x_0) = a + b\arcsin(1) = 12 + b\left(\frac{\pi}{2}\right)$ and also $S(x_0) = 24$, we must have $12 + b\left(\frac{\pi}{2}\right) = 24$ so $b\left(\frac{\pi}{2}\right) = 12$ and $b = \frac{24}{\pi} \approx 7.64$.

(c) $S(32°13') \approx 14.12$ and $S(46°4') \approx 15.58$.

(d) As the graph to the right shows, $S(x)$ is not smooth. This is because above the Arctic Circle, there cannot be more than 24 hours of sunlight in a day.

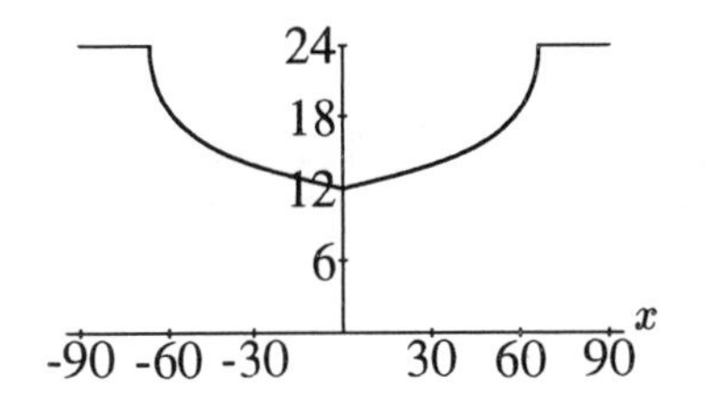

Chapter 2

2.1 Solutions

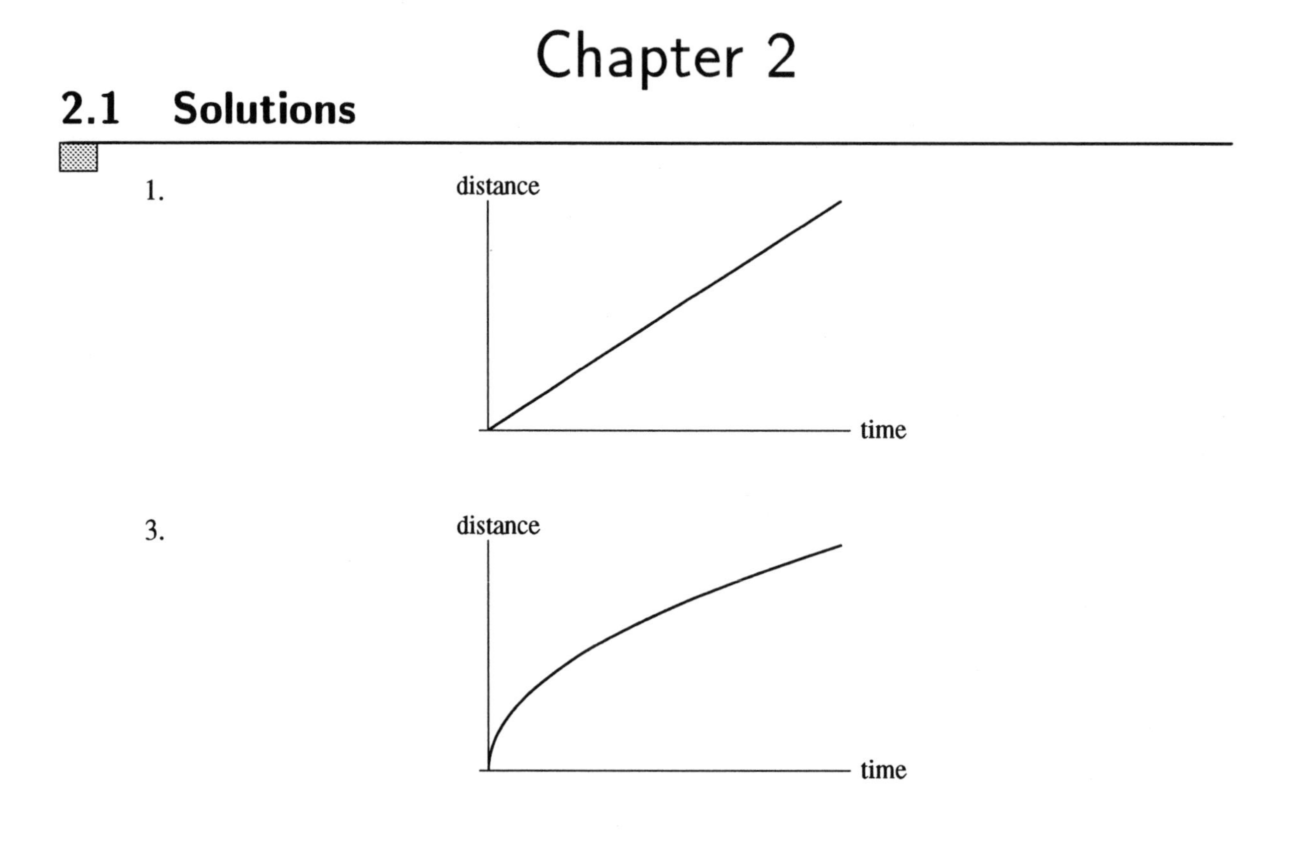

5.

$$\begin{pmatrix}\text{Average velocity}\\ 0 < t < 0.2\end{pmatrix} = \frac{s(0.2) - s(0)}{0.2 - 0} = \frac{0.5}{0.2} = 2.5 \text{ ft/sec.}$$

$$\begin{pmatrix}\text{Average velocity}\\ 0.2 < t < 0.4\end{pmatrix} = \frac{s(0.4) - s(0.2)}{0.4 - 0.2} = \frac{1.3}{0.2} = 6.5 \text{ ft/sec.}$$

A reasonable estimate of the velocity at $t = 0.2$ is the average: $\frac{1}{2}(6.5 + 2.5) = 4.5$ ft/sec.

7.

slope	point
-2	F
-1	C
0	B
$\frac{1}{2}$	E
1	D
3	A

9. $0 <$ slope at C $<$ slope at B $<$ slope of AB $< 1 <$ slope at A.

2.2 Solutions

1. (a)

x	1	1.5	2	2.5	3
$\log x$	0	0.18	0.30	0.40	0.48

(b) The average rate of change of $f(x) = \log x$ between $x = 1$ and $x = 3$ is

$$\frac{f(3)-f(1)}{3-1} = \frac{\log 3 - \log 1}{3-1} \approx \frac{0.48-0}{2} = 0.24$$

(c) First we find the average rates of change of $f(x) = \log x$ between $x = 1.5$ and $x = 2$, and between $x = 2$ and $x = 2.5$.

$$\begin{aligned} \frac{\log 2 - \log 1.5}{2-1.5} &= \frac{0.30-0.18}{0.5} \approx 0.24 \\ \frac{\log 2.5 - \log 2}{2.5-2} &= \frac{0.40-0.30}{0.5} \approx 0.20 \end{aligned}$$

Now we approximate the instantaneous rate of change at $x = 2$ by finding the average of the above rates, i.e.

$$\begin{pmatrix}\text{the instantaneous rate of change} \\ \text{of } f(x) = \log x \text{ at } x = 2\end{pmatrix} \approx \frac{0.24+0.20}{2} = 0.22.$$

3.

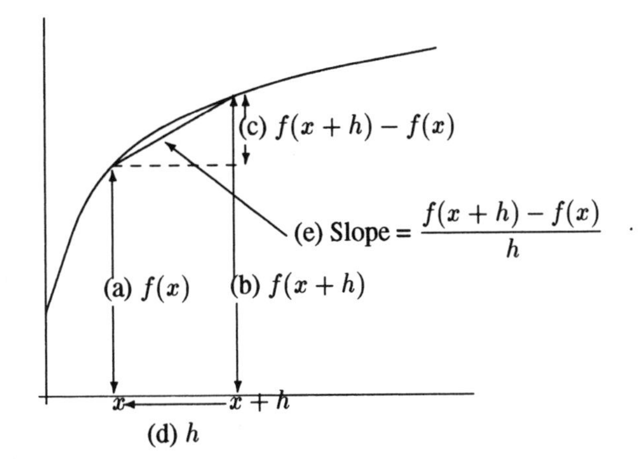

5.

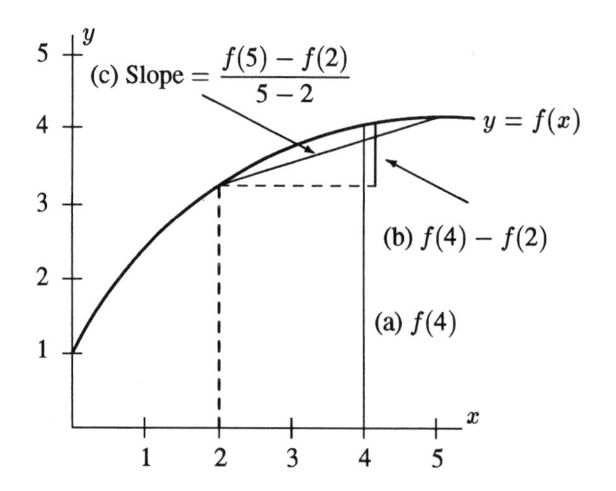

7. (a) $\frac{f(4)}{4}$ is the slope of the line connecting (0,0) to $(4, f(4))$. (See the figure to the right.)

   (b) It is clear from the picture for part (a) that the slope of $f(3)/3$ is greater than that of $f(4)/4$.

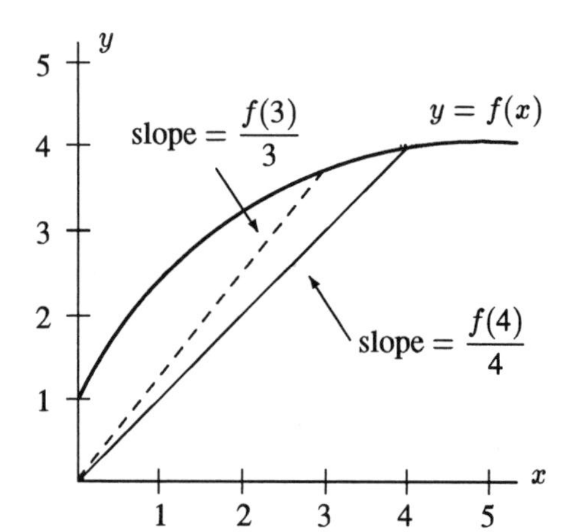

9. (a) C and D.

   (b) B and C.

   (c) A and B, and C and D.

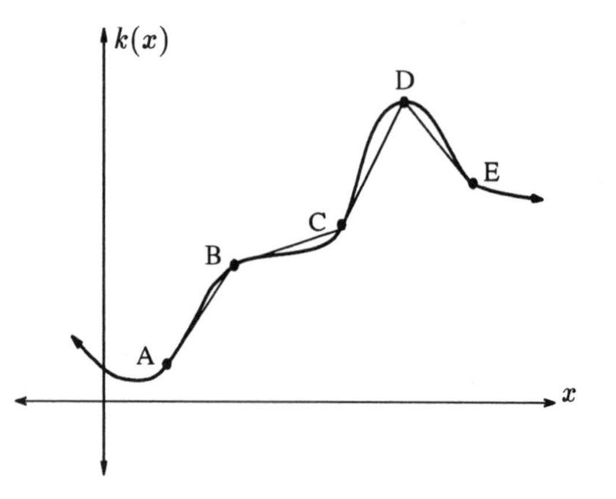

11.

x	2.998	2.999	3.000	3.001	3.002
x^3+4x	38.938	38.969	39.000	39.031	39.062

We see that each x increase of 0.001 leads to an increase in $f(x)$ by about 0.031, so $f'(3) \approx \frac{0.031}{0.001} = 31$.

13. $f'(1) = \lim_{h \to 0} \dfrac{\log(1+h) - \log 1}{h} = \lim_{h \to 0} \dfrac{\log(1+h)}{h}$

Evaluating $\frac{\log(1+h)}{h}$ for $h = 0.01, 0.001$, and 0.0001, we get $0.43214, 0.43408, 0.43427$, so $f'(1) \approx 0.43427$. The corresponding secant lines are getting steeper, because the graph of $\log x$ is concave down. We thus expect the limit to be more than 0.43427 . If we consider negative values of h, the estimates are too large. We can also see this from the graph below:

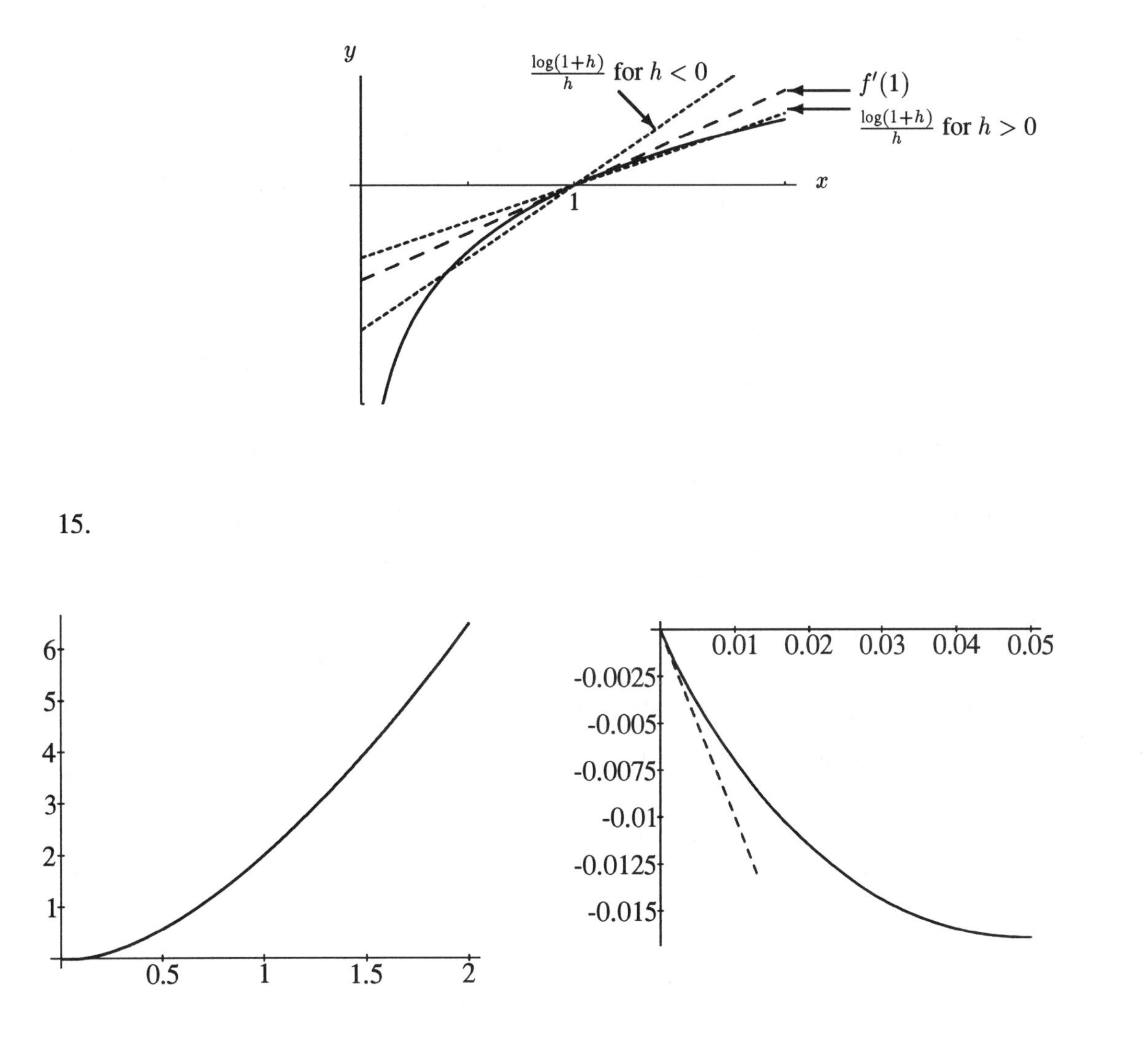

15.

Notice that we can't get all the information we want just from the graph of f for $0 \le x \le 2$, shown above to the left. Looking at this graph, it looks as if the slope at $x = 0$ is 0. But if we zoom in on the graph near $x = 0$, we get the graph of f for $0 \le x \le 0.05$, shown above on the upper right. We see that f does dip down quite a bit between $x = 0$ and $x \approx 0.11$. In

fact, it now looks like $f'(0)$ is around -1. Note that since $f(x)$ is undefined for $x < 0$, this derivative only makes sense as we approach zero from the right.
We zoom in on the graph of f near $x = 1$ to get a more accurate picture to estimate $f'(1)$. A graph of f for $0.7 \le x \le 1.3$ is shown below. [Keep in mind that the axes shown in this graph don't cross at the origin!] Here we see that $f'(1)$ looks to be around 3.5.

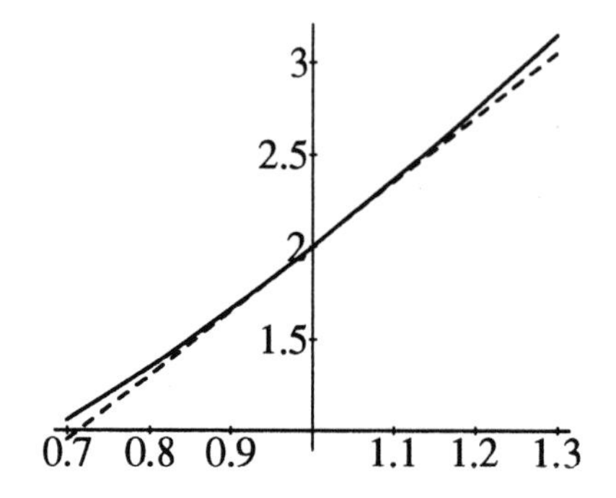

17.

$$f'(1) = \lim_{h \to 0} \frac{f(1+h) - f(1)}{h} = \lim_{h \to 0} \frac{\ln(\cos(1+h)) - \ln(\cos 1)}{h}$$

For $h = 0.001$, the difference quotient $= -1.55912$; for $h = 0.0001$, the difference quotient $= -1.55758$.
The instantaneous rate of change therefore appears to be about -1.558 at $x = 1$.
At $x = \frac{\pi}{4}$, if we try $h = 0.0001$, then

$$\text{difference quotient} = \frac{\ln[\cos(\frac{\pi}{4} + 0.0001)] - \ln(\cos \frac{\pi}{4})}{0.0001} \approx -1.0001.$$

The instantaneous rate of change appears to be about -1 at $x = \frac{\pi}{4}$.

19. A graph of one such f is shown below:

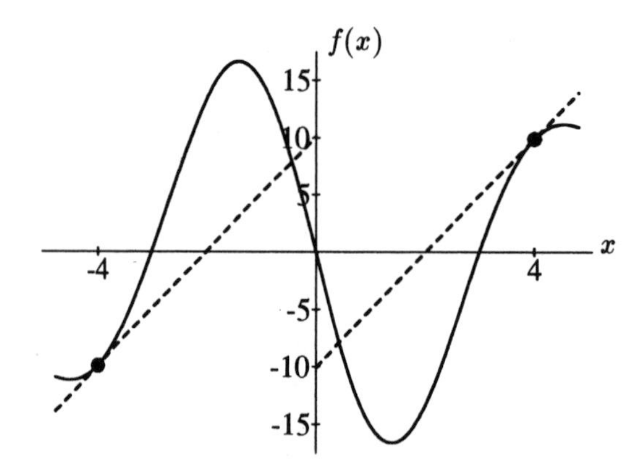

From the graph above, we can see that since f is symmetric about the origin, the tangent line at $x = -4$ is just the tangent line at $x = 4$ flipped about the origin, and so these two tangent lines have the same slope. Thus, $g'(-4) = 5$.

2.3 Solutions

1. We can see from the graph that we have the linear equation, $y = -2x + 2$. Thus the derivative is -2.

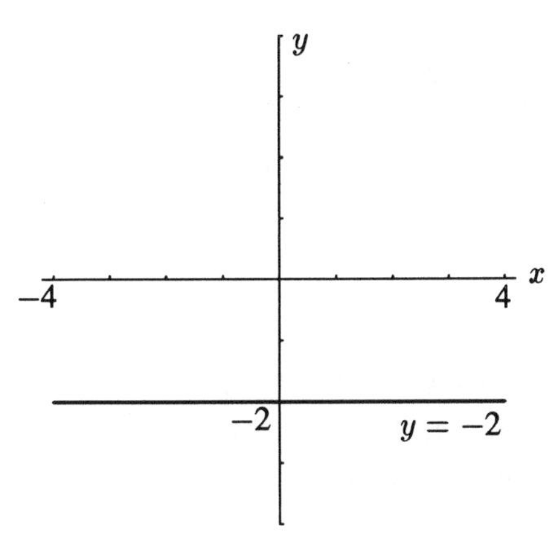

3. Note: This graph should start at $x = -4$, $y = -1$, intersect the x-axis around -2.5, the y-axis around 1, then the x-axis around 1.5, and end around $(4, -1)$.

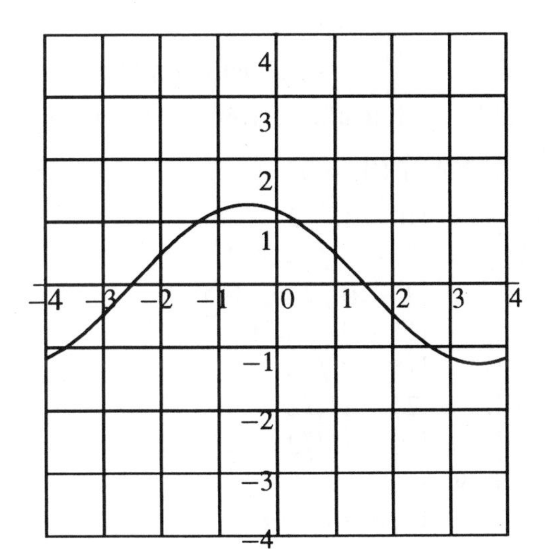

5.

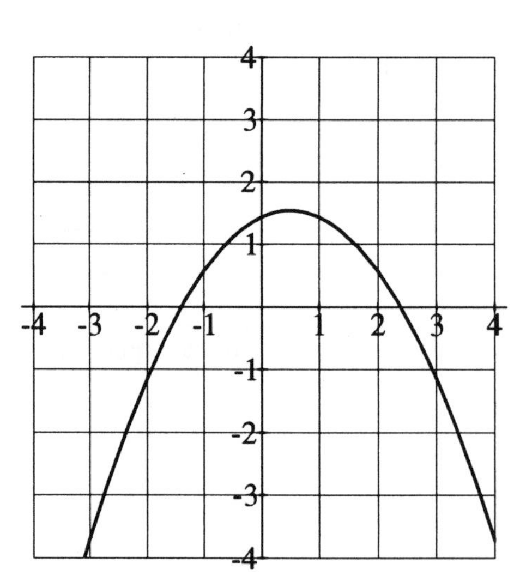

7. A possible graph for g' is shown in the figure to the right. On intervals where $g(t)$ is constant (i.e. flat), $g' = 0$. The small interval where g decreases very quickly corresponds to the downwards spike on the graph of g'. At the point where g' hits the bottom of its spike, the rate of decrease of $g(t)$ is greatest.

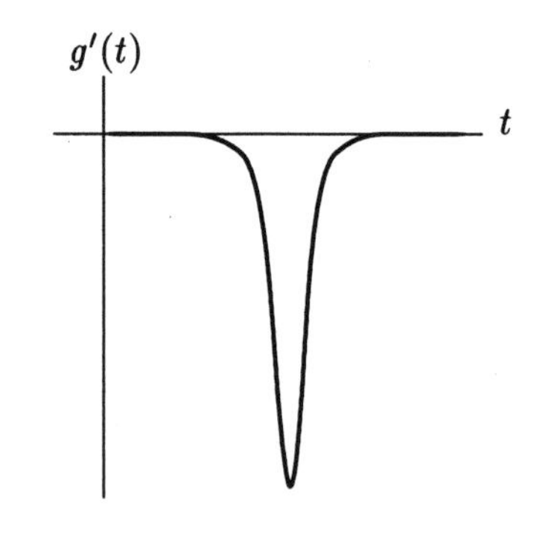

9. We know that $f'(x) \approx \dfrac{f(x+h) - f(x)}{h}$. For this problem, we'll take the average of the values obtained for $h = 1$ and $h = -1$; that's the average of $f(x+1) - f(x)$ and $f(x) - f(x-1) = \dfrac{f(x+1) - f(x-1)}{2}$. Thus, $f'(0) \approx f(1) - f(0) = 13 - 18 = -5$.
$f'(1) \approx [f(2) - f(0)]/2 = [10 - 18]/2 = -4$.
$f'(2) \approx [f(3) - f(1)]/2 = [9 - 13]/2 = -2$.
$f'(3) \approx [f(4) - f(2)]/2 = [9 - 10]/2 = -0.5$.
$f'(4) \approx [f(5) - f(3)]/2 = [11 - 9]/2 = 1$.
$f'(5) \approx [f(6) - f(4)]/2 = [15 - 9]/2 = 3$.
$f'(6) \approx [f(7) - f(5)]/2 = [21 - 11]/2 = 5$.
$f'(7) \approx [f(8) - f(6)]/2 = [30 - 15]/2 = 7.5$.
$f'(8) \approx f(8) - f(7) = 30 - 21 = 9$.
The rate of change of $f(x)$ is positive for $4 \leq x \leq 8$, negative for $0 \leq x \leq 3$. The rate of change is greatest at about $x = 8$.

11.

t	$g(t)$
0.998	1.994
0.999	1.997
1.000	2.000
1.001	2.003
1.002	2.006

t	$g(t)$
1.998	5.990
1.999	5.995
2.000	6.000
2.001	6.005
2.002	6.010

t	$g(t)$
2.998	11.986
2.999	11.993
3.000	12.000
3.001	12.007
3.002	12.014

Near 1, the values of $g(t)$ increase by 0.003 for each t increase of 0.001, so the derivative appears to be 3. Near 2, the increase is 0.005 for each step of 0.001, so the derivative appears to be 5. Near 3, the increase is 0.007 for each step of 0.001, so the derivative appears to be 7. These values seem to be increasing in a linear manner as we go to higher and higher t values, so we'll guess that the formula is $2t + 1$.

13.

(a) Since $f'(x) > 0$ for $x < -1$, $f(x)$ is increasing on this interval.

(b) Since $f'(x) < 0$ for $x > -1$, $f(x)$ is decreasing on this interval.

(c) Since $f'(x) = 0$ at $x = -1$, the tangent to $f(x)$ is horizontal at $x = -1$.

One of many possible shapes of $y = f(x)$ is shown to the right.

15. Note f' and g' are periodic, with the same period as f and g respectively. Since g oscillates between 1 and -1 more quickly, its values change faster and therefore we would expect its derivative to reach larger values (both positive and negative); g' has a larger amplitude than f.

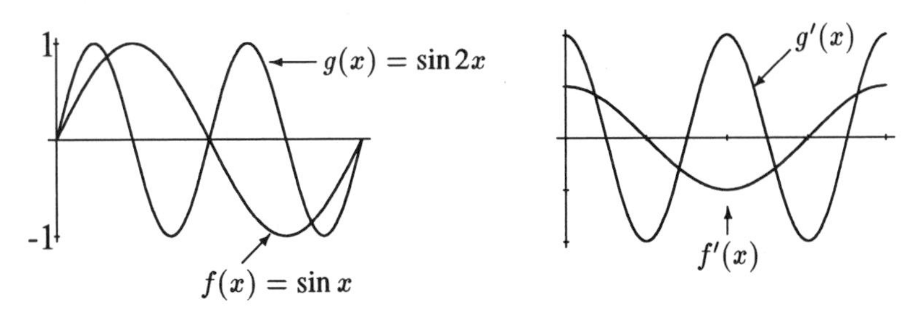

17.

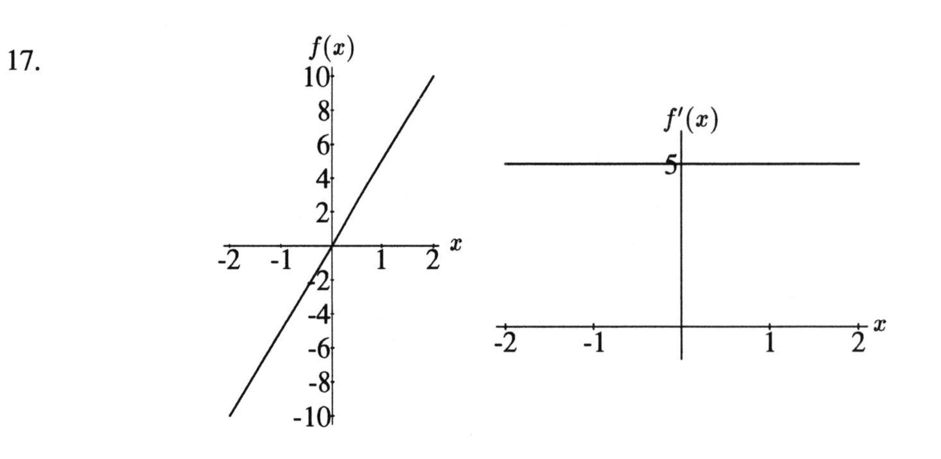

19. The graph of $f(x)$ and its derivative look the same, like the graph below:

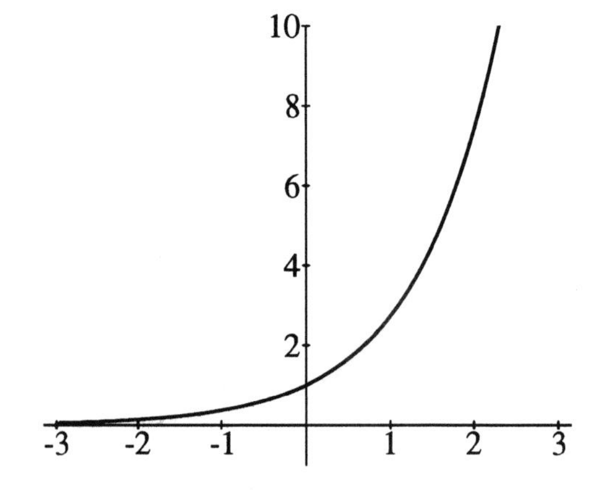

21.

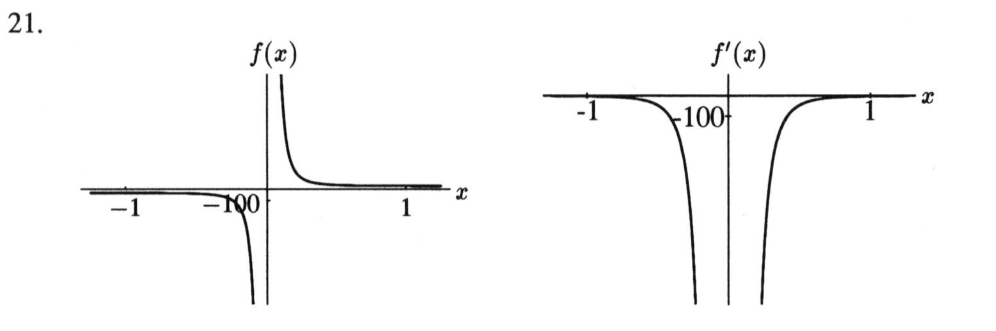

23.

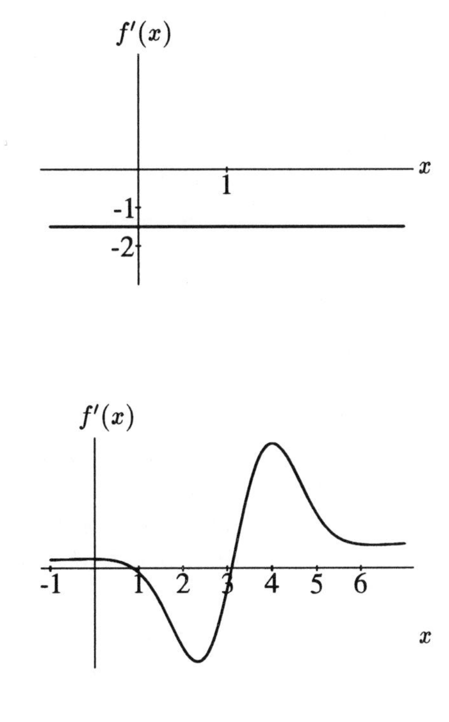

25.

27.

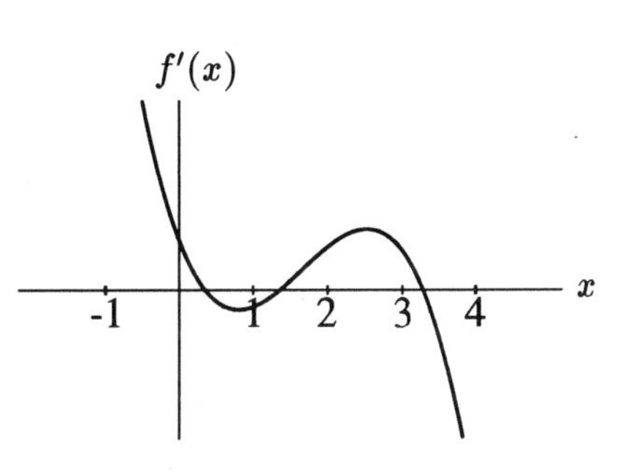

29. $f(x)$ even $\Rightarrow$ graphically, $f(x)$ symmetric about the y-axis $\Rightarrow$ the slopes of $f(x)$ and $f(-x)$ are opposite in sign $\Rightarrow$ $f'(x)$ is odd.

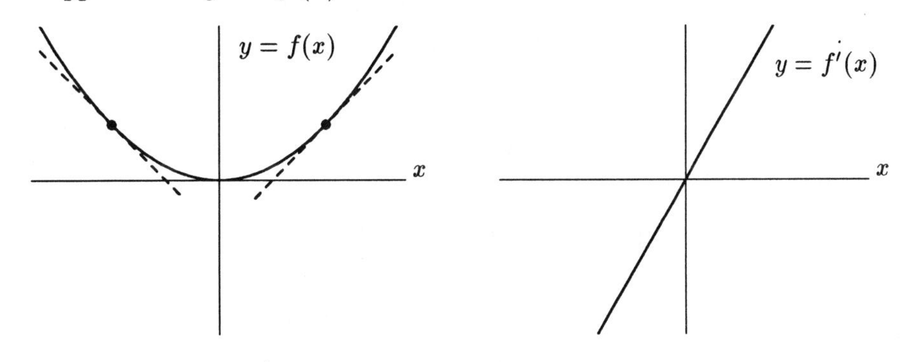

31. (a)

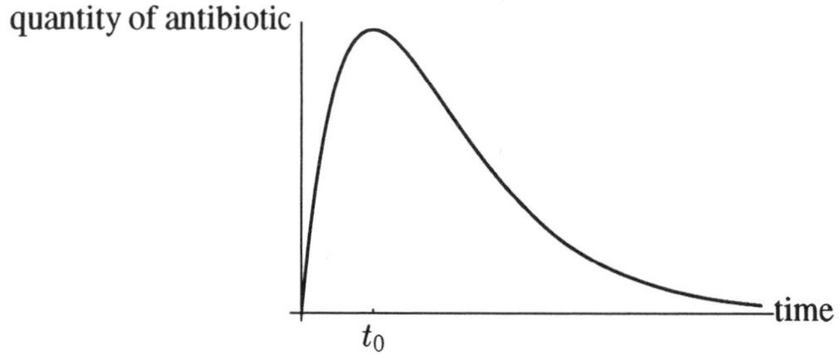

(b) Just after the injection there is a reservoir of antibiotic in the muscle which begins to diffuse into the blood. As the antibiotic diffuses into the bloodstream, it begins to leave the blood either through normal metabolic action or absorption by some other organ. For a while the diffusion rate into the blood is faster than the loss rate and the concentration rises, but as the reservoir in the muscle is drawn down, the diffusion rate into the blood decreases and eventually becomes less than the loss rate. After that, the concentration in the blood goes down and eventually all the antibiotic disappears. So we can expect the graph of concentration against time to look like the above graph.

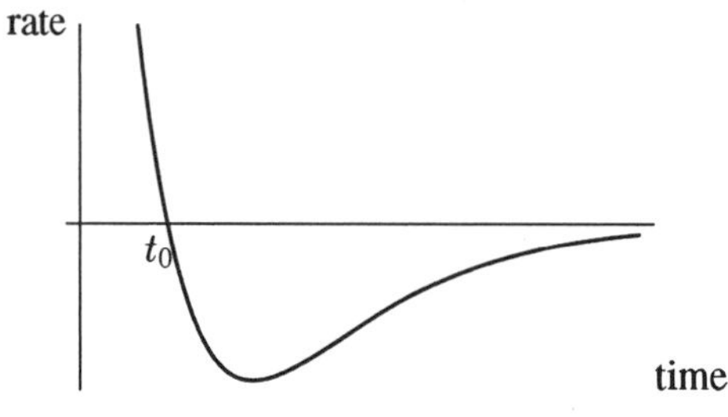

2.4 Solutions

1. (a) Velocity is zero at points A, C, F, and H.

   (b) These are points where the acceleration is zero, and hence where the particle switches from speeding up to slowing down or vice versa.

3. $\frac{dB}{dt}$ represents how fast your money is growing, in units of dollars/year. If the interest is compounded instantaneously, it is just equal to the amount of money you have at time t times the rate of interest, 7%. Otherwise, it depends on both the rate of interest and the number of times per year it is compounded.

5. Note that we are considering the average temperature of the yam, since its temperature is different at different points inside it.

   (a) It is positive, indicating that the temperature of the yam is increasing as it sits in the oven longer.

   (b) $f'(20) = 2$ means that at time $t = 20$ minutes, the temperature T is increasing by 2 degrees Fahrenheit for each additional minute in the oven.

7. $g'(55) = -0.54$ means that at 55 miles per hour the fuel efficiency (in miles per gallon, or mpg) of a car decreases as the velocity increases at a rate of approximately one half mpg for an increase of one mph.

9. (a)

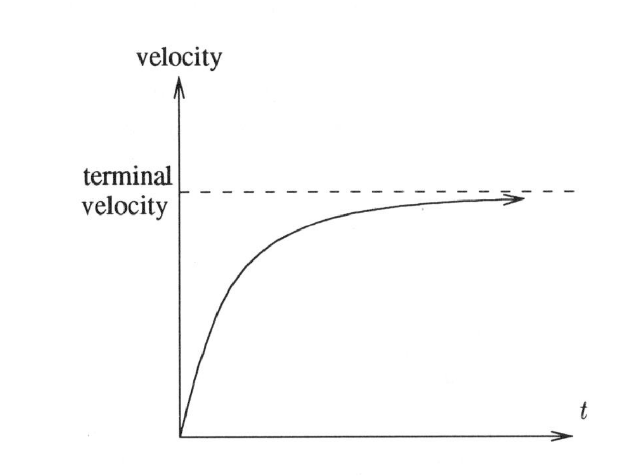

   (b) The graphs should be concave down because wind resistance decreases your acceleration as you speed up, and so the slope of the graph of velocity is decreasing.

   (c) The slope represents the acceleration due to gravity.

11. Since $\frac{P(67)-P(66)}{67-66}$ is an estimate of $P'(66)$, we may think of $P'(66)$ as an estimate of $P(67) - P(66)$, and the latter is the number of people between 66 and 67 inches tall. Since $\frac{P(66.5)-P(65.5)}{66.5-65.5}$ is a better estimate of $P'(66)$, we may regard $P'(66)$ as an estimate of the number of people of height between 65.5 and 66.5 inches. The units for P' are people per inch.

    If we assume that there are 190 million full grown persons in the U.S. whose heights are roughly evenly distributed between 60 and 75 inches, we expect $P'(x)$ to be around $\frac{190 \text{ million}}{15}$ or about 12.7 million people per inch for every x between 60 and 75. The 60 million children are probably about evenly distributed in height between 15 and 60 inches, so we expect $P'(x)$ to be about 1.33 million people per inch for x between 15 and 60 inches.

    $P'(x)$is never negative because $P(x)$ is never decreasing. To see this, let's look at an example involving a particular value of x, say $x = 80$. Now $P(80)$ represents the number of people whose height is less than 80 inches, and P(81)represents the number of people whose height is less than 81 inches. Surely everyone less than 80 inches is also less than 81 inches, so $P(80) \leq P(81)$. In general, $P(x)$ increases as x increases.

13. (a) Estimating derivatives by going to the right (but other answers are possible):

$$P'(1900) \approx \frac{P(1910) - P(1900)}{10} = \frac{92.0 - 76.0}{10} = 1.6 \text{ million people per year}$$

$$P'(1945) \approx \frac{P(1950) - P(1940)}{10} = \frac{150.7 - 131.7}{10} = 1.9 \text{ million people per year}$$

$$P'(1980) \approx \frac{P(1980) - P(1970)}{10} = \frac{226.5 - 205.0}{10} = 2.15 \text{ million people per year}$$

(b) The population growth was maximal somewhere between 1950 and 1960.

(c) $P'(1950) \approx \dfrac{P(1960) - P(1950)}{10} = \dfrac{179.0 - 150.7}{10} = 2.83$ million people per year, so $P(1956) \approx P(1950) + P'(1950)(1956 - 1950) = 150.7 + 2.83(6) \approx 167.7$ million people.

(d) If the growth rate between 1980 and 1990 was the same as the growth rate from 1970 to 1980, then the total population should be about 248 million people in 1990.

2.5 Solutions

1. If f'' is positive on an interval, then f' is <u>increasing</u> on that interval, and f is <u>concave up</u> on that interval.

   If f'' is negative on an interval, then f' is <u>decreasing</u> on that interval, and f is <u>concave down</u> on that interval.

3. (a) Such a graph is shown below. The change takes place at the origin.

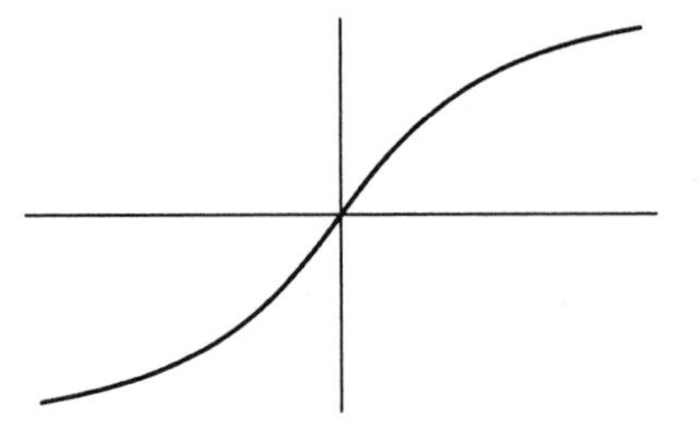

(b)

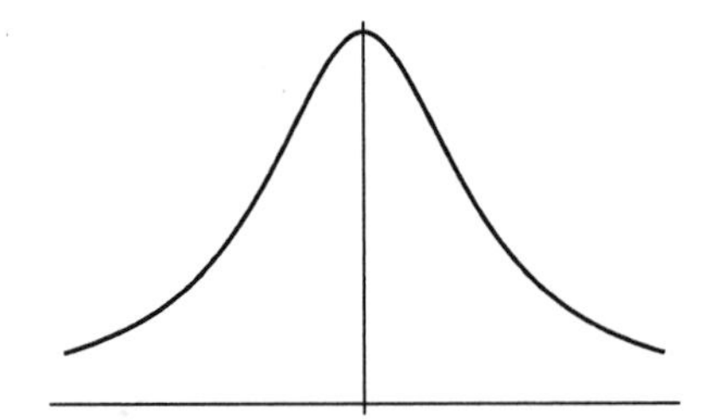

(c)

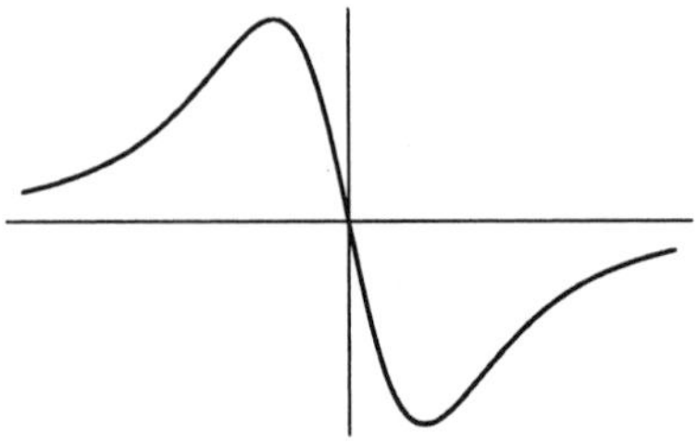

5. (a)

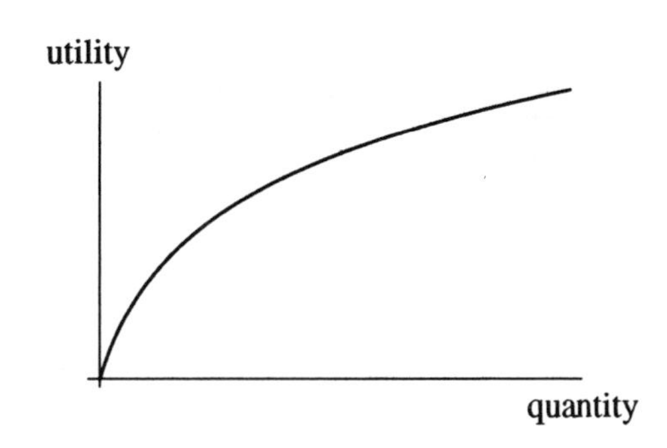

(b) As a function of quantity, utility is increasing but at a decreasing rate; the graph is increasing but concave down. So the derivative of utility is positive, but the second derivative of utility is negative.

7. (a) The EPA will say that the rate of discharge is still rising. The industry will say that the rate of discharge is increasing less quickly, and may soon level off or even start to fall.

(b) The EPA will say that the rate at which pollutants are being discharged is levelling off, but not to zero — so pollutants will continue to be dumped in the lake. The industry will say that the rate of discharge has decreased significantly.

9. (a)

$$f'(0.6) \approx \frac{f(0.8) - f(0.6)}{0.8 - 0.6} = \frac{4.0 - 3.9}{0.2} = 0.5.$$

$$f'(0.5) \approx \frac{f(0.6) - f(0.4)}{0.6 - 0.4} = \frac{0.4}{0.2} = 2$$

(b) $f''(0.6) \approx \frac{f'(0.7)-f'(0.5)}{0.7-0.5} = \frac{0.5-2}{0.2} = \frac{-1.5}{0.2} = -7.5$, where $f'(0.7) \approx \frac{f(0.8)-f(0.6)}{0.8-0.6} = \frac{4.0-3.9}{0.2} = \frac{0.1}{0.2} = 0.5$.

(c) The maximum value of f is probably near $x = 0.8$. The minimum value of f is probably near $x = 0.3$.

11. (a) B and E
(b) A and D

2.6 Solutions

1. The tangent line has slope $1/4$ and goes through the point $(4, 2)$. The equation is

$$y - 2 = \frac{1}{4}(x - 4) \qquad y = \frac{1}{4}x + 1.$$

Thus when $x = 4.006$, we have $y = 1/4(4.006) + 1 = 2.0015$, so the point (4.006, 2.0015) lies on the tangent line.

3.

x	0.80	0.81	0.82	0.83	0.84	0.85	0.86	0.87	0.88	0.89	0.90
$\tan x$	1.030	1.050	1.072	1.093	1.116	1.138	1.162	1.185	1.210	1.235	1.260

The values of $\tan x$ seem to be increasing by about 0.023 for each increase of 0.01 in x.

5. (a) $f'(6.75) \approx \frac{f(7.0)-f(6.5)}{7.0-6.5} = \frac{8.2-10.3}{0.5} = -4.2$.
$f'(7.0) \approx \frac{f(7.5)-f(6.5)}{7.5-6.5} = \frac{6.5-10.3}{1.0} = -3.8$.
$f'(8.5) \approx \frac{f(9.0)-f(8.0)}{9.0-8.0} = \frac{3.2-5.2}{1.0} = -2.0$.

(b) To estimate f'' at 7, we should have values for f' at points near 7. We know from (a) that $f'(6.75) \approx -4.2$. Next, estimate $f'(7.25) \approx \frac{6.5-8.2}{0.5} = -3.4$. Then
$f''(7) \approx \frac{f'(7.25)-f'(6.75)}{0.5} \approx \frac{-3.4-(-4.2)}{0.5} = 1.6$.

(c) $y - 8.2 = -3.8(x - 7)$ or $y = -3.8x + 34.8$.

(d) We may use the tangent line from (c) to approximate $f(6.8)$. In this case we get

$$y \approx -3.8x + 34.8 = 8.96.$$

We may also estimate $f(6.8)$ by assuming that the graph of f is straight between the given points $(6.5, 10.3)$ and $(7.0, 8.2)$. This line has the equation $y = -4.2(x - 6.5) + 10.3$ and passes through (6.8, 9.04), so we may estimate $f(6.8) \approx 9.04$. This method of estimation is called linear interpolation.
As we can see, the two estimates are fairly close.

7. At (0,0): $f'(0) = 0$, so the tangent line has slope 0 and equation $y = 0$.
At (1,1):

$$\begin{aligned} f'(1) &= \lim_{h\to0} \frac{f(1+h) - f(1)}{h} \\ &= \lim_{h\to0} \frac{(1+h)^2 - (1)^2}{h} \\ &= \lim_{h\to0} (2 + h) = 2, \end{aligned}$$

so the tangent line has slope 2 and equation $y - 1 = 2x - 2$, or $y = 2x - 1$.

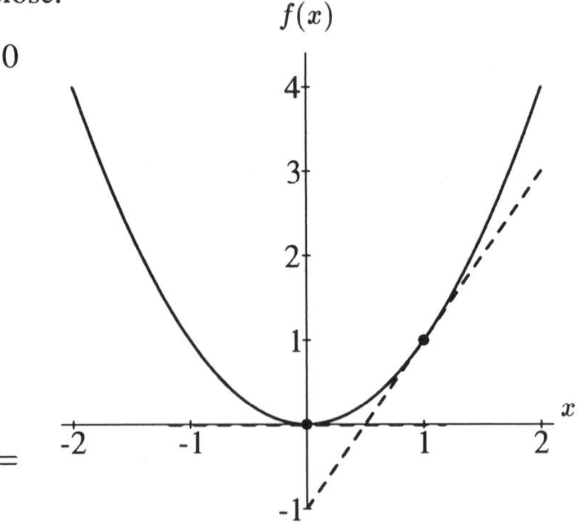

9. (a) The graph is a sine curve. It looks straight because the graph shows only a small part of the curve magnified greatly. The period of the sine curve is $\frac{2\pi}{0.0172} \approx 365$ (the number of days in a year).

   (b) The month is March: We see that about the 21st of the month there are twelve hours of daylight and hence twelve hours of night. This phenomenon (the length of the day equaling the length of the night) occurs at the equinox, midway between winter and summer. Since the length of the days is increasing, (and Madrid is in the northern hemisphere) we are looking at March, not September.

   (c) The slope of the curve is found either from the graph or the formula to be about 0.04 (the rise is about 0.8 hours in 20 days or 0.04 hours/day). This means that the amount of daylight is increasing by about 0.04 hours (about $2\frac{1}{2}$ minutes) per calendar day, or that each day is $2\frac{1}{2}$ minutes longer than its predecessor.

11. (a)

    (b)

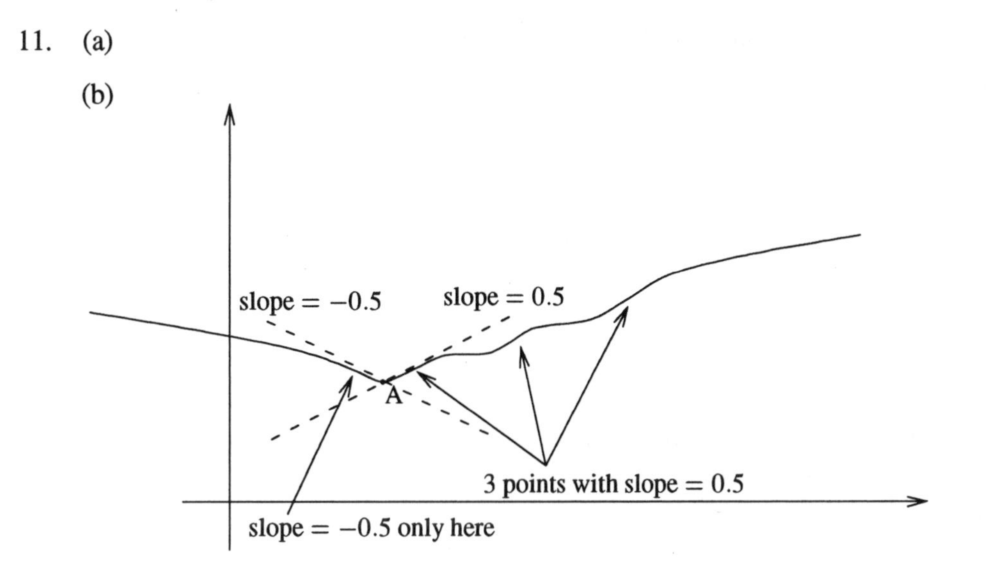

   (c) $f(a+3) = 1 + 0.5(3) = 2.5$ at most.
   $f(a+2) > 1 - 0.5(2) = 0$ at least. (It cannot be equal because, outside of one point, $f'(x) > -0.5$.)

   Note that the graph of f lies between the two lines we have drawn. It cannot cross these lines except at A.

2.7 Solutions

1. The limit appears to be 1; a graph and table of values is shown below.

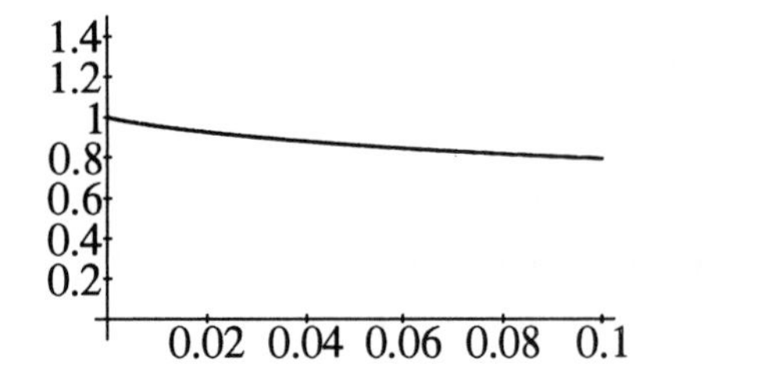

x	x^x
0.1	0.7943
0.01	0.9550
0.001	0.9931
0.0001	0.9990
0.00001	0.9999

3. For $-1 \leq \theta \leq 1$, $-1 \leq y \leq 1$, the graph of $y = \dfrac{\cos\theta - 1}{\theta}$ is shown to the right. Therefore,

$$\lim_{\theta \to 0} \frac{\cos\theta - 1}{\theta} = 0.$$

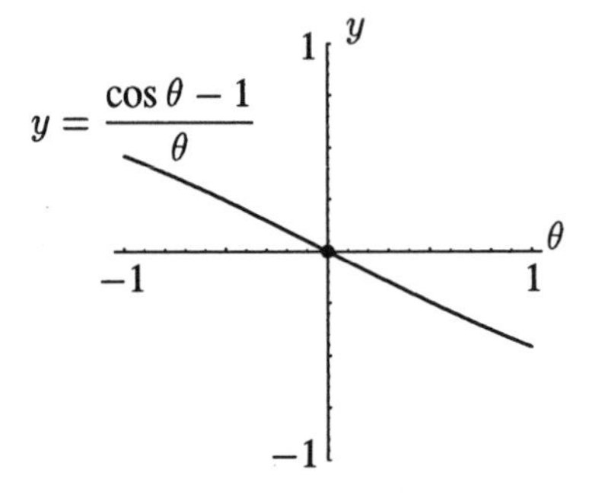

5. The answer (see the graph to the right) appears to be about 2.7; if we zoom in further, it appears to be about 2.72, which is close to the value of $e \approx 2.71828$.

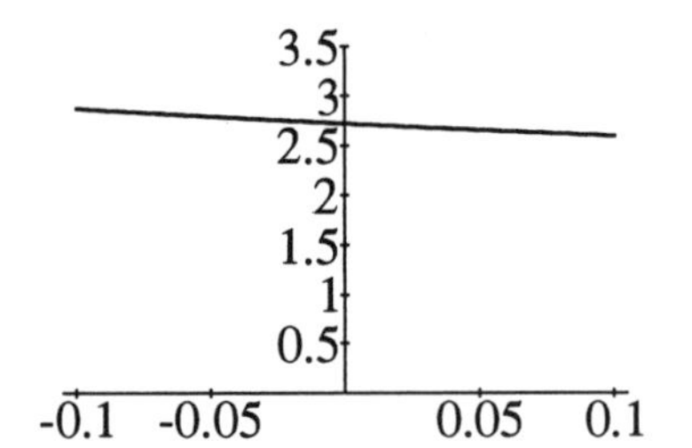

2.8 Solutions

1. Yes, this does appear to be smooth.

3. Yes, this appears to be smooth.

5. The graph of

$$f(x) = \begin{cases} 0 & \text{if } x < 0. \\ x^2 & \text{if } x \geq 0. \end{cases}$$

is shown to the right. The graph is continuous and has no vertical segments or corners, so $f(x)$ is differentiable everywhere.

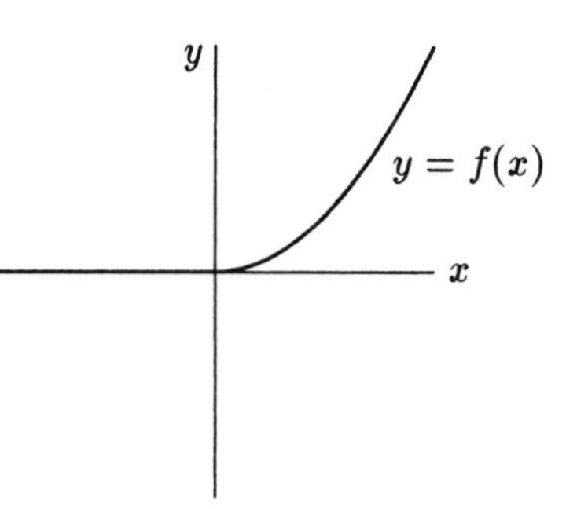

By Example 4 on page 143,

$$f'(x) = \begin{cases} 0 & \text{if } x < 0 \\ 2x & \text{if } x \geq 0 \end{cases}$$

So its graph is shown to the right.

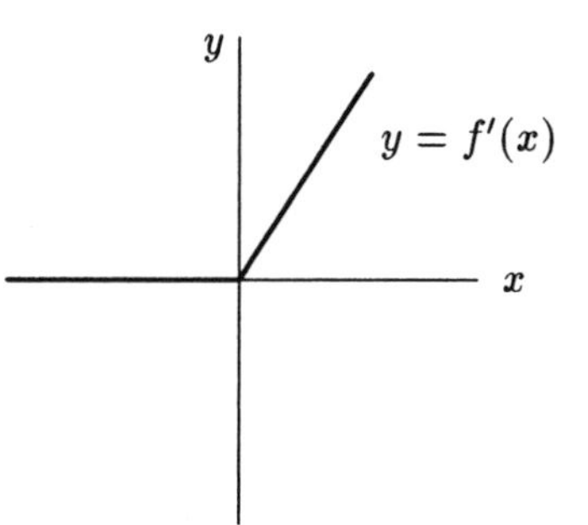

The graph of the derivative has a corner at $x = 0$ so $f'(x)$ is not differentiable at $x = 0$. The graph of

$$f''(x) = \begin{cases} 0 & \text{if } x < 0 \\ 2 & \text{if } x > 0 \end{cases}$$

looks like:

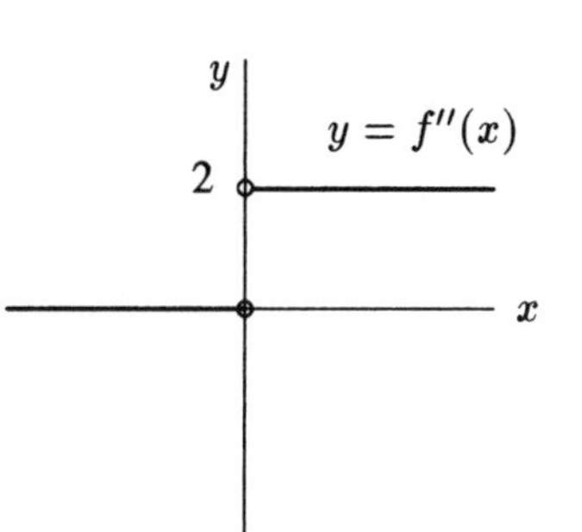

The second derivative is not defined at $x = 0$. So it is certainly not differentiable at $x = 0$.

2.9 Answers to Miscellaneous Exercises for Chapter 2

(a) (b) The graph looks like:

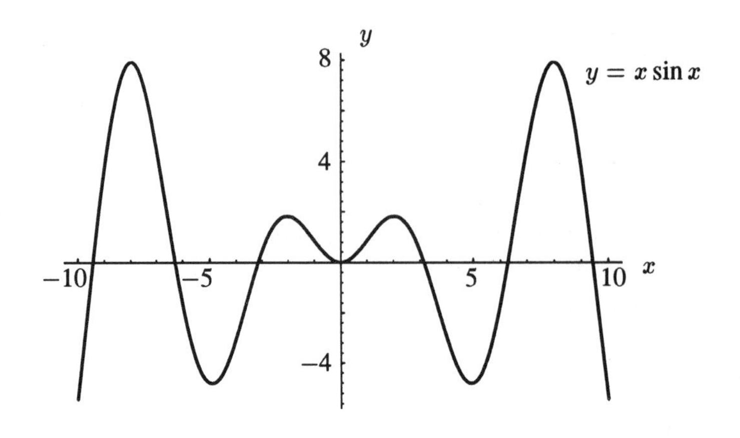

(a) Seven. $x \sin x = 0$ at $x = 0, \pm\pi, \pm 2\pi, \pm 3\pi$.

(b) $x \sin x$ is increasing at $x = 1$, decreasing at $x = 4$.

(c) $\dfrac{f(2) - f(0)}{(2-0)} = \dfrac{2\sin 2 - 0}{2} = \sin 2 \approx 0.91$
$\dfrac{f(8) - f(6)}{(8-6)} = \dfrac{8\sin 8 - 6\sin 6}{2} \approx 4.80$. So the average rate of change over $6 \leq x \leq 8$ is greater.

(d) It's greater at $x = -9$.

3.

x	x^3
0.998	0.9940
0.999	0.9970
1.000	1.0000
1.001	1.0030
1.002	1.0060

x	x^3
2.998	26.946
2.999	26.973
3.000	27.000
3.001	27.027
3.002	27.054

x	x^3
4.998	124.850
4.999	124.925
5.000	125.000
5.001	125.075
5.002	125.150

At $x = 1$, the values of x^3 are increasing by about 0.0030 for each increase of 0.001 in x, so the derivative appears to be $\frac{0.0030}{0.001} = 3$. At $x = 3$, the values increase by 0.027 over an x increase of 0.001, so the value appears to be 27. At $x = 5$, the values increase by 0.075 for a change in x of 0.001, so the derivative appears to be 75. The function $3x^2$ fits these data, so it is a good candidate for the derivative function (although it is not immediately obvious just from these calculations).

5.

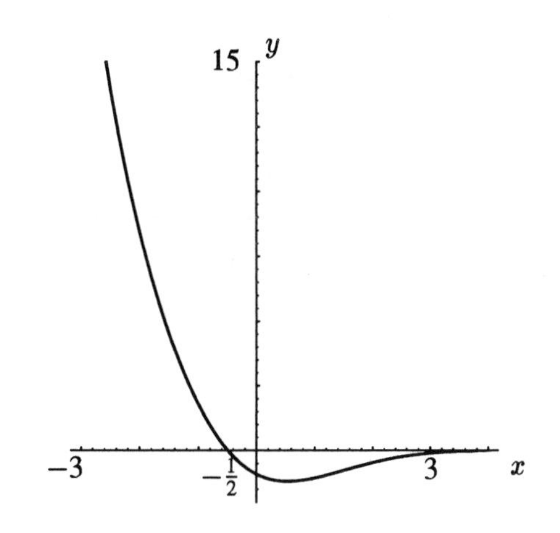

7.

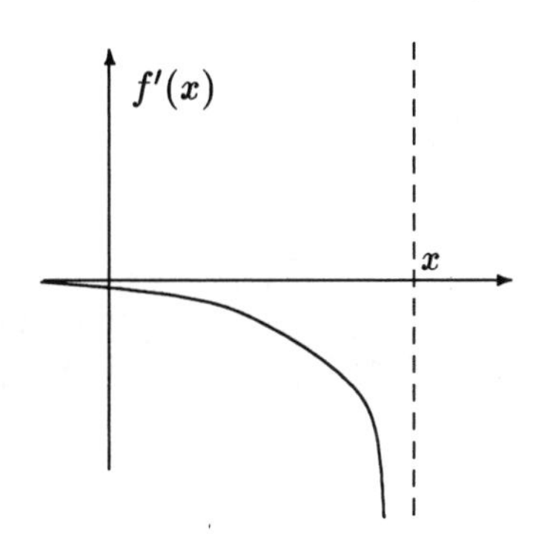

9.

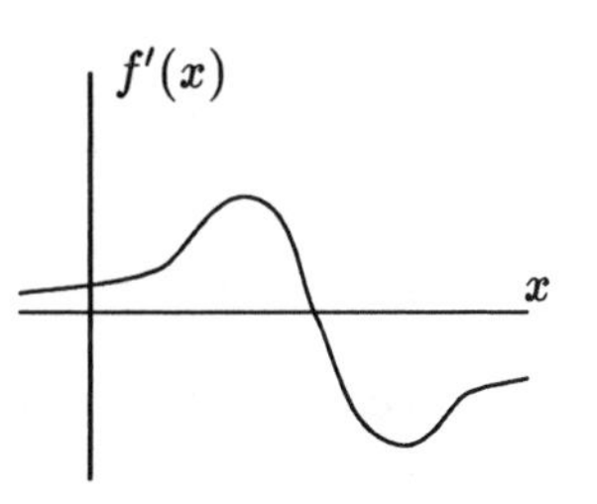

11.

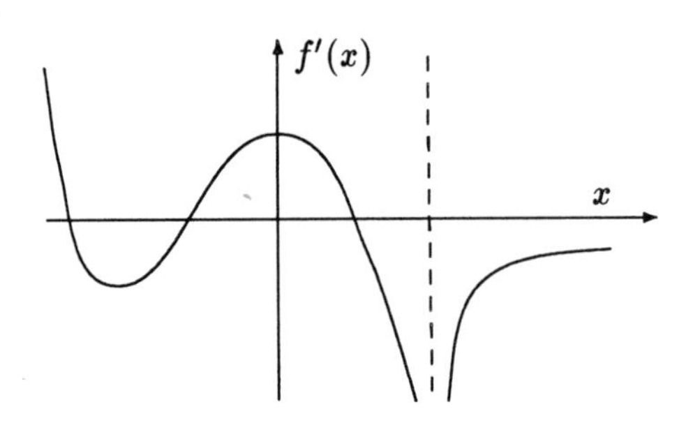

13. (a)

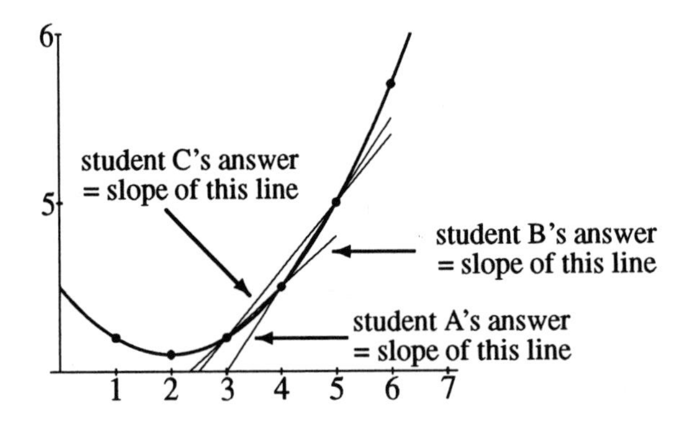

(b) The slope of f appears to be somewhere between student A's answer and student B's, so student C's answer, halfway in between, is probably the most accurate.

(c) Student A's estimate is $f'(x) \approx \frac{f(x+h)-f(x)}{h}$.

Student B's estimate is $f'(x) \approx \frac{f(x)-f(x-h)}{h}$.

Student C's estimate is the average of these two, or

$$f'(x) \approx \frac{1}{2}\left[\frac{f(x+h)-f(x)}{h} + \frac{f(x)-f(x-h)}{h}\right] = \frac{f(x+h)-f(x-h)}{2h}.$$

This estimate is the slope of the chord connecting $(x-h, f(x-h))$ to $(x+h, f(x+h))$. Thus we estimate that the tangent to a curve is nearly parallel to a chord connecting points h units to the right and left. It turns out that this estimate is always exact for functions given by a quadratic polynomial.

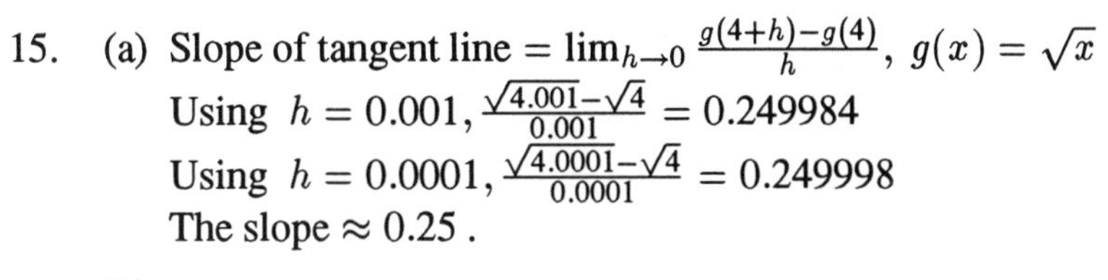

15. (a) Slope of tangent line $= \lim_{h\to 0} \frac{g(4+h)-g(4)}{h}$, $g(x) = \sqrt{x}$
Using $h = 0.001$, $\frac{\sqrt{4.001}-\sqrt{4}}{0.001} = 0.249984$
Using $h = 0.0001$, $\frac{\sqrt{4.0001}-\sqrt{4}}{0.0001} = 0.249998$
The slope ≈ 0.25 .

(b)

$$\begin{aligned} y - y_1 &= m(x - x_1) \\ y - 2 &= 0.25(x-4) \\ y - 2 &= 0.25x - 1 \\ y &= 0.25x + 1 \end{aligned}$$

(c) $f(x) = kx^2$
If $(4,2)$ is on the graph of f, then $f(4) = 2$, so $k \cdot 4^2 = 2$. Thus $k = \frac{1}{8}$, and $f(x) = \frac{1}{8}x^2$.

(d) To find where the graph of f crosses then line $y = 0.25x + 1$, we solve:

$$\begin{aligned} \frac{1}{8}x^2 &= 0.25x + 1 \\ x^2 &= 2x + 8 \\ x^2 - 2x - 8 &= 0 \\ (x-4)(x+2) &= 0 \\ x = 4 \quad &\text{or} \quad x = -2 \\ f(-2) &= \frac{1}{8}(4) = 0.5 \end{aligned}$$

Therefore, $(-2, 0.5)$ is the other point of intersection. (Of course, $(4,2)$ is a point of intersection; we know that from the start.)

17. (a)
 (b)

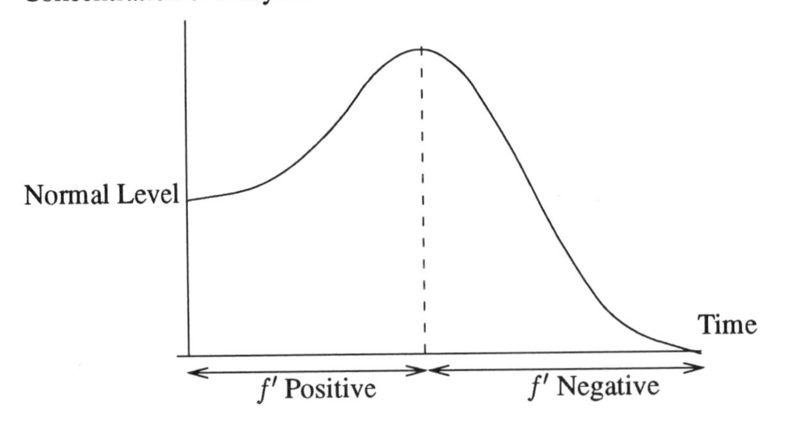

 (c) f' is the rate at which the concentration is increasing or decreasing. f' is positive at the start of the disease and negative toward the end. In practice, of course, one cannot measure f' directly. Checking the value of C in blood samples taken on consecutive days would tell us

$$\frac{f(t+1)-f(t)}{(t+1)-t},$$

which is our estimate of $f'(t)$.

19. (a)

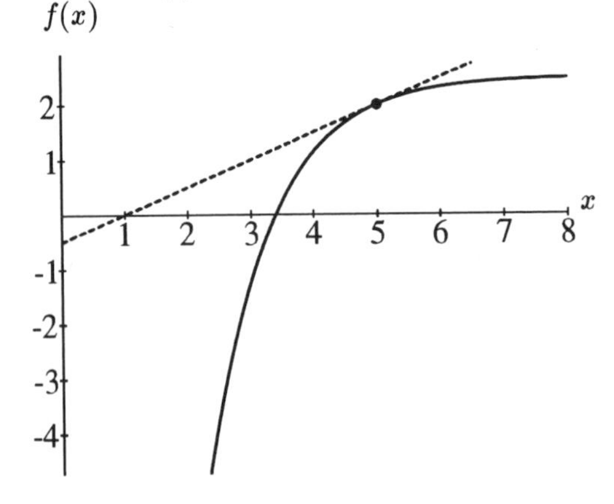

 (b) Exactly one. There can't be more than one zero because f is increasing everywhere. There does have to be one zero because f stays below its tangent line (dotted line in above graph), and therefore f must cross the x-axis.

 (c) The equation of the (dotted) tangent line is $y = \frac{1}{2}x - \frac{1}{2}$, and so it crosses the x-axis at $x = 1$. Therefore the zero of f must be between $x = 1$ and $x = 5$.

 (d) $\lim_{x\to-\infty} f(x) = -\infty$, because f is increasing and concave down. Thus, as $x \to -\infty$, $f(x)$ decreases, at a faster and faster rate.

(e) Yes.

(f) No. The slope is decreasing since f is concave down, so $f'(1) > f'(5)$, i.e. $f'(1) > \frac{1}{2}$.

21.

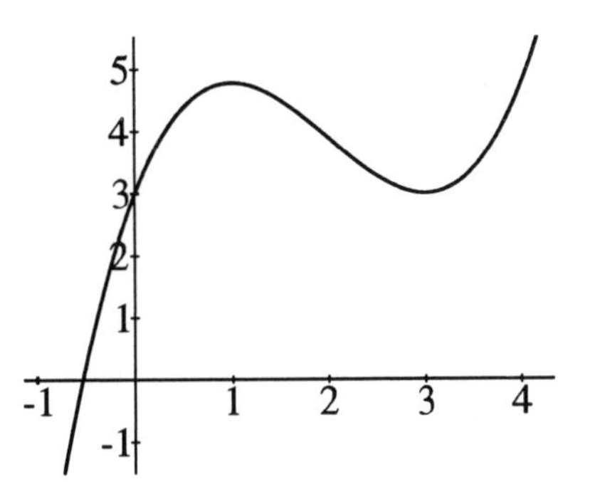

Chapter 3

3.1 Solutions

1. (a) Lower estimate $= 60+40+25+10+0 =$ 135 feet. Upper estimate $= 88+60+40+25+10 = 223$ feet.

   (b)

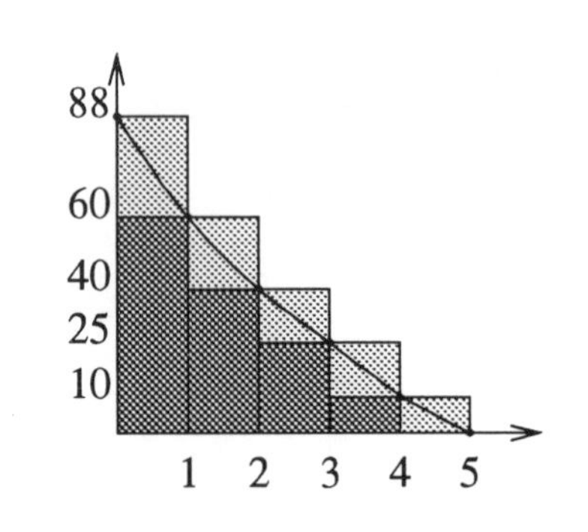

3. (a) An overestimate is 7 tons. An underestimate is 5 tons.

   (b) An overestimate is $7+8+10+13+16+20 = 74$ tons. An underestimate is $5+7+8+10+13+16 = 59$ tons.

   (c) If measurements are made every Δt months, then the error is $|f(6)-f(0)|\cdot\Delta t$. So for this to be less than 1 ton, we need $(20-5)\cdot\Delta t < 1$, or $\Delta t < 1/15$. So measurements every 2 days or so will guarantee an error of less than 1 ton.

5. (a) We want the error to be less than 0.1, so take Δx such that $|f(1)-f(0)|\Delta x < 0.1$, giving

$$\Delta x < \frac{0.1}{|e^{-\frac{1}{2}}-1|} \approx 0.25$$

   so take $\Delta x = 0.25$ or $n = 4$. Then the left sum $= 0.9016$, and the right sum $= 0.8033$, so a reasonable estimate for the area is $\frac{0.9016+0.8033}{2} = 0.8525$. Thus 0.85 is certainly within 0.1 of the actual answer.

   (b) Take Δx smaller. To have an error of at most E, you need Δx such that

$$|f(1)-f(0)|\Delta x < E$$

   This means

$$\Delta x < \frac{E}{|e^{-\frac{1}{2}}-1|} \approx \frac{E}{0.39}.$$

   Using n equal subdivisions, we have

$$\Delta x = \frac{b-a}{n} = \frac{1-0}{n} = \frac{1}{n}.$$

   Thus, to approximate the shaded area with an error $< E$ requires $n > \frac{0.39}{E}$ subdivisions.

7. First, note that, for $0 \leq t \leq 1.1$, $v(t)$ is an increasing function. The difference between the values of the function at the endpoints of the interval [0,1.1] is $(\sin 1.21 - \sin 0) \approx 0.936$. If we approximate the area under the curve by rectangles above and below the curve of base h, then the difference between upper and lower estimates of the area $\approx 0.936h$. For one decimal place accuracy, we want $0.936h < 0.1$, so $h < 0.107$. This is the same as saying that we must have more than $\frac{1.1}{0.107} \approx 10.3$ rectangles for this degree of accuracy. With 11 rectangles, our underestimate is ≈ 0.353 and the overestimate is ≈ 0.447. Hence the value of the integral is 0.400 to one decimal place accuracy.

9. (a) An upper estimate is $9.81 + 8.03 + 6.53 + 5.38 + 4.41 = 34.16$ m/sec. A lower estimate is $8.03 + 6.53 + 5.38 + 4.41 + 3.61 = 27.96$ m/sec.

   (b) The average is $\frac{1}{2}(34.16 + 27.96) = 31.06$ m/sec. Because the graph of acceleration is concave up, this estimate is too high, as can be seen in the figure to the right. The area of the shaded region is the average of the areas of the rectangles $ABFE$ and $CDFE$.

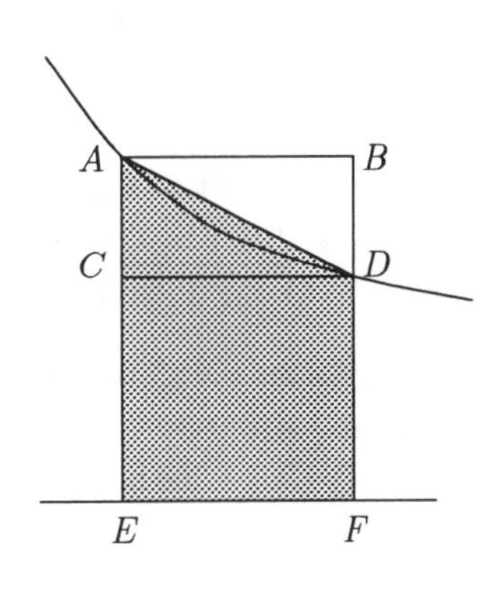

3.2 Solutions

1.

n	2	10	50	250
Left-hand Sum	0.0625	0.2025	0.2401	0.248004
Right-hand Sum	0.5625	0.3025	0.2601	0.252004

The sums seem to be converging to $\frac{1}{4}$. Since x^3 is monotone on [0,1], the true value is between 0.248004 and 0.252004 .

3.

n	2	10	50	250
Left-hand Sum	1.14201	1.38126	1.44565	1.45922
Right-hand Sum	2.00115	1.55309	1.48002	1.46610

There is no obvious guess as to what the limiting sum is. We can only observe that since e^{t^2} is monotonic on [0,1], the true value is between 1.45922 and 1.46610 .

5.

n	2	10	50	250
Left-hand Sum	−0.394991	−0.0920539	−0.0429983	−0.0335556
Right-hand Sum	0.189470	0.0248382	−0.0196199	−0.0288799

There is no obvious guess as to what the limiting sum is. Moreover, since $\sin(t^2)$ is *not* monotonic on $[2,3]$, we cannot be sure that the true value is between −0.0335556 and −0.0288799.

7. For $n = 1300$, LHS ≈ 41.618 (rounding down) and RHS ≈ 41.715 (rounding up). Since the left and right sums differ by 0.097, their average must be within 0.0485 of the true value, so $\int_0^5 x^2\,dx = 41.66$ to one decimal place.

9. For $n = 10$, LHS ≈ 0.465 (rounding down) and RHS ≈ 0.474 (rounding up). Since the left and right sums differ by 0.009, their average must be within 0.0045 of the true value, so $\int_1^{1.5} \sin x\,dx = 0.470$ to one decimal place.

11. For $n = 110$, LHS ≈ 4.810 (rounding down) and RHS ≈ 4.905 (rounding up). Since the left and right sums differ by 0.095, their average must be within 0.0475 of the true value, so $\int_1^5 (\ln x)^2\,dx = 4.858$ to one decimal place.

13. For $n = 30$, LHS ≈ 2.852 (rounding down) and RHS ≈ 2.919 (rounding up). Since the left and right sums differ by 0.067, their average must be within 0.0335 of the true value, so $\int_1^2 2^x\,dx = 2.886$ to one decimal place.

15. For $n = 10$, LHS ≈ 0.0045 (rounding down) and RHS ≈ 0.0276 (rounding up). Since the left and right sums differ by 0.0231, their average must be within 0.01155 of the true value, so $\int_{-2}^{-1} \cos^3 y\,dy = 0.016$ to one decimal place.

17. Instead of finding n by trial and error, as in the text, you can also start by calculating the value of n. Since the interval $1 \le t \le 3$ is of length 2, the length of one subdivision is $\Delta t = \frac{2}{n}$. If we choose n so that the difference between the left- and right-hand sums is less than 0.1, our result about errors tells us that

$$\text{Error} < \left| \begin{array}{c}\text{Difference between} \\ f(1) \text{ and } f(3)\end{array} \right| \cdot \Delta t < 0.1.$$

Therefore we should pick n so that

$$\begin{aligned} |f(3) - f(1)|\frac{2}{n} &< 0.1 \\ (9-1)\frac{2}{n} &< 0.1 \\ \text{so} \quad n &> 160. \end{aligned}$$

Thus $n = 5000$ used in the text was larger than we needed.

19.

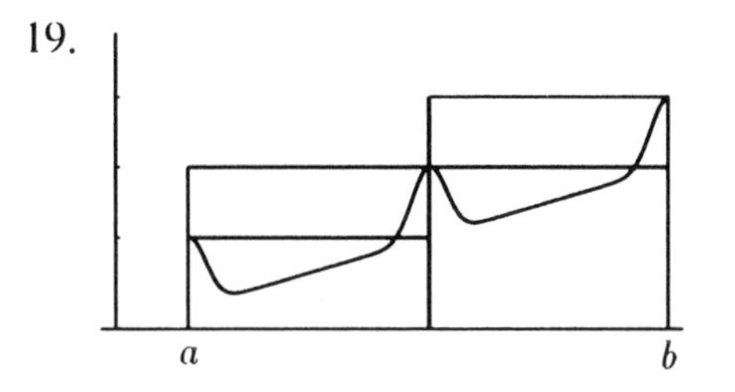

Figure 3.1: Integral vs. Left- and Right-Hand Sums

3.3 Solutions

1. (a) Clearly, the points where $x = \sqrt{\pi}, \sqrt{2\pi}, \sqrt{3\pi}, \sqrt{4\pi}$ are where the graph intersects the x-axis because $f(x) = \sin(x^2) = 0$ where $x =$ the square root of some multiple of π.

   (b) Let $f(x) = \sin(x^2)$, and let A, B, C, and D be the areas of the regions indicated in the figure to the right. Then we see that $A > B > C > D$.

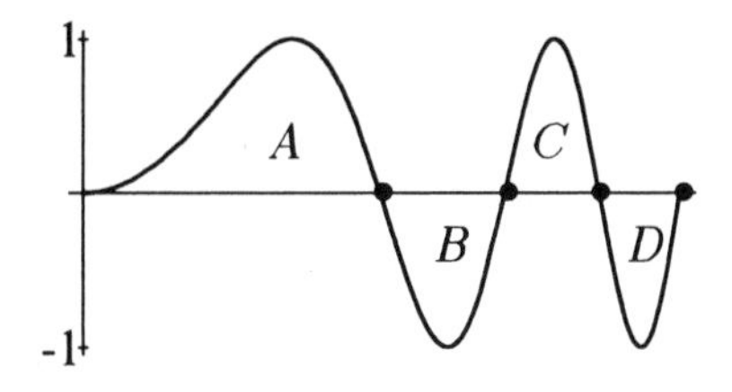

Note that

$$\int_0^{\sqrt{\pi}} f(x)\,dx = A, \int_0^{\sqrt{2\pi}} f(x)\,dx = A - B,$$

$$\int_0^{\sqrt{3\pi}} f(x)\,dx = A - B + C, \text{ and } \int_0^{\sqrt{4\pi}} f(x)\,dx = A - B + C - D.$$

It follows that

$$\int_0^{\sqrt{\pi}} f(x)\,dx = A > \int_0^{\sqrt{3\pi}} f(x)\,dx = A - (B - C) = A - B + C >$$

$$\int_0^{\sqrt{4\pi}} f(x)\,dx = A - B + C - D > \int_0^{\sqrt{2\pi}} f(x)\,dx = (A - B) > 0.$$

And thus the ordering is $n = 1$, $n = 3$, $n = 4$, and $n = 2$ from lárgest to smallest. All the numbers are positive.

3. (a) Quantity used $= \int_0^5 f(t)\,dt$.

   (b) Using a left sum, our approximation is

$$32e^{0.05(0)} + 32e^{0.05(1)} + 32e^{0.05(2)} + 32e^{0.05(3)} + 32e^{0.05(4)} = 177.27.$$

   Since f is an increasing function, this represents an underestimate.

(c) Each term is a lower estimate of one year's consumption of oil.

5. (a) Since $t = 0$ to $t = 31$ covers January:

$$\begin{pmatrix}\text{Average number of}\\ \text{daylight hours in January}\end{pmatrix} = \frac{1}{31}\int_0^{31} [12 + 2.4\sin(0.0172(t-80))]\,dt.$$

Using left and right sums with $n = 100$ gives

$$\text{Average} \approx \frac{306}{31} \approx 9.9 \text{ hours}.$$

(b) Assuming it is not a leap year, the last day of May is $t = 151 (= 31+28+31+30+31)$ and the last day of June is $t = 181 (= 151+30)$. Again finding the integral numerically:

$$\begin{pmatrix}\text{Average number of}\\ \text{daylight hours in June}\end{pmatrix} = \frac{1}{30}\int_{151}^{181} [12 + 2.4\sin(0.0172(t-80))]\,dt \approx \frac{431}{30} \approx 14.4 \text{ hours}.$$

(c)

$$(\text{Average for whole year}) = \frac{1}{365}\int_0^{365} [12 + 2.4\sin(0.0172(t-80))]\,dt \approx \frac{4381}{365} \approx 12.0 \text{ hours}.$$

(d) The average over the whole year should be 12 hours, as computed in (c). Since Madrid is in the northern hemisphere, the average for a winter month, such as January, should be less than 12 hours (it is 9.9 hours) and the average for a summer month, such as June, should be more than 12 hours (it is 14.4 hours).

7. (a) For $-2 \le x \le 2$, we have symmetry about the y-axis, so $\int_{-2}^{0} f(x)\,dx = \int_0^2 f(x)\,dx$ and $\int_{-2}^{2} f(x)\,dx = 2\int_0^2 f(x)\,dx$.

(b) For any function f, we have $\int_0^2 f(x)\,dx = \int_0^5 f(x)\,dx - \int_2^5 f(x)\,dx$.

(c) Note that $\int_{-2}^{0} f(x)\,dx = \frac{1}{2}\int_{-2}^{2} f(x)\,dx$, so we have $\int_0^5 f(x)\,dx = \int_{-2}^{5} f(x)\,dx - \int_{-2}^{0} f(x)\,dx = \int_{-2}^{5} f(x)\,dx - \frac{1}{2}\int_{-2}^{2} f(x)\,dx$.

9. (a) Average value of $f = \frac{1}{5}\int_0^5 f(x)\,dx$.

(b) Average value of $f = \frac{1}{5}\int_0^5 |f(x)|\,dx = \frac{1}{5}(\int_0^2 f(x)\,dx - \int_2^5 f(x)\,dx)$.

11. (a) The distance traveled is the integral of the velocity, so in T seconds you fall

$$\int_0^T \frac{g}{k}(1 - e^{-kt})\,dt.$$

Putting in the given values we have:

$$\text{distance fallen in } T \text{ seconds} = \int_0^T 49(1 - e^{-0.2t})\,dt \text{ meters}.$$

(b) We want the number T for which

$$\int_0^T 49(1 - e^{-0.2t})\, dt = 5000.$$

Finding T precisely is difficult, but we can show that $102 < T < 124$ as follows:

Since your velocity is always less than 49 m/s it will take more than $\frac{5000}{49}(\approx 102)$ seconds to fall 5000 meters; i.e. $T > 102$.

If we draw the graph of v versus t we see that it has an asymptote $v = 49$. This is called the *terminal velocity*. Terminal velocity is reached, in a practical sense, fairly soon. When $t = 20$ v is already more than 48.1. Hence in the next 104 seconds you would fall at least $(48.1)(104) > 5000$ meters. So your fall will surely end before 20+104 seconds, so $T < 124$.

If we interpret the integrals as areas, what we have done is to estimate $\int_0^{\frac{5000}{49}} 49(1 - e^{-0.2t})\, dt$ as less than the rectangle OABC and $\int_0^{124} 49(1 - e^{-0.2t})\, dt$ as more than the rectangle DEFG. Since these estimates are rather crude, our result that $102 < T < 124$ is also crude. Better results can be obtained by using more intervals to estimate the integrals. For example, we might use bisection: decide by a numerical method whether $\int_0^{113} 49(1 - e^{-0.2t})\, dt$ is more or less than 5000, and thus trap T between 102 and 113 or between 113 and 124.

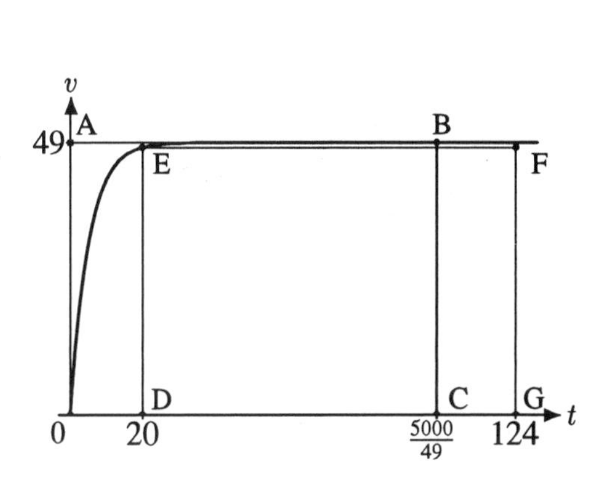

The large amount of computation required to solve for T by such methods points to the importance of finding more efficient procedures to evaluate definite integrals.

3.4 Solutions

1. Since $F(0) = 0$, $F(b) = \int_0^b f(t)\, dt$. This integral is the difference $S_+(b) - S_-(b)$ of the two areas, $S_+(b)$ and $S_-(b)$. Here $S_\pm(b)$ denotes the area between the positive (negative) portion of the graph of $f(x)$, $0 \leq x \leq b$, and the t-axis. Therefore, employing the figure from the exercise, we have:

$F(0) = S_+(0) = 0$
$F(1) = S_+(1) = 0 + 1 = 1$
$F(2) = S_+(2) = 1 + 0.5 = 1.5$
$F(3) = S_+(3) - S_-(3) = 1.5 - 0.5 = 1$
$F(4) = S_+(4) - S_-(4) = 1.5 - 1.5 = 0$
$F(5) = S_+(5) - S_-(5) = 1.5 - 2.5 = -1$
$F(6) = S_+(6) - S_-(6) = 1.5 - 3 = -1.5$

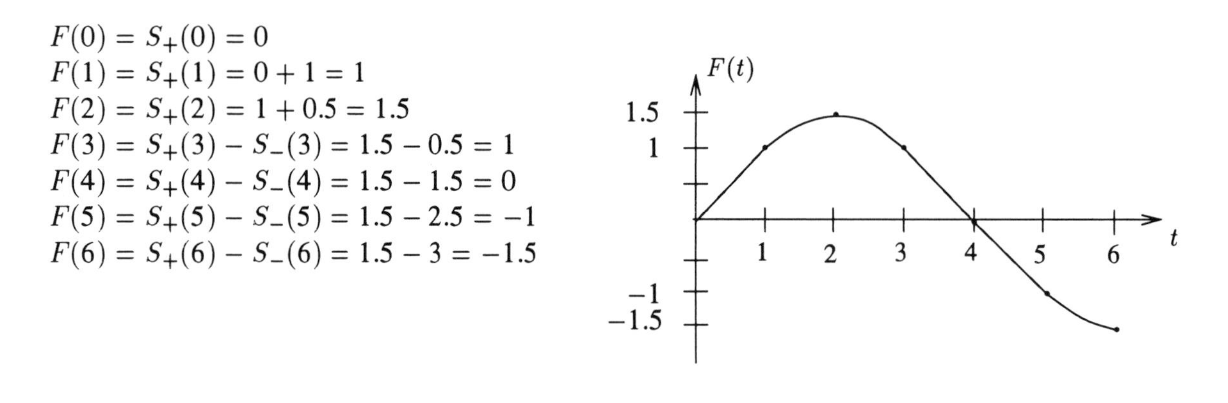

Figure 3.2: Graph of F

The graph of $F(t)$, $0 \le t \le 6$, is shown at Figure 3.2.

2. Using the table,

θ	0	0.25	0.5	0.75	1	1.25	1.5	1.75	2	2.25	2.5
$\sin\theta^2 \approx$	0	0.06	0.25	0.53	0.84	1.0	0.78	0.08	-0.76	-0.94	-0.03

the graph of $f(\theta) = F'(\theta) = \sin\theta^2$, for $0 \le \theta \le 2.5$, is in Figure 3.3:

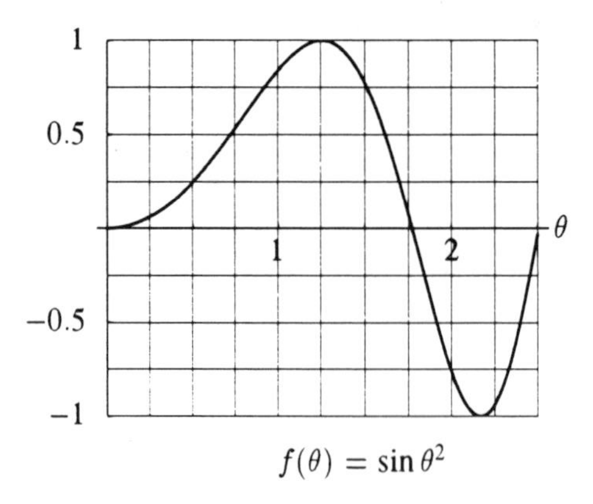

Figure 3.3: $f(\theta) = \sin\theta^2$

The area of an elementary square in the grid is $0.25 \times 0.25 \approx 0.06$. For any b, counting the number of elementary squares in the region $0 \le \theta \le b$, $0 \le y \le f(\theta)$ minus the number of squares in the region $0 \le \theta \le b$, $f(\theta) \le y \le 0$, we get:

$F(0) = 0$
$F(0.5) \approx 0.06 \times 0.7 \approx 0.04$

$F(1) \approx 0.04 + 0.06 \times 4.5 = 0.31$
$F(1.5) \approx 0.31 + 0.06 \times 7.5 = 0.76$
$F(2) \approx 0.76$ (no net change)
$F(2.5) \approx 0.76 - 0.06 \times 5.5 = 0.43$.

3. Using the graph of $f(\theta)$ in Figure 3.3 and viewing $F(\theta)$ as the difference of the two areas (as has been explained in Problem 2), we see that F is increasing for $0 \leq \theta < \sqrt{\pi} \approx 1.77$ and decreasing for $1.77 < \theta \leq 2.5$.

5. (a) For $n = 10$ subdivisions, the left sum ≈ 4.77 (rounding down) and the right sum ≈ 4.82 (rounding up). Since $\ln t$ is monotone on $[10,12]$, the actual value is between the sums. Since the left and right sums differ by 0.05, their average must be within 0.025 of the true value, so 4.795, is correct to one decimal place.

   (b) The Fundamental Theorem of Calculus tells us that we can get the exact answer by looking at $F(12) - F(10) = (12\ln 12 - 12) - (10\ln 10 - 10) \approx 4.79303$.

7. (a)

$$\begin{aligned}
F'(x) &= \lim_{h\to 0} \frac{F(x+h) - F(x)}{h} \\
&= \lim_{h\to 0} \frac{2(x+h) - 2x}{h} \\
&= \lim_{h\to 0} \frac{2x + 2h - 2x}{h} \\
&= \lim_{h\to 0} \frac{2h}{h} \\
&= \lim_{h\to 0} 2 \\
&= 2 = f(x)
\end{aligned}$$

   (b) Assuming that $b > a$, we find that the integral is simply $2(b - a)$ by recognizing the area under the curve $y = 2$ between a and b as a rectangle of height 2 and width $b - a$.

   (c) As predicted by the Fundamental Theorem of Calculus, the integral, which we calculated above as $2(b - a)$ is indeed $F(b) - F(a)$ since

$$2(b - a) = 2b - 2a = F(b) - F(a).$$

9.

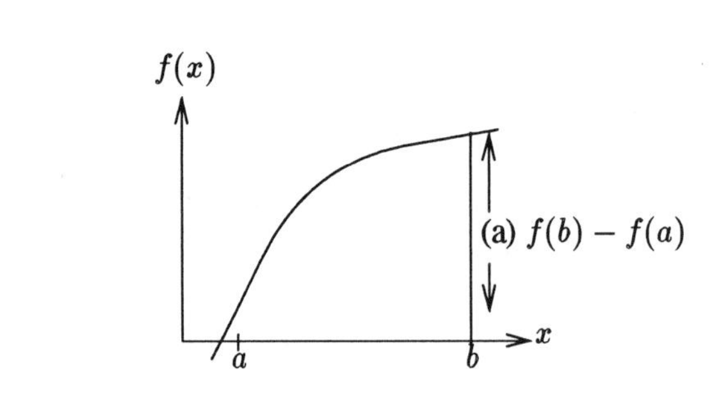
$f(x)$
(a) $f(b) - f(a)$
x
a
b

11.

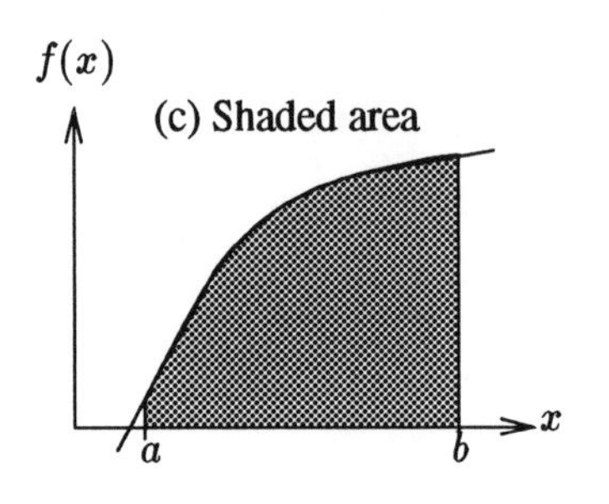
$f(x)$
(c) Shaded area
x
a
b

13.

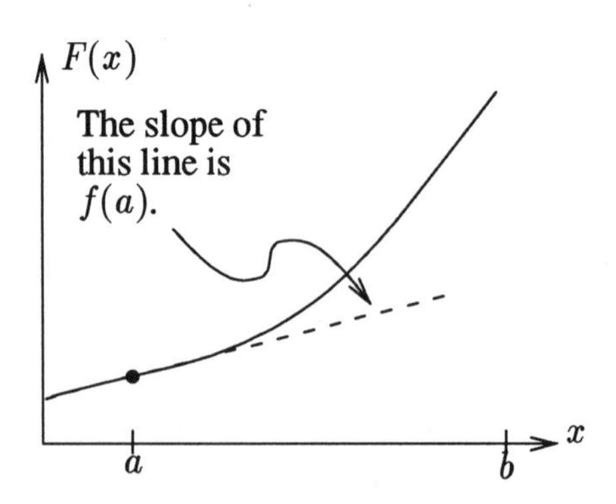
$F(x)$
The slope of
this line is
$f(a)$.
x
a
b

15.

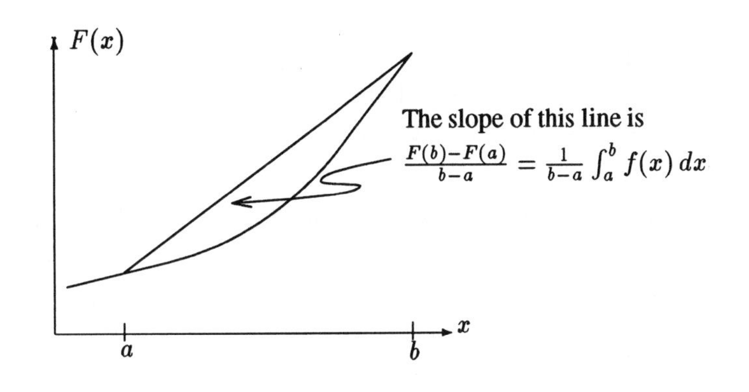

17. Let $v = f(t)$. If we can show that $\int_a^b f(t)dt$ is the change in position of the particle over the time interval $a < t < b$, then $\frac{1}{b-a}\int_a^b f(t)dt$ should give us the average velocity. But

$$\int_a^b f(t)\,dt = \int_0^b f(t)\,dt - \int_0^a f(t)\,dt = s(b) - s(a),$$

where $s(b) = \int_0^b f(t)\,dt$ and $s(a) = \int_0^a f(t)\,dt$ are positions of the particle at time $t = b$ and $t = a$, respectively. So $\int_a^b f(t)dt$ is indeed the change of position over $a < t < b$.

3.5 Solutions

1. For $0 \le t \le 10, 0 \le y \le 1$ the graph of $y = te^{-t}$ looks like the figure to the right. Therefore,

$$\lim_{t\to\infty} te^{-t} = 0.$$

Also, we are taking the limit of $\frac{t}{e^t}$ as $t \to \infty$, and as discussed in Section 1.5, powers like e^t grows much faster than t as $t \to \infty$.

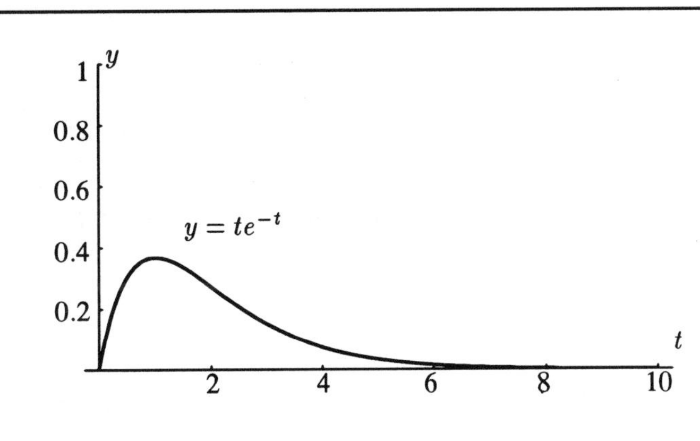

3. For $0 \leq x \leq 25$, $-1 \leq y \leq 1$, the graph of $\dfrac{\sin x}{x}$ is given to the right. As $x \to \infty$, the graph oscillates about the x-axis, but the amplitude of oscillation is constantly decreasing. By choosing x sufficiently large $y = \dfrac{\sin x}{x}$ can be brought as close to $y = 0$ as desired; we conclude that

$$\lim_{x\to\infty} \frac{\sin x}{x} = 0.$$

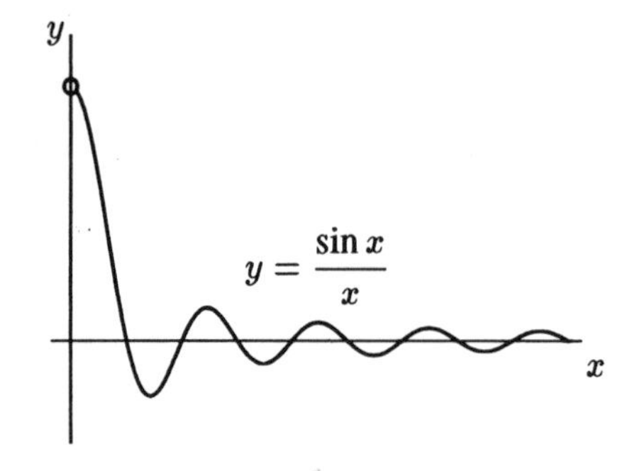

5. (a) It's true, for example, that since

$$\begin{aligned} \lim_{n\to\infty}(1+\frac{1}{n}) &= 1, \\ \lim_{n\to\infty}(1+\frac{1}{n})^3 &= (\lim_{n\to\infty}(1+\frac{1}{n}))^3 = 1^3, \end{aligned}$$

because the exponent, 3, has a fixed value, so we can bring it outside the limit. But it is not true that

$$\lim_{n\to\infty}(1+\frac{1}{n})^n = \lim_{n\to\infty} 1^n = 1$$

since the exponent is not fixed, but rather approaching infinity at the same time as the sum $1+\frac{1}{n}$ is approaching 1. Thus we are applying the limit only to part of the expression, namely the $\frac{1}{n}$, rather than taking the limit of the whole, which gives the incorrect result. In fact we can verify that

$$2.25 = (1+\frac{1}{2})^2 < (1+\frac{1}{n})^n \text{ for any } n > 2.$$

We will prove this result in Exercise 33 in the Miscellaneous Exercises for Chapter 4.

(b) Your calculator first finds $1+\frac{1}{n}$ and then raises it to the nth power. But it can only do this with a certain number of digits of accuracy, and so if n gets too large, $\frac{1}{n}$ becomes so small that the calculator rounds $1+\frac{1}{n}$ down to exactly 1, and returns the result $1^n = 1$.

3.6 Answers to Miscellaneous Exercises for Chapter 3

1. For $n = 210$, the left sum ≈ 1.466 (rounding up) and the right sum ≈ 1.417 (rounding down). Since the left and right sums differ by 0.049, their average must be within 0.245 of the true value, so the value of the integral will be 1.442 to one decimal place accuracy.

3. For $n = 5$, the left sum ≈ 1.338 (rounding up) and the right sum ≈ 1.282 (rounding down). Since the function is monotone over the interval [0,1], as can be seen with a graphing calculator, the actual value is between the two sums. Since the left and right sums differ by

0.056, their average must be within 0.28 of the true value, so the value of the integral correct to one decimal place is 1.31.

5. For $n = 10$, LHS ≈ -0.086 (rounding down) and RHS ≈ -0.080 (rounding up). Since the left and right sums differ by 0.006, their average must be within 0.003 of the true value, so $\int_2^3 -\frac{1}{(r+1)^2}\,dx = -0.083$ to one decimal place.

7. (a) $F(0) = 0$.
   (b) F increases because $F'(x) = e^{-x^2}$ is positive.
   (c) Using right and left sums, we get $F(1) \approx 0.7468$, $F(2) \approx 0.8821$, $F(3) \approx 0.8862$.

9. (a) For $n = 10$, the left sum ≈ 0.054 (rounding down), and the right sum ≈ 0.058 (rounding up). Since the left and right sums differ by 0.004, their average must be within 0.002 of the true value, so the value of the integral is 0.056 to one decimal place.
   (b) Using the Fundamental Theorem of Calculus we find that the integral is $F(0.4) - F(0.2) = \frac{1}{2}(\sin^2(0.4) - \sin^2(0.2)) \approx 0.05609$.

11. (a) If the level first becomes acceptable at time t_1, then $R_0 = 4R(t_1)$, or

$$\frac{1}{4}R_0 = R_0 e^{-0.004t_1}$$
$$\frac{1}{4} = e^{-0.004t_1}.$$

Taking natural logs on both sides,

$$\ln\frac{1}{4} = -0.004t_1$$
$$t_1 = \frac{\ln\frac{1}{4}}{-0.004} \approx 346.574 \text{ hours.}$$

   (b) The acceptable limit is 0.6 millirems/hour, so from the above, $R_0 = 4(0.6) = 2.4$. Since the rate at which radiation is emitted is $R(t) = R_0 e^{-0.004t}$,

$$\text{Total radiation emitted} = \int_0^{346.574} 2.4e^{-0.004t}\,dt.$$

Approximating the integral using left and right sums with 200 subdivisions, we find that about 450 millirems were emitted over that interval.

13. (a)

$$\text{Average population} = \frac{1}{10}\int_0^{10} 67.38(1.026)^t\,dt$$

Evaluating the integral numerically gives

$$\text{Average population} \approx 76.8 \text{ million}$$

(b) In 1980, $t = 0$, and $P = 67.38(1.026)^0 = 67.38$.
In 1990, $t = 10$, and $P = 67.38(1.026)^{10} = 87.10$.
Average= $\frac{1}{2}(67.38 + 87.10) = 77.24$ million.

(c) If P had been linear, the average value found in (a) would have been the one we found in (b). Since the population graph is concave up, it is below the secant line. Thus, the actual values of P are less than the corresponding values on the secant line, and so the average found in (a) is smaller than that in (b).

15. (a) The acceleration is positive for $0 \le t < 40$ and for a tiny period before $t = 60$, since the slope is positive over these intervals. Just to the left of $t = 40$, it looks like the acceleration is approaching 0. Between $t = 40$ and a moment just before $t = 60$, the acceleration is negative.

(b) The maximum altitude was about 500 feet, when t was a little greater than 40 (here we are estimating the area under the graph for $0 \le t \le 40$).

(c) The magnitude of the acceleration is largest when the absolute value of the slope of the velocity is greatest. This happens just after $t = 40$, where the velocity is plunging, where the direction of the acceleration is negative, or down.

(d) After the Montgolfier Brothers hit their top climbing speed (at $t = 40$), they suddenly stopped climbing and started to fall. This suggests some kind of catastrophe—the flame going out, the balloon ripping, etc. (In actual fact, in their first flight in 1783, the material covering their balloon, held together by buttons, ripped and the balloon landed in flames.)

(e) The total change in altitude for the Montgolfiers and their balloon is just the integral of their velocity, or just the total area under the given graph (counting the part after $t = 40$ as negative, of course). As mentioned before, the total area of the graph for $0 \le t \le 40$) is about 500. The area for $t > 40$ is about the area of a triangle with base 20 and height 18, or about 180. So subtracting, we see that the balloon finished 320 feet or so higher than where it began.

17. Let's just express (a)–(c) more precisely. (a) is the slope of a tangent line at $x = 1$, which looks to be about -2 (note that the scales on the x- and y-axes are not quite equal). (b) is

$$\frac{\int_0^a f(x)\,dx}{a}$$

which, estimating the area under the curve ($\int_0^a f(x)\,dx$) to be 1, and estimating a to be about 1.5, is about 0.67. As for (c), let's note that the rate of change in $y = f(x)$ at x is just $f'(x)$, and the average value of this over $0 \le x \le a$ is just

$$\frac{1}{a}\int_0^a f'(x)\,dx = \frac{(f(a) - f(0))}{a}$$

by the Fundamental Theorem. Letting $a = 1.5$ again, and using $f(a) = 0$ and $f(0) = 1$, we estimate that (c) is about -0.67. (d) is the area under the curve, which we estimated to be about 1. Putting these in order, we have

$$\text{(a)} < \text{(c)} < \text{(b)} < \text{(d)}.$$

Chapter 4

4.1 Solutions

1. (a) $f(x) = -3x + 2, g(x) = 2x + 1.$

$$\begin{aligned} k(x) &= f(x) + g(x) \\ &= (-3x + 2) + (2x + 1) \\ &= -x + 3 \\ k'(x) &= -1. \end{aligned}$$

Also, $f'(x) = -3, g'(x) = 2$, so $f'(x) + g'(x) = -3 + 2 = -1$.

(b)

$$\begin{aligned} j(x) &= f(x) - g(x) \\ &= (-3x + 2) - (2x + 1) \\ &= -5x + 1 \\ j'(x) &= -5. \end{aligned}$$

Also, $f'(x) - g'(x) = -3 - 2 = -5$.

3. (a) $f(x) = 5x - 3, g(x) = -2x + 1.$
$f[g(x)] = f(-2x + 1) = 5(-2x + 1) - 3 = -10x + 5 - 3 = -10x + 2.$
So, $\frac{d}{dx}[f[g(x)]] = -10$.

(b) Note that the derivatives of f and g are: $f'(x) = 5, g'(x) = -2$, and the derivative of $f[g(x)] = -10 = f'(x)g'(x)$. So one might speculate that if f and g are linear functions, then the derivative of $f[g(x)]$ is $f'(x)g'(x)$. This is true, and it can be proved as follows. Consider general linear functions f and g:

$$\begin{aligned} f(x) &= m_1 x + b_1 & \qquad g(x) &= m_2 x + b_2 \\ f'(x) &= m_1 & \qquad g'(x) &= m_2. \end{aligned}$$

Then

$$f[g(x)] = m_1(m_2 x + b_2) + b_1 = m_1 m_2 x + m_1 b_2 + b_1,$$

and

$$\frac{d}{dx}\Big(f[g(x)]\Big) = m_1 m_2 = f'(x)g'(x).$$

5. Say $\lim_{x\to a} f(x) = A$ and $\lim_{x\to a} g(x) = B$, i.e., as x comes arbitrarily close to a, $f(x)$ comes arbitrarily close to A and $g(x)$ comes arbitrarily close to B. So, as $x \to a$, the sum $f(x)+g(x)$ comes arbitrarily close to $A + B$, i.e.,

$$\lim_{x\to a}[f(x) + g(x)] = A + B = \lim_{x\to a} f(x) + \lim_{x\to a} g(x).$$

4.2 Solutions

1. $y' = 12x^{11}$.

2. $y' = -12x^{-13}$.

3. $y' = \frac{4}{3}x^{\frac{1}{3}}$.

4. $y' = \frac{3}{4}x^{-\frac{1}{4}}$.

5. $y' = -\frac{3}{4}x^{-\frac{7}{4}}$.

6. $f'(x) = -4x^{-5}$

7. $f'(x) = \frac{1}{4}x^{-\frac{3}{4}}$.

9. $y' = 6x^{\frac{1}{2}} - \frac{5}{2}x^{-\frac{1}{2}}$.

11. $y' = -12x^3 - 12x^2 - 6$.

13. $y' = 6t - \frac{6}{t^{3/2}} + \frac{2}{t^3}$.

15. $y = x + \frac{1}{x} \Rightarrow y' = 1 - \frac{1}{x^2}$.

17. The problems with derivatives which don't exist at $x = 0$ are: 2, 4, 5, 6, 7, and 9.

19. So far, we can only take the derivative of powers of x and the sums of constant multiples of powers of x. Since we cannot write $\sqrt{x+3}$ in this form, we cannot yet take its derivative.

21. $y' = -\frac{2}{3z^3}$. (power rule and sum rule)

23. We cannot write $\frac{1}{3x^2+4}$ as the sum of powers of x multiplied by constants.

25. $y' = 3x^2$. (power rule)

27. Once again, the x is in the exponent and we haven't learned how to handle that yet.

29. (a) Since the power of x will go down by one every time you take a derivative (until the exponent is zero after which the derivative will be zero), we can see immediately that $f^{(8)}(x) = 0$.

    (b) $f^{(7)}(x) = 7 \cdot 6 \cdot 5 \cdot 4 \cdot 3 \cdot 2 \cdot 1 \cdot x^0 = 5040$.

31.

$$\begin{aligned} y' = 3x^2 - 18x - 16 &= 5 \\ 3x^2 - 18x - 21 &= 0 \\ x^2 - 6x - 7 &= 0 \\ (x+1)(x-7) &= 0 \\ x = -1 \text{ or } x = 7. \end{aligned}$$

$f(-1) = 7, f(7) = -209.$
Thus, the two points are $(-1, 7)$ and $(7, -209)$.

33. Decreasing means $f'(x) < 0$:

$$f'(x) = 4x^2(x-3) < 0, \quad \text{so} \quad x < 3, \text{ and } x \neq 0.$$

Concave up means $f''(x) > 0$:

$$f''(0) = 12x^2 - 24x > 0 \Rightarrow 12x(x-2) > 0 \Rightarrow x < 0 \quad \text{or} \quad x > 2.$$

So, both conditions hold for $x < 0$ or $2 < x < 3$.

35. (a) $p(x) = x^2 - x$.

$$p'(x) = 2x - 1 < 0 \Rightarrow x < \frac{1}{2}.$$

So, p is decreasing when $x < \frac{1}{2}$.

(b) $p(x) = x^{\frac{1}{2}} - x$.

$$p'(x) = \frac{1}{2}x^{-\frac{1}{2}} - 1 < 0 \Rightarrow \frac{1}{2}x^{-\frac{1}{2}} < 1 \Rightarrow x^{-\frac{1}{2}} < 2 \Rightarrow x^{\frac{1}{2}} > \frac{1}{2} \Rightarrow x > \frac{1}{4}.$$

Thus, $p(x)$ is decreasing when $x > \frac{1}{4}$.

(c) $p(x) = x^{-1} - x$.

$$p'(x) = -1x^{-2} - 1 < 0 \Rightarrow -x^{-2} < 1 \Rightarrow x^{-2} > -1,$$

which is always true where x^{-2} is defined, since $x^{-2} = \frac{1}{x^2}$ is always positive. Thus $p(x)$ is always decreasing, unless $x = 0$.

37. Yes. To see why, we solve the equation $13x\dfrac{dy}{dx} = y$ for $y = x^n$. To solve the equation, we first evaluate $\dfrac{dy}{dx} = \dfrac{d}{dx}(x^n) = nx^{n-1}$. The equation becomes

$$13x(nx^{n-1}) = x^n$$

But $13x(nx^{n-1} = 13n(x \cdot x^{n-1}) = 13nx^n$, so we have

$$13n(x^n) = x^n$$

For $x \neq 0$, we divide through by x^n to get $13n = 1$, so $n = \frac{1}{13}$. Thus, $y = x^{\frac{1}{13}}$ is a solution.

39. (a) $T = 2\pi\sqrt{\frac{l}{g}} = \frac{2\pi}{\sqrt{g}}\left(l^{\frac{1}{2}}\right)$, so $\frac{dT}{dl} = \frac{2\pi}{\sqrt{g}}\left(\frac{1}{2}l^{-\frac{1}{2}}\right) = \frac{\pi}{\sqrt{gl}}$.

(b) Since $\frac{dT}{dl}$ is positive, the period T increases as the length l increases.

41. (a) $A = \pi r^2$
$\frac{dA}{dr} = 2\pi r$.

(b) This is the formula for the circumference of a circle.

(c) $A'(r) = \lim_{h\to 0} \frac{A(r+h)-A(r)}{h}$
The numerator of the difference quotient denotes the area contained between the inner circle (radius r) and the outer circle (radius $r+h$). As h approaches 0, this area can be approximated by the product of the circumference of the inner circle and the "height" of the area, i.e., h. Dividing this by the denominator, h, we get A' = the circumference of the circle with radius r.

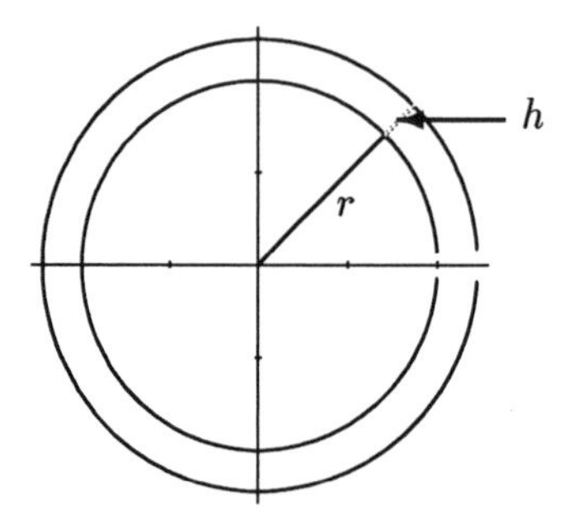

Also, if we think about the derivative of A as the rate of change of area for a small change in radius, the answer seems clear. If the radius increases by a tiny amount, the area will increase by a thin ring whose area is simply the circumference at that radius times the samll amount. To get the rate of change, we divide by the small amount and end up with the circumference as our rate of change.

43. $f(x) = \frac{1}{x} = x^{-1}$. $f'(x) = -x^{-2} = -\frac{1}{x^2}$.

The tangent line at $x = 1$ will have slope $f'(1) = -1$. Using the point $(1, 1)$ which lies on the line, we obtain the equation $y = -x + 2$. We approximate $f(2)$ by using the y value corresponding to $x = 2$, so $f(2) \approx 0$.

Similarly, the tangent line at $x = 100$ will have slope $f'(100) = \frac{-1}{(100)^2} = -0.0001$. The equation of the line is then $y = -0.0001x + 0.02$. The approximate value of $f(2)$ predicted by this tangent line is $f(2) \approx 0.0198$.

The actual value of $f(2)$ is $\frac{1}{2}$, so the approximation from $x = 100$ is better than that from $x = 1$. This is because the slope changes less between $x = 2$ and $x = 100$ than it does between $x = 1$ and $x = 2$.

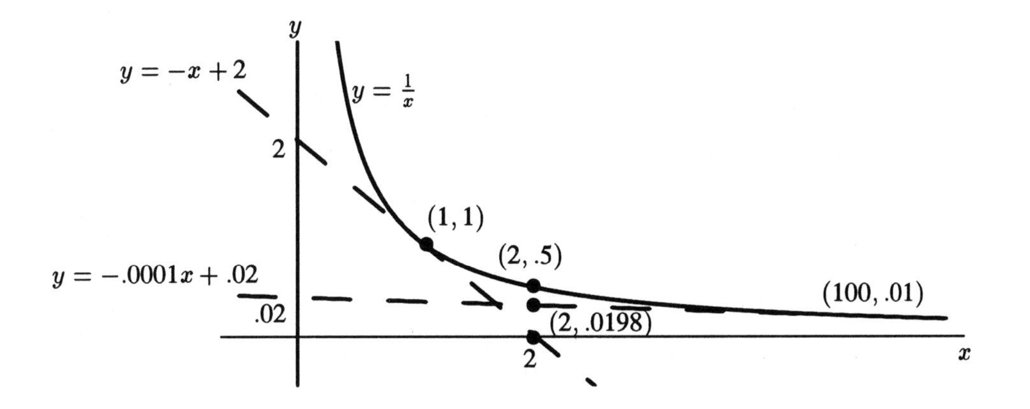

4.3 Solutions

1. $y' = 10t + 4e^t$.

3. $f'(x) = (\ln 2)2^x + 2(\ln 3)3^x$.

5. $\dfrac{dy}{dx} = 3 - 2(\ln 4)4^x$.

7. $f'(x) = ex^{e-1}$.

9. $f(t) = e^t \cdot e^2$. Then, since e^2 is just a constant, $f'(t) = \frac{d}{dt}(e^t e^2) = e^2 \frac{d}{dt} e^t = e^2 e^t = e^{t+2}$.

11. $z' = (\ln 4)e^x$.

13. $f'(z) = (2\ln 3)z + (\ln 4)e^z$.

15. $y' = 2x + (\ln 2)2^x$.

17. We can take the derivative of the sum $x^2 + 2^x$, but not the product.

19. $y = e^5 e^x \Rightarrow y' = e^5 e^x = e^{x+5}$.

21. $f(s) = 5^s e^s = (5e)^s \Rightarrow f'(s) = \ln(5e) \cdot (5e)^s = (\ln(5e))5^s e^s = (\ln 5 + 1)5^s e^s$.

23. $f'(z) = (\ln\sqrt{4})(\sqrt{4})^z = (\ln 2)2^z$.

25. (a) $f(x) = 1 - e^x$ crosses the x-axis where $0 = 1 - e^x \Rightarrow e^x = 1 \Rightarrow x = 0$. $f'(x) = -e^x$ so $f'(0) = -e^0 = -1$.

    (b) $y = -x$

    (c) The negative reciprocal of -1 is 1, so the equation of the normal line is $y = x$.

27.
$$P = 1 \cdot (1.05)^t \Rightarrow \frac{dP}{dt} = \ln(1.05)1.05^t.$$

When $t = 10$,

$$\frac{dP}{dt} = (\ln 1.05)(1.05)^{10} \approx 0.07947\ldots\frac{\text{dollars}}{\text{year}} \approx 7.95\frac{\text{cents}}{\text{year}}.$$

29. (a) $P = 4.1(1 + 0.02)^t = 4.1(1.02)^t$.

    (b)

$$\begin{aligned}\frac{dP}{dt} &= 4.1\frac{d}{dt}(1.02)^t = 4.1(1.02)^t \ln 1.02\\ \left.\frac{dP}{dt}\right|_{t=0} &= 4.1(1.02)^0 \ln 1.02 \approx 8.12 \times 10^{-2}\\ \left.\frac{dP}{dt}\right|_{t=15} &= 4.1(1.02)^{15} \ln 1.02 \approx 0.11\,.\end{aligned}$$

$\frac{dP}{dt}$ is the rate of growth of the world's population; $\left.\frac{dP}{dt}\right|_{t=0}$ and $\left.\frac{dP}{dt}\right|_{t=15}$ are the rates of growth in the years 1975 and 1990, respectively.

31. The tangent line has slope $\frac{d}{dx}(e^x) = e^x$. At $x = 0$, this gives us a slope of 1, and the tangent line is $y = x + 1$. A function which is always concave up will always stay above any of its tangent lines. Thus $e^x \geq x + 1$ for all x.

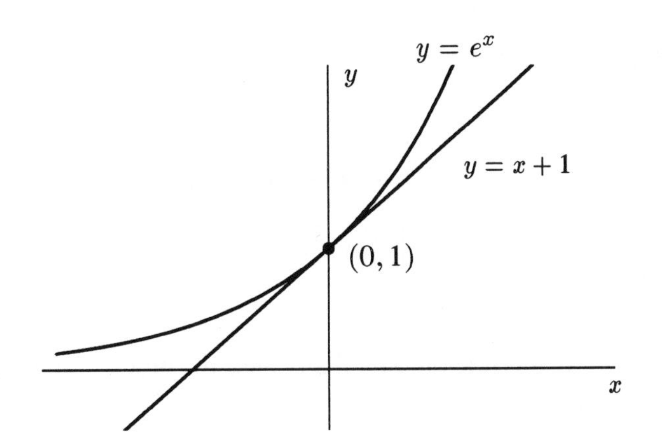

33.

x	3^x	Difference Quotient $= \frac{3^{x+h}-3^x}{h}$	Difference Quotient / 3^x
0	1	1.09921	1.09921
0.1	1.11612	1.22686	1.09921
0.2	1.24573	1.36932	1.09921
0.3	1.39038	1.52833	1.09921
0.4	1.55184	1.70581	1.09921
0.5	1.73205	1.90389	1.09921

We conclude that $\frac{d}{dx}(3^x) \approx (1.09921) \cdot 3^x$.

4.4 Solutions

1. By the product rule,

$$f'(x) = 2x(x^3 + 5) + x^2(3x^2) = 2x^4 + 3x^4 + 10x = 5x^4 + 10x.$$

Alternatively,

$$f'(x) = (x^5 + 5x^2)' = 5x^4 + 10x.$$

The two answers should, and do, match.

3. $f'(x) = x \cdot e^x + e^x \cdot 1 = e^x(x+1)$.

5. $y' = 2^x + x(\ln 2)2^x = 2^x(1 + x\ln 2)$.

7.

$$\begin{aligned} f'(x) &= (x^2 - x^{\frac{1}{2}}) \cdot 3^x(\ln 3) + 3^x\left(2x - \frac{1}{2}x^{-\frac{1}{2}}\right) \\ &= 3^x\left[(\ln 3)(x^2 - x^{\frac{1}{2}}) + \left(2x - \frac{1}{2\sqrt{x}}\right)\right]. \end{aligned}$$

9. It is easier to multiply this one out first than to use the product rule:
$z = s^4 - s \quad z' = 4s^3 - 1$.

11. $y' = \dfrac{(2t+5)(t+3) - (t^2+5t+2)}{(t+3)^2}$.

13.

$$\begin{aligned} f'(x) &= \frac{x(2x) - (x^2+3)(1)}{x^2} \\ &= \frac{2x^2 - x^2 - 3}{x^2} \\ &= \frac{x^2 - 3}{x^2}. \end{aligned}$$

15.

$$\begin{aligned} f'(x) &= \frac{(2+3x+4x^2)(1) - (1+x)(3+8x)}{(2+3x+4x^2)^2} \\ &= \frac{2+3x+4x^2 - 3 - 11x - 8x^2}{(2+3x+4x^2)^2} \\ &= \frac{-4x^2 - 8x - 1}{(2+3x+4x^2)^2}. \end{aligned}$$

16. Notice that you can cancel a z out of the numerator and denominator to get

$$\begin{aligned} f(z) &= \frac{3z}{5z+7}, z \neq 0 \\ \text{then } f'(z) &= \frac{(5z+7)3 - 3z(5)}{(5z+7)^2} \\ &= \frac{15z + 21 - 15z}{(5z+7)^2} \\ &= \frac{21}{(5z+7)^2}, z \neq 0. \end{aligned}$$

[If you used the quotient rule correctly without cancelling the z out first, your answer should simplify to this one, but it is usually a good idea to simplify as much as possible before taking derivatives.]

17.

$$f'(x) = 3(2x-5) + 2(3x+8) = 12x + 1$$

$$f''(x) = 12.$$

19. $f(x) = e^x \cdot e^x$
$f'(x) = e^x \cdot e^x + e^x \cdot e^x = 2e^{2x}$.

21. Since $\frac{d}{dx}e^{2x} = 2e^{2x}$ and $\frac{d}{dx}e^{3x} = 3e^{3x}$, we might guess that $\frac{d}{dx}e^{4x} = 4e^{4x}$.

23. This is the same function we were asked to differentiate in Problem 16, so we know that, if $x \neq 0$,

$$f'(x) = \frac{21}{(5x+7)^2}.$$

So at $x = 1$,

$$y = f(1) = \frac{3}{12} = \frac{1}{4},$$

$$y' = \frac{21}{144} = \frac{7}{48}.$$

So,

$$y - \frac{1}{4} = \frac{7}{48}(x-1).$$

$$y = \frac{7}{48}x + \frac{5}{48}.$$

25. (a) $f(x) = \frac{x}{x^2-1}$.
$x = 1$ or -1 makes $f(x)$ undefined. All other values of x define f.

$$f'(x) = \frac{x^2 - 1 - (x)(2x)}{(x^2-1)^2} = \frac{-x^2-1}{(x^2-1)^2}.$$

As x gets closer to one, the denominator of f approaches zero while the numerator goes to 1. Thus, $f(x)$ approaches infinity, and is large in magnitude at $x \approx 1.01$. Now, observe that $f'(x)$ has $(x^2-1)^2$ in the denominator. Since $\frac{1}{(x^2-1)^2}$ is much larger than $\frac{1}{x^2-1}$ as x gets closer to one, $f'(x)$ will be larger in magnitude than $f(x)$ as x goes to 1 (though $f'(x)$ will be negative because its numerator approaches -2).

(b) $g(x) = \frac{x^2+3x-4}{x^2-1}$.
$x = 1$ or -1 makes $g(x)$ undefined. All other values of x define g.

$$\begin{aligned} g'(x) = \frac{(2x+3)(x^2-1)-(2x)(x^2+3x-4)}{(x^2-1)^2} &= \frac{-3x^2+6x-3}{(x^2-1)^2} \\ &= \frac{-3(x-1)^2}{(x^2-1)^2} = \frac{-3}{(x+1)^2} \end{aligned}$$

You can factor the numerator and denominator of g to get this formula:

$$g(x) = \frac{(x+4)(x-1)}{(x+1)(x-1)} = \frac{x+4}{x+1}, x \neq 1.$$

At $x = 1$, g is not defined. But it does not go to infinity there. A graph of g would look like a graph of $\frac{x+4}{x+1}$ with a hole at $x = 1$. One can also say the numerator and denominator of g both go to zero at the same rate as x gets close to 1, so there is no reason for g to be large at 1.01. For $g'(x)$, as x goes to 1, the denominator does not go to 0 but rather to 4 (while the numerator is -3), so $g'(x)$ will not be large.

27. (a) $f'(x) = (x-2)+(x-1)$.

(b) Think of f as the product of two factors, with the first as $(x-1)(x-2)$. (The reason for this is that we have already differentiated $(x-1)(x-2)$).

$$f(x) = [(x-1)(x-2)](x-3).$$

Now $f'(x) = [(x-1)(x-2)]'(x-3) + [(x-1)(x-2)](x-3)'$
Using the result of a):

$$\begin{aligned} f'(x) &= [(x-2)+(x-1)](x-3) + [(x-1)(x-2)] \cdot 1 \\ &= (x-2)(x-3)+(x-1)(x-3)+(x-1)(x-2). \end{aligned}$$

(c) Because we have already differentiated $(x-1)(x-2)(x-3)$, rewrite f as the product of two factors, the first being $(x-1)(x-2)(x-3)$:

$$f(x) = [(x-1)(x-2)(x-3)](x-4)$$

Now $f'(x) = [(x-1)(x-2)(x-3)]'(x-4) + [(x-1)(x-2)(x-3)](x-4)'$.

$$\begin{aligned} f'(x) &= [(x-2)(x-3)+(x-1)(x-3)+(x-1)(x-2)](x-4) \\ &\quad +[(x-1)(x-2)(x-3)] \cdot 1 \\ &= (x-2)(x-3)(x-4)+(x-1)(x-3)(x-4) \\ &\quad +(x-1)(x-2)(x-4)+(x-1)(x-2)(x-3). \end{aligned}$$

From the solutions above, one can observe that when f is a product, its derivative is obtained by differentiating each factor in turn (leaving the other factors alone), and adding the results.

29. Assume for $g(x) \neq f(x)$, $g'(x) = g(x)$ and $g(0) = 1$. Then for

$$h(x) = \frac{g(x)}{e^x}$$

$$h'(x) = \frac{g'(x)e^x - g(x)e^x}{(e^x)^2} = \frac{e^x(g'(x) - g(x))}{(e^x)^2} = \frac{g'(x) - g(x)}{e^x}.$$

But, since $g(x) = g'(x)$, $h'(x) = 0$, so $h(x)$ is constant. Thus, the ratio of $g(x)$ to e^x is constant. Since $\frac{g(0)}{e^0} = \frac{1}{1} = 1$, $\frac{g(x)}{e^x}$ must equal 1 for all x. Thus $g(x) = e^x = f(x)$ for all x, so f and g are the same function.

31. Note first that $f(v) = \frac{\text{liters}}{\text{km}}$, and $v = \frac{\text{km}}{\text{hour}}$.

(a) $g(v) = \frac{1}{f(v)}$. (This is $\frac{\text{km}}{\text{liter}}$.)

$$g'(v) = \frac{-f'(v)}{[f(v)]^2}.$$

So,

$$g(80) = \frac{1}{0.05} = 20\frac{\text{km}}{\text{liter}}.$$

$$g'(80) = \frac{-0.0005}{0.0025} = -\frac{1}{5}\frac{\text{km}}{\text{liter}}.$$

(b) $h(v) = v \cdot f(v)$. (This is $\frac{\text{km}}{\text{hour}} \cdot \frac{\text{liters}}{\text{km}} = \frac{\text{liters}}{\text{hour}}$.) $h'(v) = f(v) + v \cdot f'(v)$, so

$$h(80) = 80(0.05) = 4\frac{\text{liters}}{\text{hr}}.$$

$$h'(80) = 0.05 + 80(0.0005) = 0.09\frac{\text{liters}}{\text{hr}}.$$

(c) Part (a) tells us that at 80 km/hr, the car can go 20 km on 1 liter. Since the first derivative evaluated at this velocity is negative, this implies that as velocity increases, fuel efficiency decreases, i.e., at higher velocities the car will not go as far on 1 liter of gas. Part (b) tells us that at 80 km/hr, the car uses 4 liters in an hour. Since the first derivative evaluated at this velocity is positive, this means that at higher velocities, the car will use more gas per hour.

4.5 Solutions

1. $f'(x) = 99(x+1)^{98} \cdot 1 = 99(x+1)^{98}$.

3. $w' = 100(t^2+1)^{99}(2t) = 200t(t^2+1)^{99}$.

5. $w' = 100(\sqrt{t}+1)^{99}\left(\frac{1}{2\sqrt{t}}\right) = \frac{50}{\sqrt{t}}(\sqrt{t}+1)^{99}$.

7. $y' = \frac{3}{2}e^{\frac{3}{2}w}$.

9. $y' = \dfrac{3s^2}{2\sqrt{s^3+1}}$.

11. $y' = 1 \cdot e^{-t^2} + te^{-t^2}(-2t) = e^{-t^2} - 2t^2e^{-t^2}$.

12. $f'(x) = \dfrac{1}{2\sqrt{z}}e^{-z} - \sqrt{z}e^{-z}$.

13. We can write this as $f(z) = \sqrt{z}e^{-z}$, in which case it is the same as problem 12. So $f'(z) = \dfrac{1}{2\sqrt{z}}e^{-z} - \sqrt{z}e^{-z}$.

15. $f'(z) = -2(e^z+1)^{-3} \cdot e^z = \dfrac{-2e^z}{(e^z+1)^3}$.

17. $f'(x) = 6(e^{5x})(5) + (e^{-x^2})(-2x) = 30e^{5x} - 2xe^{-x^2}$.

19. $w' = (2t+3)(1-e^{-2t}) + (t^2+3t)(2e^{-2t})$.

21. $f'(x) = e^{-(x-1)^2} \cdot (-2)(x-1)$.

23. $f'(t) = 2(e^{-2e^{2t}})(-2e^{2t})2 = -8(e^{-2e^{2t}+2t})$.

25. The graph is concave down when $f''(x) < 0$.

$$
\begin{aligned}
f'(x) &= e^{-x^2}(-2x) \\
f''(x) &= \left[e^{-x^2}(-2x)\right](-2x) + e^{-x^2}(-2) \\
&= \frac{4x^2}{e^{x^2}} - \frac{2}{e^{x^2}} \\
&= \frac{4x^2-2}{e^{x^2}} < 0
\end{aligned}
$$

when

$$4x^2 - 2 < 0 \Rightarrow 4x^2 < 2 \Rightarrow x^2 < \frac{1}{2} \Rightarrow -\frac{1}{\sqrt{2}} < x < \frac{1}{\sqrt{2}}$$

27.

$$\begin{aligned} f'(x) &= [10(2x+1)^9(2)][(3x-1)^7] + [(2x+1)^{10}][7(3x-1)^6(3)] \\ &= (2x+1)^9(3x-1)^6[20(3x-1)+21(2x+1)] \\ &= [(2x+1)^9(3x-1)^6](102x+1) \\ f''(x) &= [9(2x+1)^8(2)(3x-1)^6 + (2x+1)^9(6)(3x-1)^5(3)](102x+1) + (2x+1)^9(3x-1)^6(102). \end{aligned}$$

29.

$$\begin{aligned} \frac{dQ}{dt} &= \frac{d}{dt}e^{-0.000121t} \\ &= -0.000121e^{-0.000121t} \end{aligned}$$

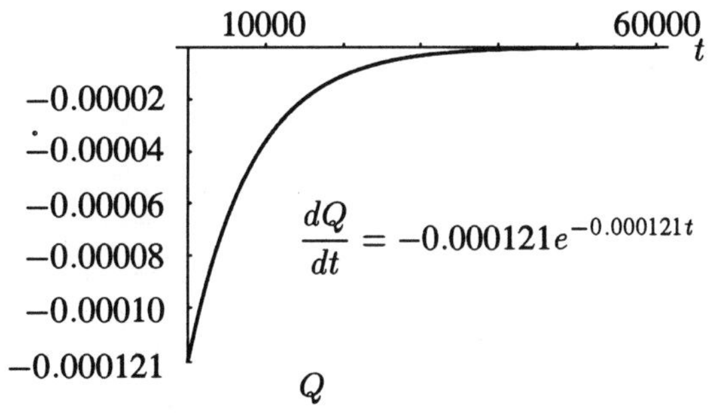

10000
60000
t
−0.00002
−0.00004
−0.00006
−0.00008
−0.00010
−0.000121
$\frac{dQ}{dt} = -0.000121e^{-0.000121t}$
Q

(a) (b)

31. (a)

$$\begin{aligned} \frac{dm}{dv} &= \frac{d}{dv}\left[m_0\left(1-\frac{v^2}{c^2}\right)^{-1/2}\right] \\ &= m_0\left(-\frac{1}{2}\right)\left(1-\frac{v^2}{c^2}\right)^{-3/2}\left(-\frac{2v}{c^2}\right) \\ &= \frac{m_0 v}{c^2}\frac{1}{\sqrt{\left(1-\frac{v^2}{c^2}\right)^3}}. \end{aligned}$$

(b) $\frac{dm}{dv}$ represents the rate of change of mass with respect to the speed v.

4.6 Solutions

1.

$$\begin{aligned} f(x) &= (1-\cos x)^{\frac{1}{2}} \\ f'(x) &= \frac{1}{2}(1-\cos x)^{-\frac{1}{2}}(-(-\sin x)) \\ &= \frac{\sin x}{2\sqrt{1-\cos x}}. \end{aligned}$$

3. $f'(x) = \cos(3x) \cdot 3 = 3\cos(3x)$.

5. $w' = e^t \cos(e^t)$.

7. $f'(x) = (e^{\cos x})(-\sin x) = -\sin x e^{\cos x}$.

9. $z' = e^{\cos\theta} - \theta(\sin\theta)e^{\cos\theta}$.

11. $f'(x) = 2\cos(2x)\sin(3x) + 3\sin(2x)\cos(3x)$.

13.

$$\begin{aligned} f'(x) &= (e^{-2x})(-2)(\sin x) + (e^{-2x})(\cos x) \\ &= -2\sin x(e^{-2x}) + (e^{-2x})(\cos x) \\ &= e^{-2x}[\cos x - 2\sin x]. \end{aligned}$$

15. $y' = 5\sin^4\theta\cos\theta$.

17. $z' = \dfrac{-3e^{-3\theta}}{\cos^2(e^{-3\theta})}$.

19. $h'(t) = 1 \cdot (\cos t) + t(-\sin t) + \frac{1}{\cos^2 t} = \cos t - t\sin t + \frac{1}{\cos^2 t}$.

21.

x	$\cos x$	Difference Quotient	$-\sin x$
0	1.0	−0.0005	0.0
0.1	0.995	−0.10033	−0.099833
0.2	0.98007	−0.19916	−0.19867
0.3	0.95534	−0.296	−0.29552
0.4	0.92106	−0.38988	−0.38942
0.5	0.87758	−0.47986	−0.47943
0.6	0.82534	−0.56506	−0.56464

23. (a) $\dfrac{dy}{dt} = -\dfrac{4.9\pi}{6}\sin\left(\dfrac{\pi}{6}t\right)$. It represents the rate of change of the depth of the water.

(b) $\dfrac{dy}{dt}$ is zero where the tangent line to the curve $y(t)$ is horizontal. $\dfrac{dy}{dt} = 0$ occurs when $\sin(\frac{\pi}{6}t) = 0$, or at $t =$ 6am, 12 noon, 6pm and 12 midnight. When $\dfrac{dy}{dt} = 0$, the depth of the water is no longer changing. Therefore, it has either just finished rising or just finished falling, and we know that the harbor's level is at a maximum or a minimum.

25. Even though sine is a periodic function, this one is not. Since e^x grows more rapidly for larger numbers and increases from near zero (for very negative numbers) to near infinity (for very positive numbers), this graph will look like the standard sine curve for angles from zero to infinity but stretched and squashed. For example, the part of the standard sine curve, $\sin z$, between $z = 0$ and $z = 1$ is in this case stretched out on the negative x-axis. On the positive x-axis, the sine curve is squashed more and more as the numbers get larger. $f(x)$ still has 1 as a max and -1 as a min. f has many zeros and can be negative. A graph of the function looks like this:

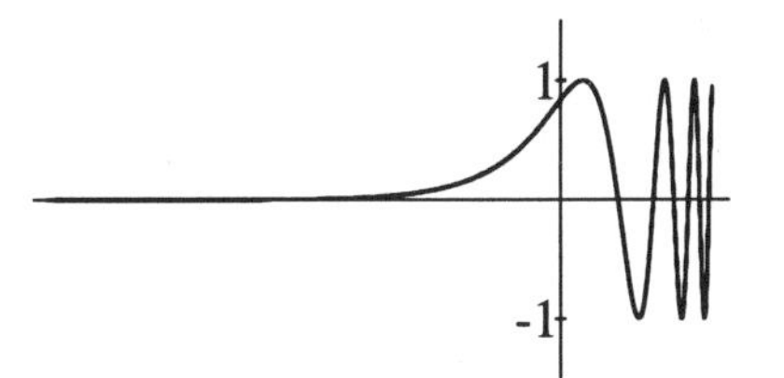

27.

$$\frac{d}{dx}(\sin x)|_{x=35^\circ} = \lim_{h\to 0} \frac{\sin(35^\circ + h) - \sin 35^\circ}{h}$$

Using $h = \pm 0.1, \pm 0.01, \pm 0.001$, we get $\frac{d}{dx}(\sin 35^\circ) \approx 0.014$.

29. The answer would be 0 no matter what $u'(-1)$ is. Algebraically, you can see this by noting that whatever $u'(-1)$ is, it gets multiplied by zero. Intuitively, the fact that the derivative of cosine is 0 at π means that it is not changing at π. Then, if you are looking at the value of $u(\cos t)$ at π, its value can't be changing either since $\cos t$ is not changing.

31. The tangent lines to $f(x) = \sin x$ have slope $\frac{d}{dx}(\sin x) = \cos x$. The tangent line at $x = 0$ has slope $f'(0) = \cos 0 = 1$ and goes through the point $(0, 0)$. Consequently, its equation is $y = g(x) = x$. The approximate value of $\sin \frac{\pi}{6}$ given by this equation is then $g(\frac{\pi}{6}) = \frac{\pi}{6} \approx 0.524$.

Similarly, the tangent line at $x = \frac{\pi}{3}$ has slope $f'(\frac{\pi}{3}) = \cos \frac{\pi}{3} = \frac{1}{2}$ and goes through the point $(\frac{\pi}{3}, \frac{\sqrt{3}}{2})$. Consequently, its equation is $y = h(x) = \frac{1}{2}x + \frac{3\sqrt{3}-\pi}{6}$. The approximate value of $\sin \frac{\pi}{6}$ given by this equation is then $h(\frac{\pi}{6}) = \frac{6\sqrt{3}-\pi}{12} \approx 0.604$.

The actual value of $\sin \frac{\pi}{6}$ is $\frac{1}{2}$, so the approximation from 0 is better than that from $\frac{\pi}{3}$. This is because the slope of the function changes less between $x = 0$ and $x = \frac{\pi}{6}$ than it does between $x = \frac{\pi}{6}$ and $x = \frac{\pi}{3}$.

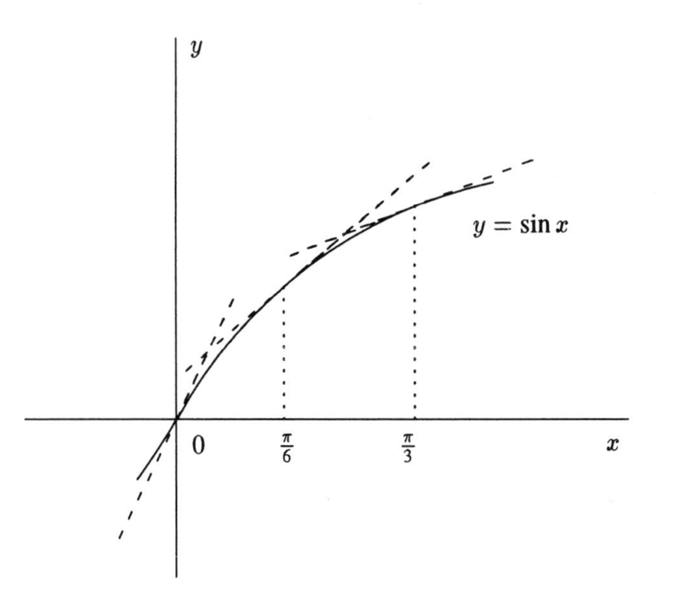

33. (a) Sector OAQ is a sector of a circle with radius $\frac{1}{\cos\theta}$ and angle $\Delta\theta$. Thus its area is the left side of the inequality. Similarly, the area of Sector OBR is the right side of the equality. The area of the triangle OQR is $\frac{1}{2}\Delta\tan\theta$ since it is a triangle with base $\Delta\tan\theta$ (the segment QR) and height 1 (if you turn it sideways, it is easier to see this). Thus, using the given fact about areas (which is also clear from looking at the picture), we have

$$\frac{\Delta\theta}{2\pi}\cdot\pi\left(\frac{1}{\cos\theta}\right)^2 \leq \frac{1}{2}\cdot\Delta(\tan\theta) \leq \frac{\Delta\theta}{2\pi}\cdot\pi\left(\frac{1}{\cos(\theta+\Delta\theta)}\right)^2.$$

(b) Dividing the inequality through by $\frac{\Delta\theta}{2}$ and cancelling the π's gives:

$$\left(\frac{1}{\cos\theta}\right)^2 \leq \frac{\Delta\tan\theta}{\Delta\theta} \leq \left(\frac{1}{\cos(\theta+\Delta\theta)}\right)^2$$

Then as $\Delta\theta \to 0$, the right and left sides both tend towards $\left(\frac{1}{\cos\theta}\right)^2$ while the middle (which is the difference quotient for tangent) tends to $(\tan\theta)'$. Thus, the derivative of tangent is "squeezed" between two values heading towards the same thing and must, itself, also tend to that value. Therefore, $(\tan\theta)' = \left(\frac{1}{\cos\theta}\right)^2$.

(c) Take the identity $\sin^2\theta + \cos^2\theta = 1$ and divide through by $\cos^2\theta$ to get $(\tan\theta)^2 + 1 = (\frac{1}{\cos\theta})^2$. Differentiating with respect to θ yields:

$$\begin{aligned} 2(\tan\theta)\cdot(\tan\theta)' &= 2\left(\frac{1}{\cos\theta}\right)\cdot\left(\frac{1}{\cos\theta}\right)' \\ 2\left(\frac{\sin\theta}{\cos\theta}\right)\cdot\left(\frac{1}{\cos\theta}\right)^2 &= 2\left(\frac{1}{\cos\theta}\right)\cdot(-1)\left(\frac{1}{\cos\theta}\right)^2(\cos\theta)' \end{aligned}$$

$$\begin{aligned} 2\frac{\sin\theta}{\cos^3\theta} &= (-1)2\frac{1}{\cos^3\theta}(\cos\theta)' \\ -\sin\theta &= (\cos\theta)'. \end{aligned}$$

(d)

$$\begin{aligned} \frac{d}{d\theta}\left(\sin^2\theta+\cos^2\theta\right) &= \frac{d}{d\theta}(1) \\ 2\sin\theta\cdot(\sin\theta)'+2\cos\theta\cdot(\cos\theta)' &= 0 \\ 2\sin\theta\cdot(\sin\theta)'+2\cos\theta\cdot(-\sin\theta) &= 0 \\ (\sin\theta)'-\cos\theta &= 0 \\ (\sin\theta)' &= \cos\theta. \end{aligned}$$

4.7 Solutions

1. $f'(x) = \frac{-1}{1-x} = \frac{1}{x-1}$.

3. $f'(z) = -1(\ln z)^{-2}\cdot\frac{1}{z} = \frac{-1}{z(\ln z)^2}$.

5. $f'(x) = \frac{1}{1-e^{-x}}\cdot -e^{-x}(-1) = \frac{e^{-x}}{1-e^{-x}}$.

7. $f'(x) = \frac{1}{e^x+1}\cdot e^x$.

9. Note that $f(x) = e^{\ln x}\cdot e^1 = x\cdot e = ex$. So $f'(x) = e$. (Remember, e is just a constant.)

   Or you might use the chain rule to get:

   $f'(x) = e^{(\ln x)+1}\cdot\frac{1}{x}$. [Are the two answers the same? Of course they are, since $e^{(\ln x)+1}\left(\frac{1}{x}\right) = e^{\ln x}\cdot e\left(\frac{1}{x}\right) = xe\left(\frac{1}{x}\right) = e$.]

11. $f(t) = \ln t$ (because $\ln e^x = x$ or because $e^{\ln t} = t$) $\Rightarrow f'(t) = \frac{1}{t}$.

13. $g'(t) = \dfrac{3}{(3t-4)^2+1}$.

15. $g(\alpha) = \alpha$, so $g'(\alpha) = 1$.

17.

$$\begin{aligned} \text{Let } g(x) &= \log x \\ \text{Then } 10^{g(x)} &= x \\ \text{Differentiating,}\quad (\ln 10)[10^{g(x)}]g'(x) &= 1 \\ g'(x) &= \frac{1}{(\ln 10)[10^{g(x)}]} \\ g'(x) &= \frac{1}{(\ln 10)x}. \end{aligned}$$

19. (a) $y = \ln x$
$y' = \frac{1}{x}$
$f'(1) = \frac{1}{1} = 1$

$y - y_1 = m(x - x_1)$
$y - 0 = 1(x - 1)$
$y = g(x) = x - 1$

(b) $g(1.1) = 1.1 - 1 = 0.1$
$g(2) = 2 - 1 = 1$

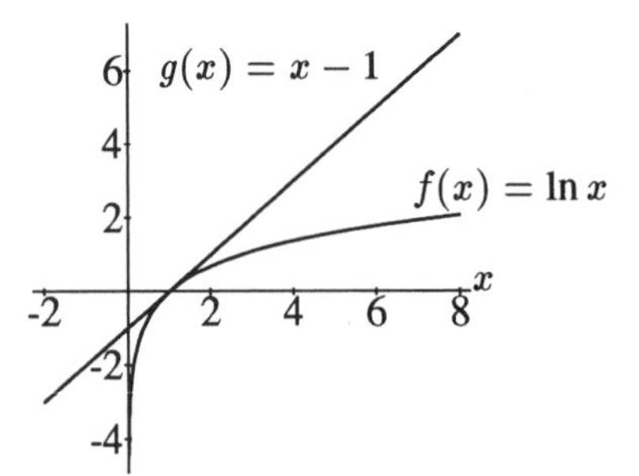

(c) $f(1.1)$ and $f(2)$ are below $g(x) = x - 1$. $f(0.9)$ and $f(0.5)$ are also below $g(x)$. This would be true for any approximation of this function by a tangent line since f is concave down ($f''(x) = -\frac{1}{x^2} < 0$ for all $x \neq 0$.) Thus, for a given x value, the y value given by the function is always below the value given by the tangent line.

21. (a)

$$\begin{aligned} f'(x) &= \frac{1}{1+x^2} + \frac{1}{1+\frac{1}{x^2}} \cdot (-\frac{1}{x^2}) \\ &= \frac{1}{1+x^2} + \left(-\frac{1}{x^2+1}\right) \\ &= \frac{1}{1+x^2} - \frac{1}{1+x^2} \\ &= 0 \end{aligned}$$

(b) f is a constant function. Checking at a few values of x,

x	$\arctan x$	$\arctan \frac{1}{x}$	$f(x) = \arctan x + \arctan \frac{1}{x}$
1	0.785392	0.7853982	1.5707963
2	1.1071487	0.4636476	1.5707963
3	1.2490458	0.3217506	1.5707963

4.8 Solutions

1. We differentiate implicitly both sides of the equation with respect to x.

$$2x + \left(y + x\frac{dy}{dx}\right) - 3y^2\frac{dy}{dx} = y^2 + x(2y)\frac{dy}{dx}\,,$$

$$x\frac{dy}{dx} - 3y^2\frac{dy}{dx} - 2xy\frac{dy}{dx} = y^2 - y - 2x\,,$$

$$\frac{dy}{dx} = \frac{y^2 - y - 2x}{x - 3y^2 - 2xy}$$

3. We differentiate implicitly both sides of the equation with respect to x.

$$x^{\frac{1}{2}} + y^{\frac{1}{2}} = 25\,,$$

$$\frac{1}{2}x^{-\frac{1}{2}} + \frac{1}{2}y^{-\frac{1}{2}}\frac{dy}{dx} = 0\,,$$

$$\frac{dy}{dx} = -\frac{\frac{1}{2}x^{-\frac{1}{2}}}{\frac{1}{2}y^{-\frac{1}{2}}} = -\frac{x^{-\frac{1}{2}}}{y^{-\frac{1}{2}}} = -\frac{\sqrt{y}}{\sqrt{x}} = -\sqrt{\frac{y}{x}}$$

5. We differentiate implicitly both sides of the equation with respect to x.

$$\ln y + x\frac{1}{y}\frac{dy}{dx} + 3y^2\frac{dy}{dx} = \frac{1}{x}\,,$$

$$\frac{x}{y}\frac{dy}{dx} + 3y^2\frac{dy}{dx} = \frac{1}{x} - \ln y\,,$$

$$\frac{dy}{dx}\left(\frac{x}{y} + 3y^2\right) = \frac{1 - x\ln y}{x}\,,$$

$$\frac{dy}{dx}\left(\frac{x + 3y^3}{y}\right) = \frac{1 - x\ln y}{x}\,,$$

$$\frac{dy}{dx} = \frac{(1 - x\ln y)}{x} \cdot \frac{y}{(x + 3y^3)}$$

7. Using the relation $\cos^2 y + \sin^2 y = 1$, the equation becomes:
$1 = y + 2$ or $y = -1$. Hence, $\dfrac{dy}{dx} = 0$.

9. First, we must find the slope of the tangent, $\left.\dfrac{dy}{dx}\right|_{(4,2)}$. Implicit differentiation yields:

$$2y\frac{dy}{dx} = \frac{2x(xy-4) - x^2\left(x\frac{dy}{dx} + y\right)}{(xy-4)^2}$$

Given the complexity of the above equation, we first want to substitute 4 for x and 2 for y (the coordinates of the point where we are constructing our tangent line), then solve for $\dfrac{dy}{dx}$.
Substitution yields:

$$2\cdot 2\frac{dy}{dx} = \frac{(2\cdot 4)(4\cdot 2-4) - (4)^2\left(4\frac{dy}{dx} + 2\right)}{(4\cdot 2-4)^2} = \frac{8(4) - 16(4\frac{dy}{dx} + 2)}{16} = -4\frac{dy}{dx}\,.$$

$$4\frac{dy}{dx} = -4\frac{dy}{dx}\,,$$

Solving for $\dfrac{dy}{dx}$, we have:

$$\frac{dy}{dx} = 0\,.$$

The tangent is a horizontal line,through $(4,2)$, hence its equation is $y = 2$.

11. Taking derivatives implicitly, we find

$$\begin{aligned}\frac{dy}{dx} + \cos y\frac{dy}{dx} + 2x &= 0\\ \frac{dy}{dx} &= \frac{-2x}{1+\cos y}\end{aligned}$$

So, at the point $x = 3, y = 0$,

$$\frac{dy}{dx} = \frac{(-2)(3)}{1+\cos 0} = \frac{-6}{2} = -3.$$

13. (a) If $x = 4$ then $16 + y^2 = 25 \Rightarrow y = \pm 3$. Finding $\dfrac{dy}{dx}$ implicitly:

$$\begin{aligned}2x + 2y\frac{dy}{dx} &= 0\\ \frac{dy}{dx} &= -\frac{x}{y}\end{aligned}$$

So the slope at $(4,3)$ is $-\frac{4}{3}$ and at $(4,-3)$ is $\frac{4}{3}$. The tangent lines are:

$$(y-3) = -\frac{4}{3}(x-4) \quad \text{and} \quad (y+3) = \frac{4}{3}(x-4)$$

(b) The normal lines have slopes that are the negative reciprocal of the slopes of the tangent lines. Thus,

$$(y-3) = \frac{3}{4}(x-4) \Rightarrow y = \frac{3}{4}x$$

and

$$(y+3) = -\frac{3}{4}(x-4) \Rightarrow y = -\frac{3}{4}x$$

are the normal lines.

(c) These lines meet at the origin, which is the center of the circle.

15. Solving the equation for x we get

$$x = \frac{6}{y} + y^2$$

whose graph looks like:

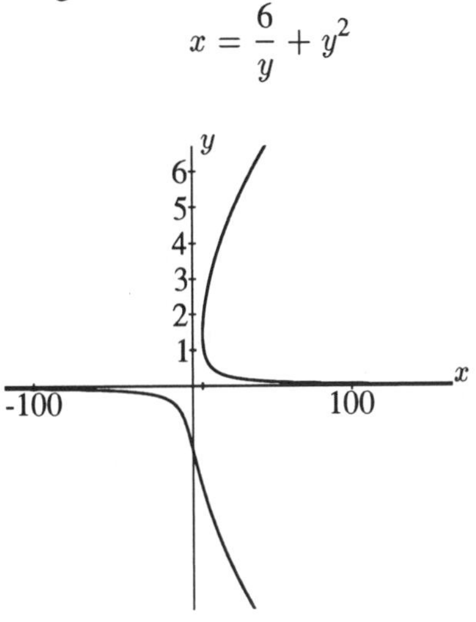

There appears to be one point where the tangent line is vertical. (The exact x-value of this point is indicated by a tick mark on the x-axis.) From the example, we see that this value is $x \approx 6.240$. For each x less than this value, there is one corresponding y value. For $x \approx 6.240$, there are two y values, and for each x greater than 6.240, there are three corresponding y values.

17. (a) Taking derivatives implicitly yields

$$e^{5y} + 5xe^{5y}\frac{dy}{dx} = 3\frac{dy}{dx} \Rightarrow \frac{dy}{dx} = \frac{e^{5y}}{3-5xe^{5y}}.$$

(b) Thus the slope at $(0,0)$ is $\dfrac{e^{5\cdot 0}}{3-(5\cdot 0)e^{5\cdot 0}} = \dfrac{1}{3}$. Then the equation of the tangent line is $y = \frac{1}{3}x$.

(c) When $x = 0.1$, $y \approx \frac{1}{3}\cdot 0.1 = 0.0333$.

4.9 Solutions

1. Let $f(x) = e^{-x}$. Then $f'(x) = -e^{-x}$. So $f(0) = 1$, $f'(0) = -e^0 = -1$. Therefore, $e^{-x} \approx f(0) + f'(0)x = 1 - x$.

3. With $f(x) = 1/x$, we see that the tangent line approximation to f near $x = 1$ is

$$f(x) \approx f(1) + f'(1)(x-1),$$

which becomes

$$\frac{1}{x} \approx 1 + f'(1)(x-1).$$

Since $f'(x) = -1/x^2$, $f'(1) = -1$. Thus our formula reduces to

$$\frac{1}{x} \approx 1 - (x-1) = 2 - x.$$

This is the local linearization of $1/x$ near $x = 1$.

5. The function e^{rt} would be the balance after t years if \$1 were compounded continuously at the nominal rate r (see Section 4.4). The function $1 + rt$ would be the balance after t years if \$1 were invested at simple interest at the nominal rate r. The local linearization $e^{rt} \approx 1 + rt$ tells us that in the beginning (when t is near zero), compounding continuously is approximately the same as simple interest.

7. (a) Since $\sin x \approx x$ for small x, we have

$$\cos x = 1 - 2\sin^2\frac{x}{2} \approx 1 - 2\left(\frac{x}{2}\right)^2 = 1 - \frac{x^2}{2}.$$

(b) i. $\displaystyle \lim_{h\to 0}\frac{\cos h - 1}{h} = \lim_{h\to 0}\frac{\left(1-\frac{h^2}{2}\right)-1}{h} = \lim_{h\to 0}\frac{-h^2/2}{h} = \lim_{h\to 0} -\frac{h}{2} = 0.$

ii. $\displaystyle \lim_{h\to 0}\frac{\cos h - 1}{h^2} = \lim_{h\to 0}\frac{\left(1-\frac{h^2}{2}\right)-1}{h^2} = \lim_{h\to 0}\frac{-h^2/2}{h^2} = \lim_{h\to 0} -\frac{1}{2} = -\frac{1}{2}.$

9. (a) Let $f(x) = 1/(1+x)$. Then $f'(x) = -1/(1+x)^2$ by the Chain rule. So $f(0) = 1$, and $f'(0) = -1$. Therefore, for x near 0, $1/(1+x) \approx f(0) + f'(0)x = 1 - x$.

(b) We know that for small y, $1/(1+y) \approx 1 - y$. Let $y = x^2$; when x is small, so is $y = x^2$. Hence, for small x, $1/(1+x^2) \approx 1 - x^2$.

(c) Since the linearization of $1/(1+x^2)$ is the line $y = 1$, and this line has a slope of 0, the derivative of $1/(1+x^2)$ is zero at $x = 0$.

11. Note that

$$[f(x)g(x)]' = \lim_{h\to 0}\frac{f(x+h)g(x+h) - f(x)g(x)}{h}.$$

We use the hint: For small h, $f(x+h) \approx f(x) + f'(x)h$, and $g(x+h) \approx g(x) + g'(x)h$. Therefore

$$\begin{aligned} f(x+h)g(x+h) - f(x)g(x) &\approx [f(x) + hf'(x)][g(x) + hg'(x)] - f(x)g(x) \\ &= f(x)g(x) + hf'(x)g(x) + hf(x)g'(x) \\ &\quad + h^2 f'(x)g'(x) - f(x)g(x) \\ &= hf'(x)g(x) + hf(x)g'(x) + h^2 f'(x)g'(x). \end{aligned}$$

Therefore

$$\begin{aligned} \lim_{h\to 0} \frac{f(x+h)g(x+h) - f(x)g(x)}{h} &= \lim_{h\to 0} \frac{hf'(x)g(x) + hf(x)g'(x) + h^2 f'(x)g'(x)}{h} \\ &= \lim_{h\to 0} \frac{h\left(f'(x)g(x) + f(x)g'(x) + hf'(x)g'(x)\right)}{h} \\ &= \lim_{h\to 0} \left(f'(x)g(x) + f(x)g'(x) + hf'(x)g'(x)\right) \\ &= f'(x)g(x) + f(x)g'(x). \end{aligned}$$

4.10 Answers to Miscellaneous Exercises for Chapter 4

1. $f'(x) = 6x(e^x - 4) + (3x^2 + \pi)e^x = 6xe^x - 24x + 3x^2e^x + \pi e^x.$

3. $\dfrac{d}{dz}\left(\dfrac{z^2+1}{\sqrt{z}}\right) = \dfrac{d}{dz}(z^{\frac{3}{2}} + z^{-\frac{1}{2}}) = \dfrac{3}{2}z^{\frac{1}{2}} - \dfrac{1}{2}z^{-\frac{3}{2}} = \dfrac{\sqrt{z}}{2}(3 - z^{-2}).$

5. $\dfrac{d}{dt}e^{(1+3t)^2} = e^{(1+3t)^2} \cdot 2(1+3t) \cdot 3.$

7. $\dfrac{d}{dx}xe^{\tan x} = e^{\tan x} + xe^{\tan x}\dfrac{1}{\cos^2 x}.$

9. $\dfrac{d}{dy}\ln\ln(2y^3) = \dfrac{1}{\ln(2y^3)}\dfrac{1}{2y^3}6y^2.$

11. $r'(\theta) = \dfrac{d}{d\theta}\sin[(3\theta - \pi)^2] = \cos[(3\theta-\pi)^2] \cdot 2(3\theta - \pi) \cdot 3 = 6(3\theta - \pi)\cos[(3\theta-\pi)^2].$

13. $\dfrac{d}{d\theta}\sqrt{a^2 - \sin^2\theta} = \dfrac{1}{2\sqrt{a^2 - \sin^2\theta}}(-2\sin\theta\cos\theta) = -\dfrac{\sin\theta\cos\theta}{\sqrt{a^2-\sin^2\theta}}.$

15. $w'(\theta) = \dfrac{d}{d\theta}\left(\theta(\sin\theta)^{-2}\right) = (\sin\theta)^{-2} + \theta\left(-2(\sin\theta)^{-3}\cos\theta\right) = \dfrac{1}{\sin^2\theta} - \dfrac{2\theta\cos\theta}{\sin^3\theta}.$

17. $h'(t) = \dfrac{d}{dt}\left(\ln\left(e^{-t} - t\right)\right) = \dfrac{1}{e^{-t} - t}\left(-e^{-t} - 1\right).$

19. $s'(x) = \frac{d}{dx}(\arctan(2-x)) = \frac{-1}{1+(2-x)^2}.$

21. $\frac{d}{dy}\left(\frac{y}{\cos y + a}\right) = \frac{\cos y + a - y(-\sin y)}{(\cos y + a)^2} = \frac{\cos y + a + y\sin y}{(\cos y + a)^2}.$

23. When we zoom in on the origin, we find that two functions are not defined there . The other functions all look like straight lines through the origin. The only way we can tell them apart is their slope.

    The following functions all have slope 0 and are therefore indistinguishable: $\sin x - \tan x$, $\frac{x^2}{x^2+1}$, $x - \sin x$, and $\frac{1-\cos x}{\cos x}$.

    These functions all have slope 1 at the origin, and are thus indistinguishable: $\arcsin x$, $\frac{\sin x}{1+\sin x}$, $\arctan x$, $e^x - 1$, $\frac{x}{x+1}$, and $\frac{x}{x^2+1}$.

    Now, $\frac{\sin x}{x} - 1$ and $-x\ln x$ both are undefined at the origin, so they are distinguishable from the other functions. In addition, while $\frac{\sin x}{x} - 1$ has a slope that approaches zero near the origin, $-x\ln x$ is becomes vertical near the origin, so they are distinguishable from each other.

    Finally, $x^{10} + \sqrt[10]{x}$ is the only function defined at the origin and with a vertical tangent there, so it is distinguishable from the others.

25. (a) $\frac{dg}{dr} = GM\frac{d}{dr}\left(\frac{1}{r^2}\right) = GM\frac{d}{dr}\left(r^{-2}\right) = GM(-2)r^{-3} = -\frac{2GM}{r^3}.$

    (b) $\frac{dg}{dr}$ is the rate of change of acceleration due to the pull of gravity. The further away from the center of the Earth, the weaker the pull of gravity is. So g is decreasing and therefore its derivative, $\frac{dg}{dr}$, is negative.

27. The population of Mexico is given by the formula

$$M = 84(1+0.026)^t = 84(1.026)^t \text{ millions}$$

and that of the U.S. by

$$U = 250(1+0.007)^t = 250(1.007)^t \text{ millions},$$

where t is measured in years ($t = 0$ corresponds to the year 1975). So,

$$\left.\frac{dM}{dt}\right|_{t=0} = \left.84\frac{d}{dt}(1.026)^t\right|_{t=0} = \left.84(1.026)^t\ln(1.026)\right|_{t=0} \approx 2.156$$

and

$$\left.\frac{dU}{dt}\right|_{t=0} = \left.250\frac{d}{dt}(1.007)^t\right|_{t=0} = \left.250(1.007)^t\ln(1.007)\right|_{t=0} \approx 1.744$$

Since $\left.\frac{dM}{dt}\right|_{t=0} > \left.\frac{dU}{dt}\right|_{t=0}$, the population of Mexico was growing faster in 1975.

29. (a)

$$\frac{dB}{dt} = P\left(1+\frac{r}{100}\right)^t \ln\left(1+\frac{r}{100}\right).$$

$\frac{dB}{dt}$ tells us how fast the amount of money in the bank is changing with respect to time for fixed initial investment P and interest rate r.

(b)

$$\frac{dB}{dr} = Pt\left(1+\frac{r}{100}\right)^{t-1}\frac{1}{100}.$$

$\frac{dB}{dr}$ indicates how fast the amount of money changes with respect to the interest rate r, assuming fixed initial investment P and time t.

31. (a)

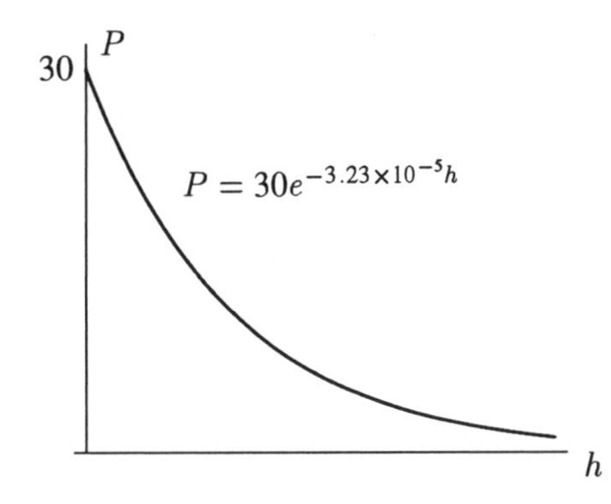

(b)

$$\frac{dP}{dh} = 30e^{-3.23\times10^{-5}h}(-3.23\times10^{-5})$$

so

$$\left.\frac{dP}{dh}\right|_{h=0} = -30(3.23\times10^{-5}) = -9.69\times10^{-4}$$

Hence, at $h = 0$, the slope of the tangent line is -9.69×10^{-4}, so the equation of the tangent line is

$$y - 30 = (-9.69\times10^{-4})(h-0)$$

or

$$y = (-9.69\times10^{-4})h + 30.$$

(c) The rule of thumb says

$$\begin{pmatrix}\text{Drop in pressure from}\\ \text{sea level to height } h\end{pmatrix} = \frac{h}{1000}$$

But since the pressure at sea level is 30, this drop in pressure is also $(30 - P)$, so

$$30 - P = \frac{h}{1000}$$

giving

$$P = 30 - 0.001h.$$

(d) The equations in (b) and (c) are almost the same: both have P intercepts of 30, and the slopes are almost the same $(9.69 \times 10^{-4} \approx 0.001)$. The rule of thumb calculates values of P which are very close to the tangent lines, and therefore yields values very close to the curve.

(e) The tangent line is slightly below the curve, and the rule of thumb line, having a slightly more negative slope, is slightly below the tangent line (for $h > 0$). Thus, the rule of thumb values are slightly smaller.

33. (a) From the figure, the slope of the secant to $y = \ln x$ between $x = 1$ and $x = 1 + \frac{1}{t}$ is given by

$$\text{slope} = \frac{\ln(1+\frac{1}{t}) - 0}{(1+\frac{1}{t}) - 1} = \frac{\ln(1+\frac{1}{t})}{\frac{1}{t}} = t\ln(1+\frac{1}{t}).$$

As $t \to \infty$, this slope approaches the value of the slope of the tangent line to $y = \ln x$ at $x = 1$. We know that the derivative of $\ln x$ is $\frac{1}{x}$, so $f'(1) = \frac{1}{1} = 1$, and we have $\lim_{t\to\infty} t\ln(1+\frac{1}{t}) = 1$. Thus,

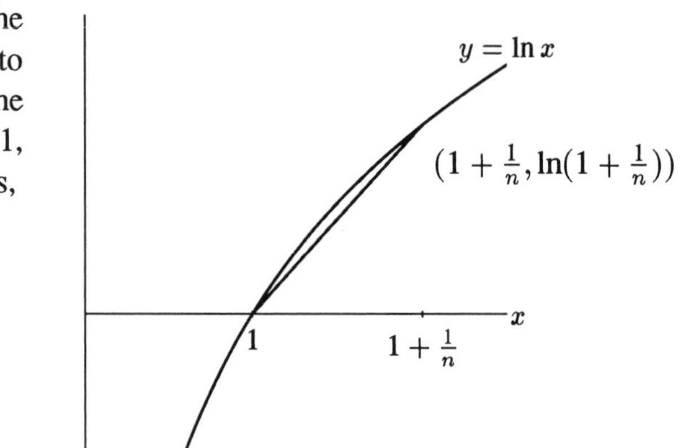

$$\begin{aligned}\lim_{n\to\infty}(1+\frac{1}{n})^n &= \lim_{n\to\infty} e^{n\ln(1+\frac{1}{n})} \\ &= e^{[\lim_{n\to\infty} n\ln(1+\frac{1}{n})]} \\ &= e^1 = e.\end{aligned}$$

(b) We have $\lim_{x\to\infty}(1+\frac{1}{x})^x = e$ from the above, replacing t by x. Now suppose r is a positive constant, and consider $\lim_{x\to\infty}(1+\frac{r}{x})^x$. As $x \to \infty$, $rx \to \infty$ as well. Thus replacing x with rx, we have

$$\begin{aligned}\lim_{x\to\infty}\left(1+\frac{r}{x}\right)^x &= \lim_{rx\to\infty}\left(1+\frac{r}{rx}\right)^{rx} = \lim_{rx\to\infty}\left(1+\frac{1}{x}\right)^{rx} \\ &= \lim_{x\to\infty}\left(1+\frac{1}{x}\right)^{rx} = \left[\lim_{x\to\infty}\left(1+\frac{1}{x}\right)^x\right]^r = e^r.\end{aligned}$$

Suppose one deposits P dollars in an account that gives an interest r. If this is compounded n times a year, at the end of that year the balance becomes

$$P\left(1+\frac{r}{n}\right)^n.$$

By the preceding, as the number of times interest is compounded goes to infinity, the balance approaches

$$\lim_{n\to\infty} P\left(1+\frac{r}{n}\right)^n = Pe^r.$$

One might expect that if the interest were compounded a large number of times, the balance after one year would be a sizable fortune. This argument shows that this is not the case.

For example, if an initial sum of \$100 is placed into an account which compounds interest twice annually at a rate $r = 0.07$, then after one year there is

$$\$100\left(1+\frac{0.07}{2}\right)^2 = \$107.12$$

in the account. On the other hand, no matter how many times the interest is compounded, the balance could be no greater than

$$\lim_{n\to\infty} \$100\left(1+\frac{0.07}{n}\right)^n = \$100e^{0.07} = \$107.25.$$

Thus the number of times a balance is compounded has very little effect on the overall interest.

(c) Suppose an initial balance is compounded n times annually at a rate of $r = 1$ (or 100%). The balance after one year is $P(1+\frac{1}{n})^n$. Clearly if P is compounded $n+1$ times, the balance will be greater, i.e.

$$P\left(1+\frac{1}{n+1}\right)^{n+1} > P\left(1+\frac{1}{n}\right)^n$$

or,

$$\left(1+\frac{1}{n+1}\right)^{n+1} > \left(1+\frac{1}{n}\right)^n.$$

(d) Using the argument in (a) we have:

slope of secant line 1 = $n\ln(1+\frac{1}{n})$

and

slope of secant line 2 = $n+1\ln(1+\frac{1}{n+1})$

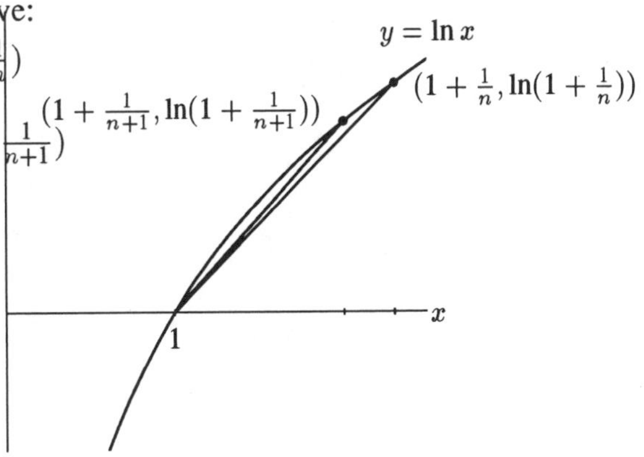

35. Since we're given the instantaneous rate of change T at $t = 30$ is 2, we want to choose a and b so that the derivative of T agrees with this value. Differentiating, $T'(t) = ab\cdot e^{-bt}$. Then we have

$$2 = T'(30) = abe^{-30b} \text{ or } e^{-30b} = \frac{2}{ab}$$

We also know that at $t = 30$, $T = 120$, so

$$120 = T(30) = 200 - ae^{-30b} \text{ or } e^{-30b} = \frac{80}{a}$$

Thus $\frac{80}{a} = e^{-30b} = \frac{2}{ab}$, so $b = \frac{1}{40} = 0.025$ and $a = 169.36$.

37. The radius r is related to the volume by the formula $V = \frac{4}{3}\pi r^3$. By implicit differentiation, we have

$$\frac{dV}{dt} = \frac{4}{3}\pi 3r^2 \frac{dr}{dt} = 4\pi r^2 \frac{dr}{dt}$$

The surface area of a sphere is $4\pi r^2$, so if $\frac{dV}{dt} = \frac{1}{3} \cdot S$, then

$$\frac{dV}{dt} = (4\pi r^2)\frac{dr}{dt} = S\frac{dr}{dt} \text{ so } \frac{dr}{dt} = \frac{1}{3}\mu\text{m}^3/\text{day}.$$

39. (a) We can approximate $\frac{d}{dx}[F(x)G(x)H(x)]$ using the large rectangular solids by which our original cube is increased:

Volume of whole − volume of original solid = change in volume.

$F(x+h)G(x+h)H(x+h) - F(x)G(x)H(x)$ = change in volume.

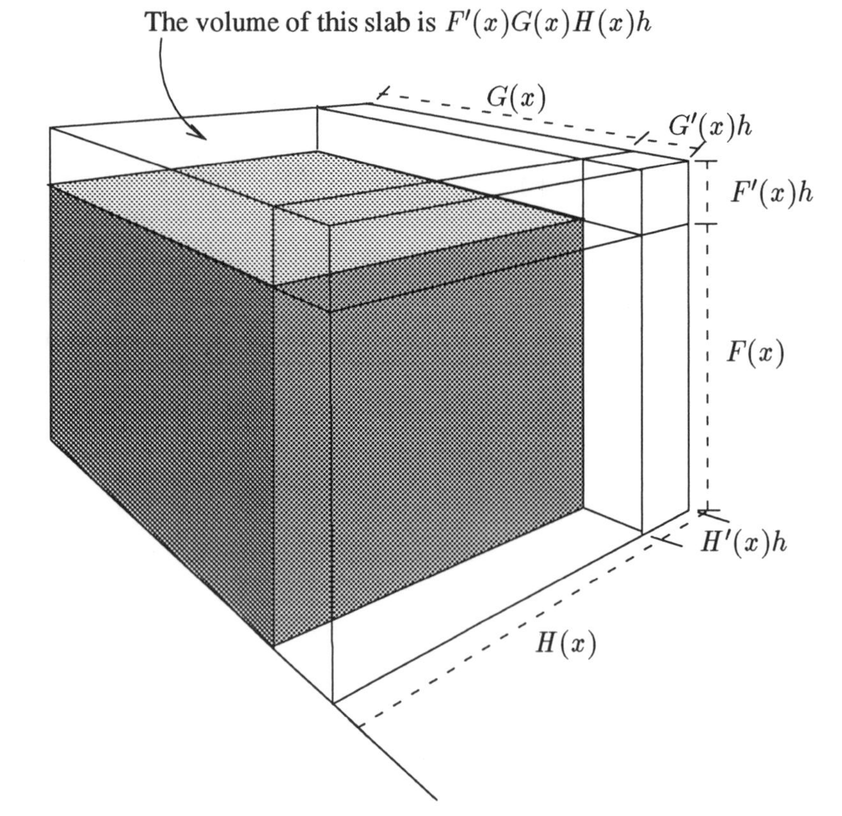

As in the book, we will ignore the smaller regions which are added (the long, thin rectangular boxes and the small cube in the corner.) This can be justified by recognizing that as $h \to 0$, these volumes will shrink much faster than the volumes of the big slabs and will therefore be insignificant. (Note that these smaller regions have an h^2 or h^3 in the formulas of their volumes.) Then we can approximate the change in volume above by:

$$\begin{aligned} F(x+h)G(x+h)H(x+h) - F(x)G(x)H(x) &\approx F'(x)G(x)H(x)h \quad \text{(the top slab)} \\ &+ F(x)G'(x)H(x)h \quad \text{(the front slab)} \\ &+ F(x)G(x)H'(x)h \quad \text{(the other slab)}, \end{aligned}$$

dividing by h gives

$$\begin{aligned} \frac{F(x+h)G(x+h)H(x+h) - F(x)G(x)H(x)}{h} &\approx F'(x)G(x)H(x) \\ &+ F(x)G'(x)H(x) + F(x)G(x)H'(x), \end{aligned}$$

letting $h \to 0$

$$(FGH)' = F'GH + FG'H + FGH'.$$

(b) Verifying,

$$\begin{aligned} &\frac{d}{dx}[(F(x) \cdot G(x)) \cdot H(x)] \\ &= (F \cdot G)'(H) + (F \cdot G)(H)' \\ &= [F'G + FG']H + FGH' \\ &= F'GH + FG'H + FGH' \end{aligned}$$

as before.

(c) From the answer to (b), we observe that the derivative of a product is obtained by differentiating each factor in turn (leaving the other factors alone), and adding the results. So, in general,

$$(f_1 \cdot f_2 \cdot f_3 \cdot \ldots \cdot f_n)' = f_1' f_2 f_3 \cdots f_n + f_1 f_2' f_3 \cdots f_n + \cdots + f_1 \cdots f_{n-1} f_n'.$$

Chapter 5

5.1 Solutions

1.

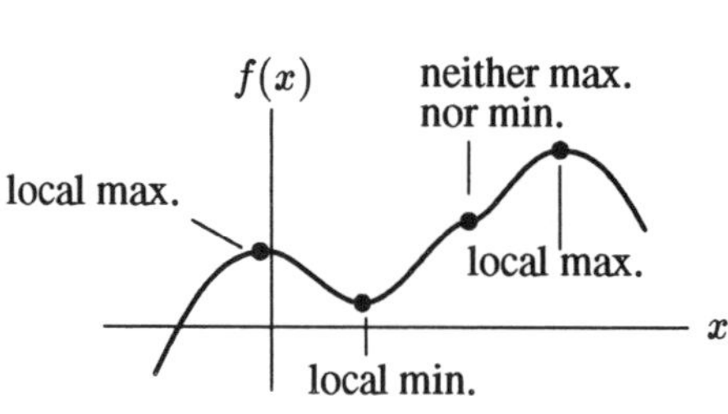

3. Let us examine the critical points, namely where $f' = 0$. Just before the first critical point $f' > 0$, meaning that the function f is increasing. Immediately after the critical point $f' < 0$, meaning that the function f is decreasing. Thus, the first point must be a maximum. To the left of the second critical point, $f' < 0$, and to its right, $f' > 0$; hence it is a minimum. On either side of the last critical point, $f' > 0$, so it is neither a maximum nor a minimum.

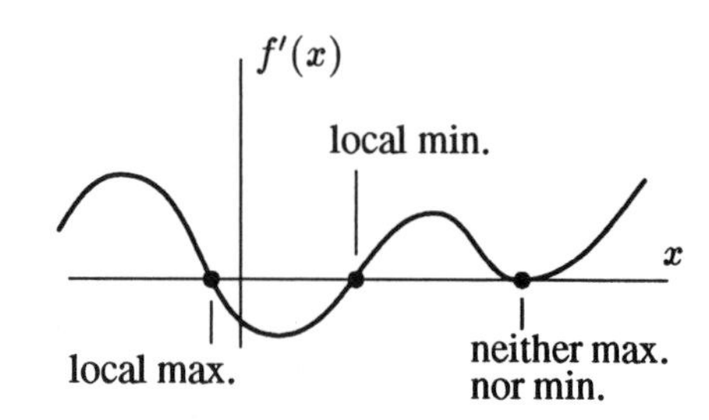

5. $f'(x) = 12x^3 - 12x^2$. To find critical points, we set $f'(x) = 0$. This implies $12x^2(x-1) = 0$. So the critical points of f are $x = 0$ and $x = 1$. To the left of $x = 0$, $f'(x) < 0$. Between $x = 0$ and $x = 1$, $f'(x) < 0$. To the right of $x = 1$, $f'(x) > 0$. Therefore, $f(1)$ is a local minimum, but $f(0)$ is not a local extremum.

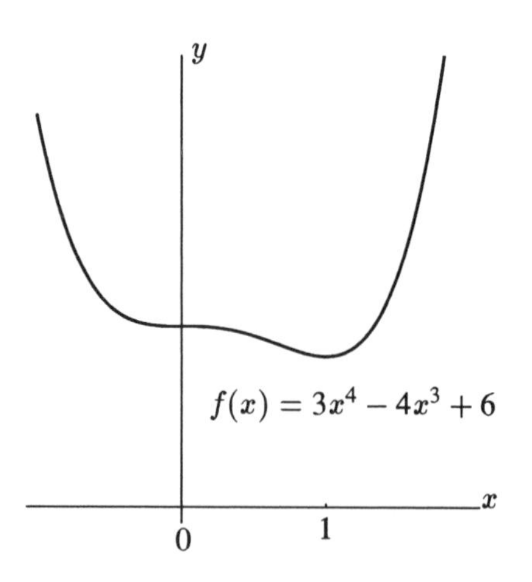

7. $f'(x) = 4(x^3 - 8)^3 3x^2$
$= 12x^2(x-2)^3(x^2+2x+4)^3$.
So the critical points are $x = 0$ and $x = 2$.
To the left of $x = 0$, $f'(x) < 0$.
Between $x = 0$ and $x = 2$, $f'(x) < 0$.
To the right of $x = 2$, $f'(x) > 0$.
Thus, $f(2)$ is a local minimum, whereas $f(0)$ is not a local extremum.

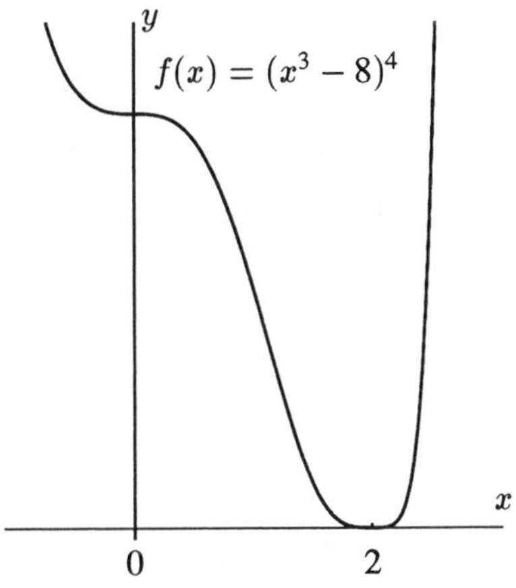

y
$f(x) = (x^3 - 8)^4$
x
0
2

9. $f'(x) = 4xe^{5x} + 2x^2e^{5x} \cdot 5 = 2e^{5x}x(5x+2)$. Notice that $e^{5x} > 0$ for all x. So, the critical points are $x = 0$ and $x = -2/5$.
To the left of $x = -2/5$, $f'(x) > 0$.
Between $x = -2/5$ and $x = 0$, $f'(x) < 0$.
To the right of $x = 0$, $f'(x) > 0$.
So, $f(-2/5)$ is a local maximum, $f(0)$ a local minimum. Notice that in the figure, you can barely discern the local maximum and minimum.

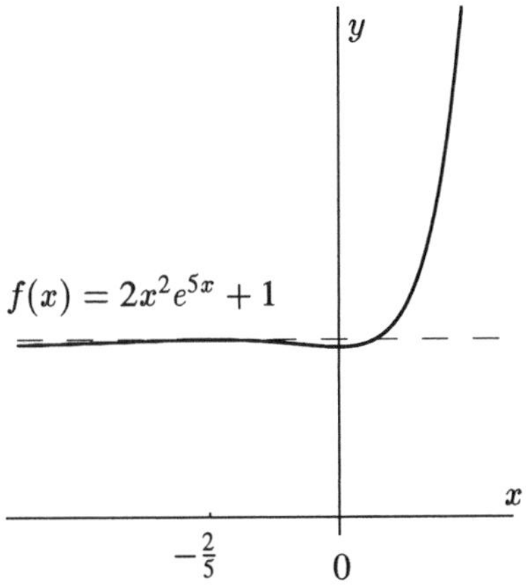

y
$f(x) = 2x^2e^{5x} + 1$
x
$-\frac{2}{5}$
0

11. (a) We have $f'(x) = 10x^9 - 10 = 10(x^9 - 1)$. This is zero when $x = 1$, so $x = 1$ is a critical point of f. For values of x less than 1, x^9 is less than 1, and thus $f'(x)$ is negative when $x < 1$. Similarly, $f'(x)$ is positive for $x > 1$. Thus $f(1) = -9$ is a local minimum.
(b) We have, by looking at the endpoints and the critical point, $f(0) = 0$, $f(1) = -9$, and $f(2) = 1004$. Thus the global minimum is at $x = 1$, and the global maximum is at $x = 2$.

13. (a)

$$\begin{aligned} f(x) &= \sin^2 x - \cos x \quad \text{for } 0 \le x \le \pi \\ f'(x) &= 2\sin x\cos x + \sin x = (\sin x)(2\cos x + 1) \end{aligned}$$

$f'(x) = 0$ when $\sin x = 0$ or when $2\cos x + 1 = 0$. Now, $\sin x = 0$ when $x = 0$ or when $x = \pi$. On the other hand, $2\cos x + 1 = 0$ when $\cos x = \frac{-1}{2}$, which happens when $x = \frac{2\pi}{3}$. So the critical points are $x = 0$, $x = \frac{2\pi}{3}$, and $x = \pi$.
Note that $\sin x > 0$ for $0 < x < \pi$. Also, $2\cos x + 1 < 0$ if $\frac{2\pi}{3} < x \le \pi$ and $2\cos x + 1 > 0$ if $0 < x < \frac{2\pi}{3}$. Therefore,

$$f'(x) \; < \; 0 \quad \text{for} \quad \frac{2\pi}{3} < x < \pi$$

$$f'(x) \;>\; 0 \quad \text{for} \quad 0 < x < \frac{2\pi}{3}.$$

Thus f has a local maximum at $x = \frac{2\pi}{3}$ and local minima at $x = 0$ and $x = \pi$.

(b) We have

$$\begin{aligned} f(0) &= [\sin(0)]^2 - \cos(0) = -1 \\ f\left(\frac{2\pi}{3}\right) &= \left[\sin\left(\frac{2\pi}{3}\right)\right]^2 - \cos\frac{2\pi}{3} = 1.25 \\ f(\pi) &= [\sin(\pi)]^2 - \cos(\pi) = 1. \end{aligned}$$

Thus the global maximum is at $x = \frac{2\pi}{3}$, and the global minimum is at $x = 0$.

15. (a)

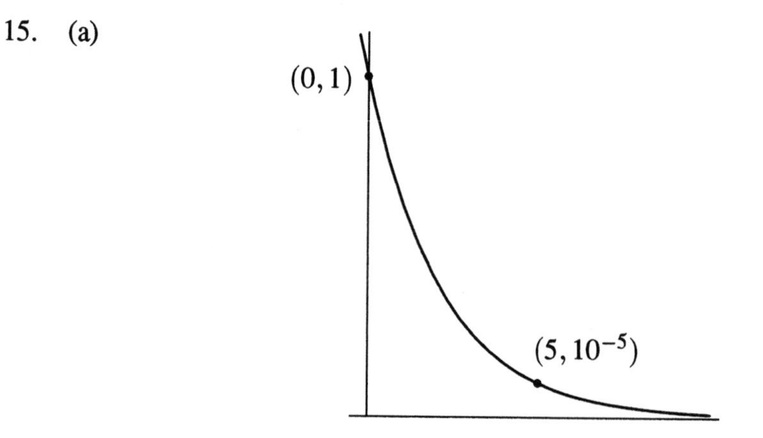

WARNING: This graph is not drawn to scale. 10^{-5} is so much smaller than 1 that you really couldn't see it on this graph if it had been drawn to scale.

(b)

$$f(0) = B = 1$$
$$f(5) = A = 10^{-5} = 0.00001$$

so $0.00001 \le 10^{-x} \le 1$.

17. Using the product rule on the function $f(x) = axe^{bx}$, we have $f'(x) = ae^{bx} + abxe^{bx}$. We want $f(\frac{1}{3}) = 1$, and since this is to be a maximum, we require $f'(\frac{1}{3}) = 0$. These conditions give

$$\begin{aligned} f(\frac{1}{3}) &= a\left(\frac{1}{3}\right)e^{\left(\frac{1}{3}\right)b} = 1, \\ f'(\frac{1}{3}) &= ae^{\left(\frac{1}{3}\right)b} + ab\left(\frac{1}{3}\right)e^{\left(\frac{1}{3}\right)b} = 0. \end{aligned}$$

Since $ae^{\left(\frac{1}{3}\right)b}$ is non-zero, we can divide both sides of the second equation by $ae^{\left(\frac{1}{3}\right)b}$ to obtain $0 = 1 + \frac{b}{3}$. This implies $b = -3$. Plugging $b = -3$ into the first equation gives us

$a(\frac{1}{3})e^{-1} = 1$, or $a = 3e$. How do we know we have a maximum at $x = \frac{1}{3}$ and not a minimum? Since $f'(x) = ae^{bx}(1 + bx) = (3e)e^{-3x}(1 - 3x)$, and $(3e)e^{-3x}$ is always positive, it follows that $f'(x) > 0$ when $x < \frac{1}{3}$ and $f'(x) < 0$ when $x > \frac{1}{3}$. Since f' is positive to the left of $x = \frac{1}{3}$ and negative to the right of $x = \frac{1}{3}$, $f(\frac{1}{3})$ is a local maximum.

19. Since $f(x) = 2x^3 - 9x^2 + 12x + 1$, $f'(x) = 6x^2 - 18x + 12 = 6(x^2 - 3x + 2) = 6(x - 2)(x - 1)$. Thus there are critical points at $x = 2$, $x = 1$. Using the formula, we find that $f(1) = 6$ and $f(2) = 5$. Using a graphing calculator and the above information as a guide, we can construct the following graph. We see that $x = 1$ gives a local max while $x = 2$ gives a local min. Furthermore, the only interval on which f is decreasing is $1 < x < 2$. Thus $f(x) = 10$ can have only one solution, $f(x) = 5$ has two solutions, $f(x) = 0$ has only one solution, and since $5 < 2e < 6$, $f(x) = 2e$ has 3 solutions.

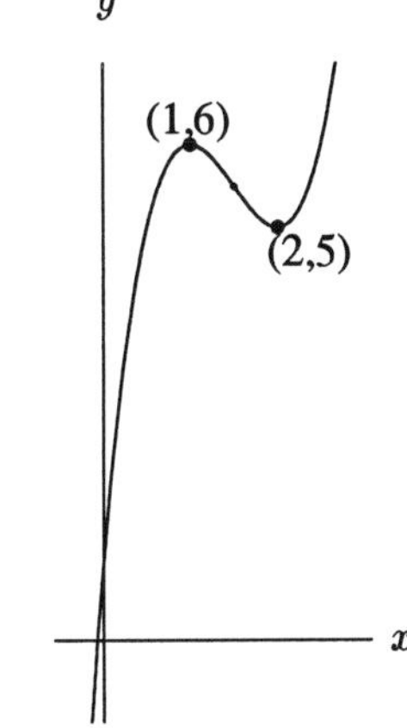

20. This is one of many possible functions which satisfy the conditions of the question:

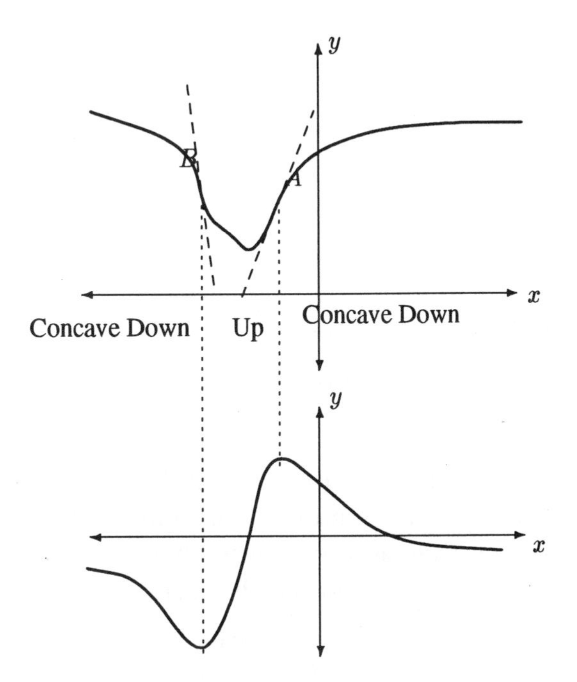

21. See mark of point A in problem 20. The slope of tangent lines to the left or right of A would have slopes smaller than that of the line through A. To the left of A, the graph is concave up; to the right, the graph is concave down.

23. Let $f(x) = \sin x$ and $g(x) = x$. Then $f(0) = 0$ and $g(0) = 0$. Also $f'(x) = \cos x$ and $g'(x) = 1$, so for all $x \geq 0$: $f'(x) \leq g'(x)$. So the graphs of f and g both go through the

origin and the graph of f climbs slower than the graph of g. Thus the graph of f is below the graph of g for $x \geq 0$. Thus $\sin x < x$ for $x > 0$ and $\sin x \leq x$ for $x \geq 0$.

25.

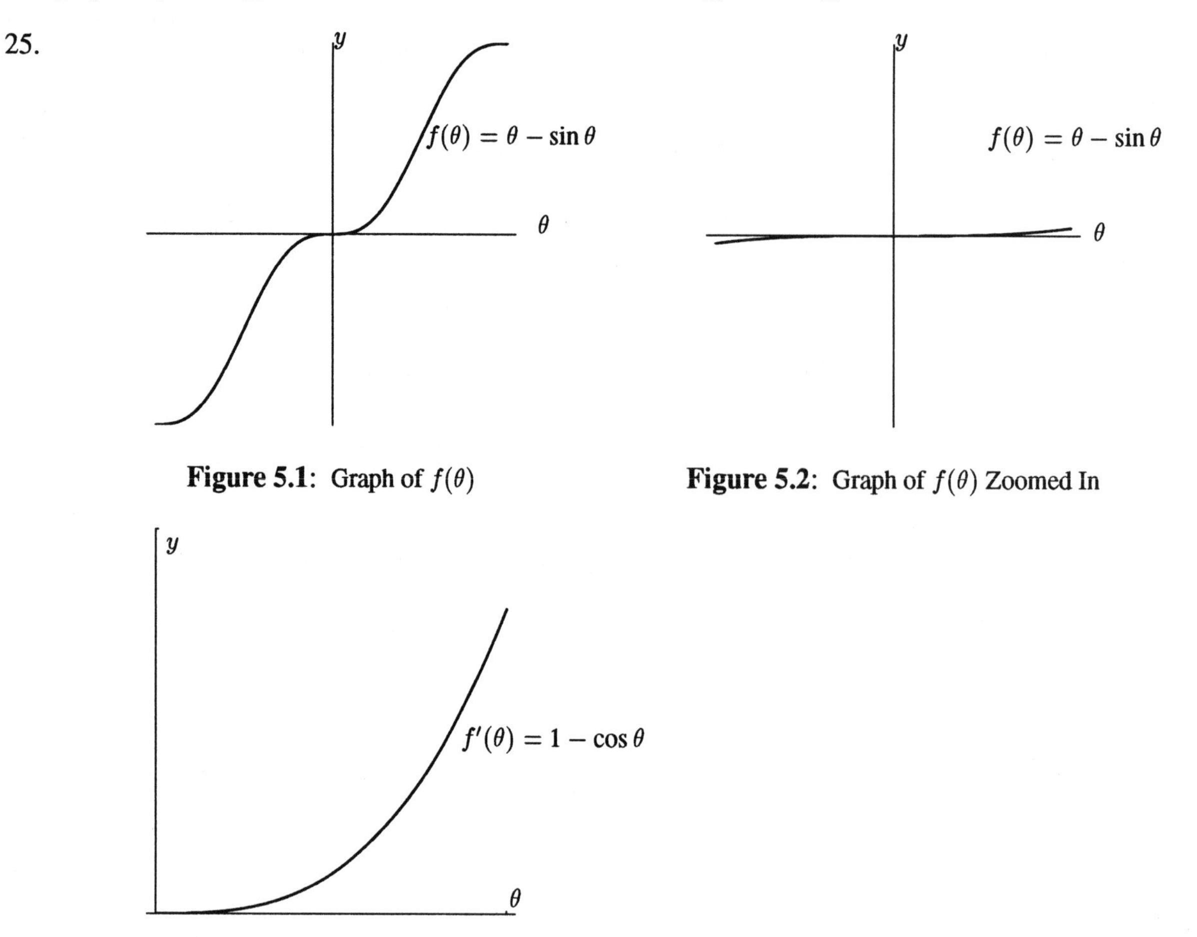

Figure 5.1: Graph of $f(\theta)$

Figure 5.2: Graph of $f(\theta)$ Zoomed In

Figure 5.3: Graph of $f'(\theta)$

(a) In Figure 5.1, we see that $f(\theta) = \theta - \sin\theta$ definitely has a zero at $\theta = 0$. To see if it has any other zeros near the origin, we use our calculator to zoom in. (See Figure 5.2.) No extra root seems to appear no matter how close to the origin we zoom. However, zooming can never tell you for sure that there is not a root that you have not found yet.

(b) Using the derivative we can argue that for sure that there is no other zero. $f'(\theta) = 1 - \cos\theta$. Since $\cos\theta < 1$ for $0 < \theta \leq 1$, $f'(\theta) > 0$ for $0 < \theta \leq 1$. Thus, f increases for $0 < \theta \leq 1$. Consequently, we conclude that the only zero of f is the one at the origin. If f had another zero at x_0, $x_0 > 0$, f would have to "turn around", and recross the x-axis at x_0. But if this were the case, f' would be nonpositive somewhere, which we know to be impossible.

27. (a) We want to find where $x > 2\ln x$, which is the same as solving $x - 2\ln x > 0$. Let $f(x) = x - 2\ln x$. Then $f'(x) = 1 - \frac{2}{x}$, which implies that $x = 2$ is the only critical

point of f. Since $f'(x) < 0$ for $x < 2$ and $f'(x) > 0$ for $x > 2$, by the first derivative test we see that f has a local minimum at $x = 2$. Since $f(2) = 2 - 2\ln 2 \approx 0.61$, then for all $x > 0$, $f(x) \geq f(2) > 0$. Thus $f(x)$ is always positive, which means $x > 2\ln x$ for any $x > 0$.

(b) We've shown that $x > 2\ln x$ for all $x > 0$, which is the same as saying $x > \ln x^2$. Thus $e^x > e^{\ln x^2} = x^2$, so $e^x > x^2$ for all $x > 0$.

(c) Let $f(x) = x - 3\ln x$. Then $f'(x) = 1 - \frac{3}{x} = 0$ at $x = 3$. By the first derivative test, f has a local minimum at $x = 3$. But, $f(3) \approx -0.295$, which is less than zero. Thus $3\ln x > x$ at $x = 3$. So, x is not less than $3\ln x$ for all $x > 0$.

(One could also see this by plugging in $x = e$: since $3\ln e = 3, x < 3\ln x$ when $x = e$.)

29. (a)

$$P(t) = \frac{2000}{1 + e^{(5.3 - 0.4t)}}$$

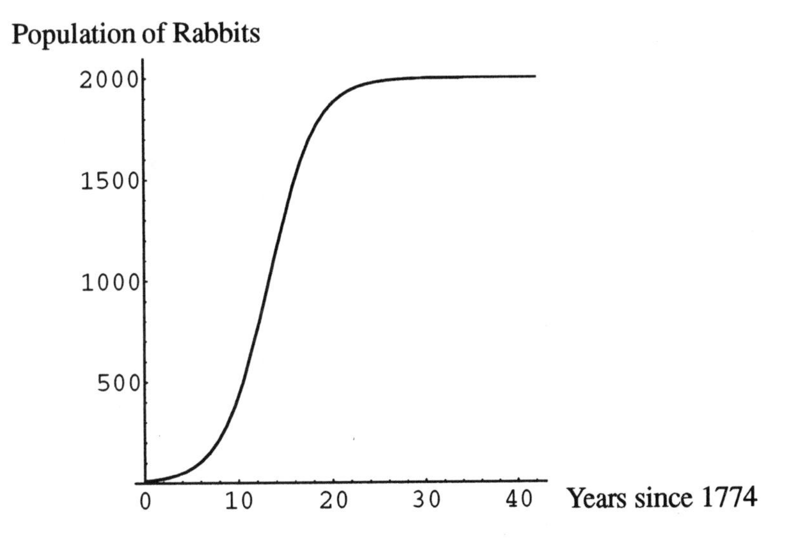

(b) The population appears to have been growing fastest when there were about 500 rabbits, just over 10 years after Captain Cook left them there.

(c) The rabbits reproduce quickly, so their population initially grew very rapidly. Limited food and space availability or perhaps predators on the island probably accounts for the population being unable to grow past 2000.

5.2 Solutions

1. To find inflection points of the function f we must find points where f'' changes sign. However, because f'' is a derivative of f', any point where f'' changes sign will be a local maximum or minimum on the graph of f'.

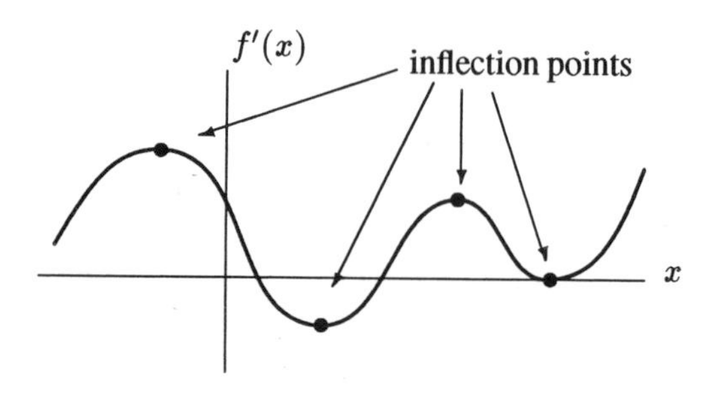

3. First, check that $x = 0$ is a critical point.

$$f'(0) = \cos 0 + 2 \cdot 0 - 1 = 1 + 0 - 1 = 0$$

so $x = 0$ is a critical point.

Now, $f''(x) = -2x \sin(x^2) + 2$. So $f''(0) = 0 + 2 = 2 > 0$.
Thus $x = 0$ is a local minimum.

5.

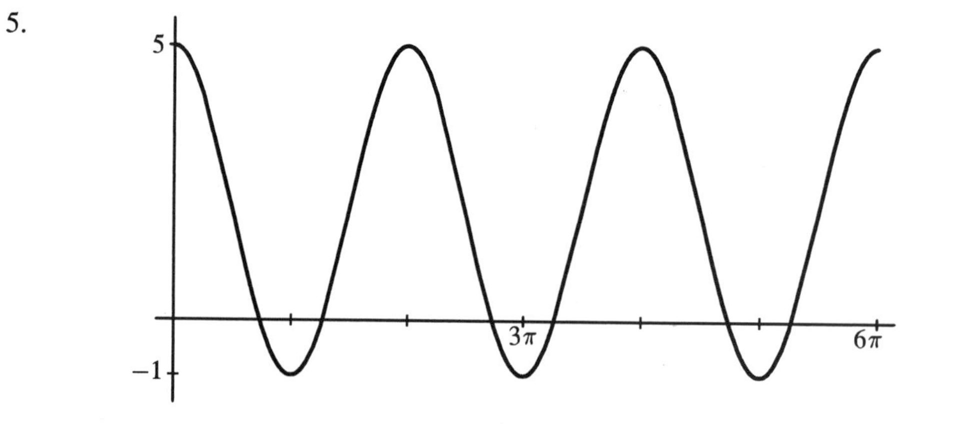

$f(x) = 2 + 3\cos x$, so $f'(x) = -3\sin x$, which is zero when $x = n\pi$ for $n = 0, 1, \ldots 5$.

The local maxima are where n is even, at $x = 0$, $x = 2\pi$, $x = 4\pi$, $x = 6\pi$. The local minima are where n is odd, at $x = \pi$, $x = 3\pi$, $x = 5\pi$. This is because when n is even, $\cos n\pi = 1$, the maximum of cosine, and so $f(x) = 5$ here, maximizing $f(x)$. When n is odd, $\cos n\pi = -1$, the minimum of cosine, and so $f(x) = -1$ here, minimizing $f(x)$..

$f''(x) = -3\cos x$, which is zero when $x = (2n+1)\pi/2$ for $n = 0, 1, \ldots 5$, so there are inflection points at $x = \pi/2, 3\pi/2, 5\pi/2, 7\pi/2, 9\pi/2$, and $11\pi/2$. These are the points with the steepest slopes, since $f'(x)$ is either maximized or minimized here.

7. (If you read Chapter 8, you'll see that this curve is just $\sqrt{2}\sin(x+\pi/4)$, making this problem fairly simple. We'll proceed without that helpful fact.)

   Since both $\sin x$ and $\cos x$ have period 2π, the same is true for $f(x)$. Therefore we need only graph $f(x)$ for $0 \le x \le 2\pi$ and repeat the graph two more times.

   We have $f'(x) = \cos x - \sin x = 0$ when $\cos x = \sin x$, or when $1 = \frac{\sin x}{\cos x} = \tan x$. For $0 \le x \le 2\pi$, this has the solutions $x = \pi/4$ and $x = 5\pi/4$. Thus $f(x)$ has critical points at $x = \frac{\pi}{4}$ and $x = \frac{5\pi}{4}$.

   We have (checking endpoints and critical points):

$$\begin{aligned} f(0) = f(2\pi) &= \sin(0) + \cos(0) = 1, \\ f\left(\frac{\pi}{4}\right) &= \sin\left(\frac{\pi}{4}\right) + \cos\left(\frac{\pi}{4}\right) \approx 1.414, \\ \text{and } f\left(\frac{5\pi}{4}\right) &= \sin\left(\frac{5\pi}{4}\right) + \cos\left(\frac{5\pi}{4}\right) \approx -1.414. \end{aligned}$$

   Since $f(0) < f(\frac{\pi}{4})$, $f(x)$ is increasing for $0 < x < \pi/4$. Since $f(\frac{\pi}{4}) > f(\frac{5\pi}{4})$, $f(x)$ is decreasing for $\pi/4 < x < 5\pi/4$. And since $f(\frac{5\pi}{4}) < f(2\pi)$, $f(x)$ is again increasing for $5\pi/4 < x < 2\pi$.

   Thus f has a local maximum at $x = \pi/4$, and a local minimum at $x = 5\pi/4$.

   We have $f''(x) = -\sin x - \cos x = 0$ when $-\sin x = \cos x$, i.e. when $\tan x = -1$. This has solutions $x = 3\pi/4$ and $x = 7\pi/4$. Furthermore, $f''(x) < 0$ for $0 \le x < 3\pi/4$, $f''(x) > 0$ for $3\pi/4 < x < 7\pi/4$, and $f''(x) < 0$ for $7\pi/4 < x \le 2\pi$. Therefore, f is concave down for $0 < x < \frac{3\pi}{4}$, concave up for $\frac{3\pi}{4} < x < \frac{7\pi}{4}$, and concave down for $\frac{7\pi}{4} < x < 2\pi$. Thus, the inflection points are $x = \frac{3\pi}{4}, \frac{7\pi}{4}$.

   Combining this with our results for $f'(x)$, we have this diagram:

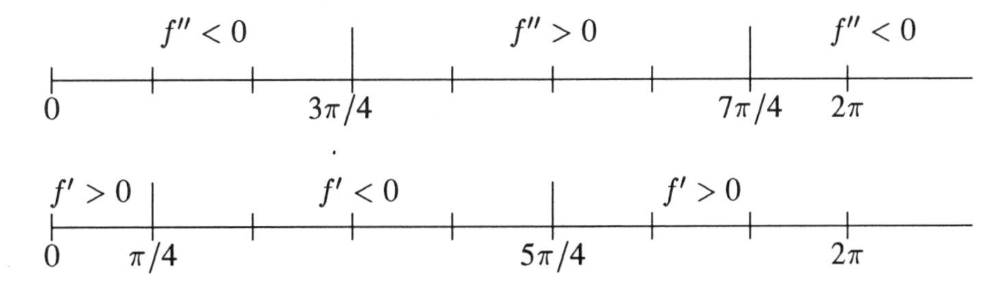

   and the graph:

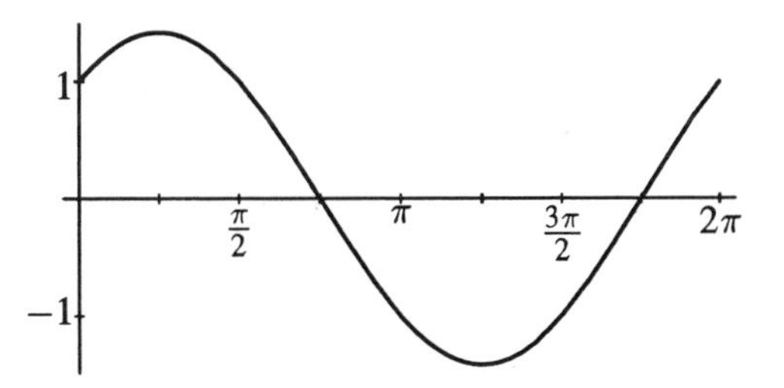

Thus the completed graph for $f(x) = \sin x + \cos x$ with $0 \le x \le 6\pi$ is:

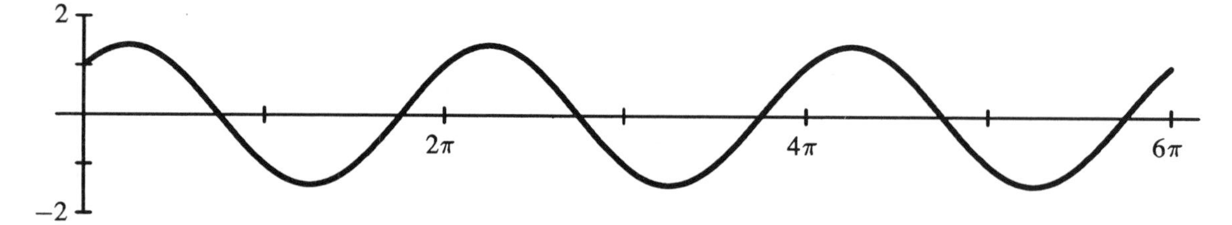

The coordinates of the local and global maxima are $(\pi/4, \sqrt{2})$, $(9\pi/4, \sqrt{2})$, $(17\pi/4, \sqrt{2})$.

The coordinates of the local and global minima are $(5\pi/4, \sqrt{2})$, $(13\pi/4, \sqrt{2})$, $(21\pi/4, \sqrt{2})$.

The inflection points are $x = \frac{3\pi}{4}, \frac{7\pi}{4}, \frac{11\pi}{4}, \frac{15\pi}{4}, \frac{19\pi}{4}$, and $\frac{23\pi}{4}$.

9. We have
$$f'(x) = 10x^9 - 10$$
and
$$f''(x) = 90x^8.$$

Since $90x^8 \ge 0$ for all x, this shows that $f'(x)$ is increasing, in particular for $0 \le x \le 2$.

Thus $f'(x)$ is greatest on the given interval when $x = 2$. Since $f'(2) = 10(2^9) - 10 = 5110 > 0$, f is increasing most rapidly when $x = 2$.

Similarly $f'(x)$ is least on the given interval when $x = 0$. $f'(0) = -10$, so this is where $f(x)$ is decreasing most rapidly.

11. (a) To find the critical points, set $f' = 0$. Since $f'(x) = 1 + \sin x$, we want $\sin x = -1$. This is never possible for $0 \le x \le \pi$. Thus there are no local minima or maxima, so the global maximum and minimum must be at the endpoints. If we check them, we see that $f(0) = -1$ and $f(\pi) = \pi + 1$. So $f(x)$ is least at $x = 0$ and greatest at $x = \pi$.

(b) $f''(x) = \cos x$. $\cos x = 0$ at $x = \frac{\pi}{2}$. Thus $x = \frac{\pi}{2}$ is a critical point for $f'(x)$. Since $f'(\frac{\pi}{2}) = 2$, and since $f'(0) = f'(\pi) = 1$, $f'(\frac{\pi}{2}) = 2$ is a local maximum of $f'(x)$, and $f'(x)$ is never less than 1. So, $f(x)$ is increasing most rapidly at $x = \frac{\pi}{2}$ and decreasing nowhere.

(c) Since $f'(x)$ gives the slope of the line tangent to f at x, we want $f'(x)$ to be increasing most rapidly. This is when $f''(x)$ is largest. To find the critical points of f'', we set $f'''(x) = 0$. But $f'''(x) = -\sin x = 0$ at $x = 0, \pi$. Now, $f''(0) = 1$, and $f''(\pi) = -1$. So $f'(x)$, the slope of the tangent lines, is decreasing most rapidly at $x = \pi$ and increasing most rapidly at $x = 0$.

13. Since $f'(x) = (\ln x)^2 - 2(\sin x)^4$, we have
$$f''(x) = 2(\ln x)\frac{1}{x} - 2 \cdot 4(\sin x)^3(\cos x) = \frac{2\ln x}{x} - 8(\sin x)^3 \cos x.$$

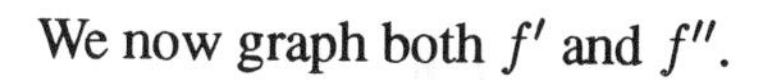

We now graph both f' and f''.

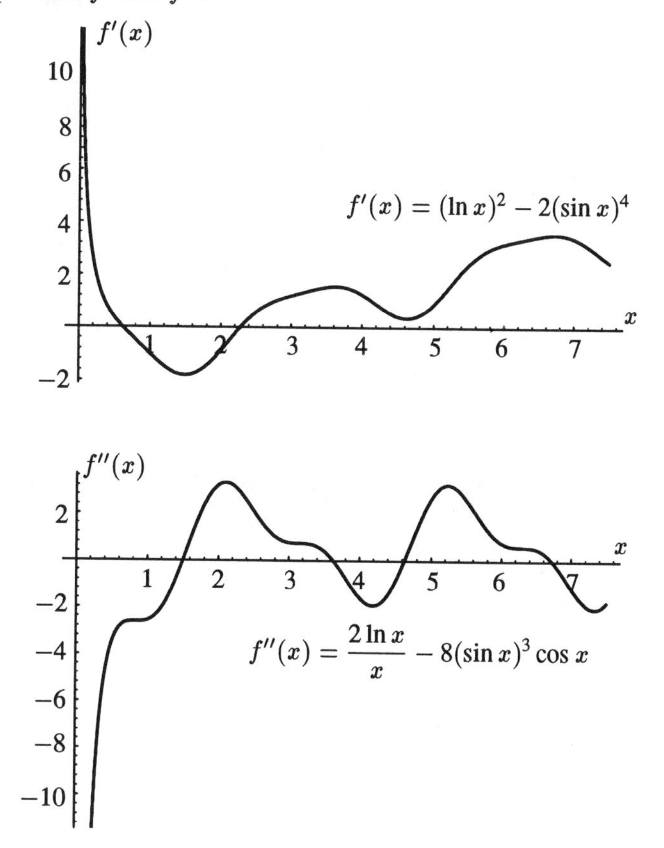

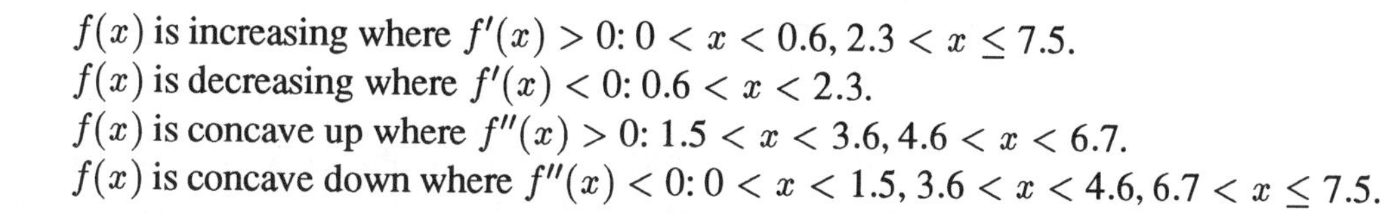

$f(x)$ is increasing where $f'(x) > 0$: $0 < x < 0.6, 2.3 < x \leq 7.5$.
$f(x)$ is decreasing where $f'(x) < 0$: $0.6 < x < 2.3$.
$f(x)$ is concave up where $f''(x) > 0$: $1.5 < x < 3.6, 4.6 < x < 6.7$.
$f(x)$ is concave down where $f''(x) < 0$: $0 < x < 1.5, 3.6 < x < 4.6, 6.7 < x \leq 7.5$.

Using this information, we sketch $f(x)$:

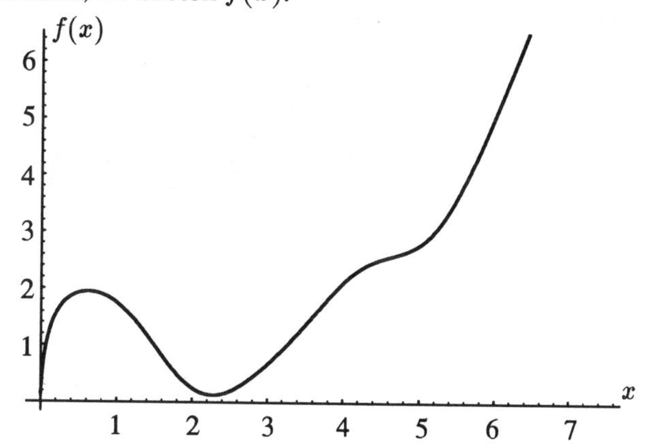

15. Remember that if a function is concave down on some interval, it will be below any line tangent to it on that interval. If we let $f(x) = \ln x$, then $f'(x) = \frac{1}{x}$ and $f''(x) = -\frac{1}{x^2}$. Since $f''x$ is everywhere negative, $f(x)$ is everywhere concave down, and will thus remain below any of its tangent lines forever. Now, the tangent line to $f(x) = \ln x$ at $x = 1$ is $y = x - 1$. This is because $f'(1) = 1$, and $f(1) = 0$. Thus $f(x)$ stays below this line, which is the same as saying $\ln x \leq x - 1$.

17. (a) As $x \to \infty$, $f(x) \to 1$. As $x \to -\infty$, $f(x) \to 1$. As $x \to 0^+$, $f(x) \to \infty$. As $x \to 0^-$, $f(x) \to 0$.

    (b) $f'(x) = (\frac{-1}{x^2})(e^{\frac{1}{x}})$. Thus $f'(x) < 0$ for all $x \neq 0$, which means $f(x)$ is decreasing everywhere it is defined.

    (c) $f''(x) = \frac{1}{x^4}e^{\frac{1}{x}} + \frac{2}{x^3}e^{\frac{1}{x}} = \frac{(2x+1)}{x^4}e^{\frac{1}{x}}$.
    $f''(x) = 0$ when $x = -1/2$.
    $f''(x) < 0$ for $x < -1/2$ and $f''(x) > 0$ for $-1/2 < x < 0$ and $x > 0$. So, $f(x)$ is concave up for $x > 0$ and $-1/2 < x < 0$, and concave down for $x < -1/2$.

    (d)

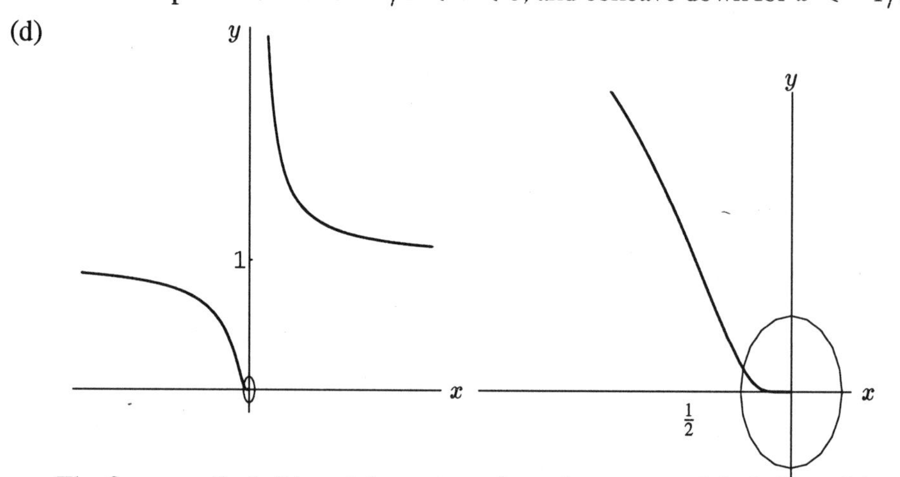

The figure on the left is not drawn to scale so that a more global view of the graph can be seen. The figure on the right is a blow-up of the graph on the left for values of x between -1 and 0. In this picture, we have a better view of the inflection point at $x = -1/2$.

19. (a) This is one of many possible graphs.

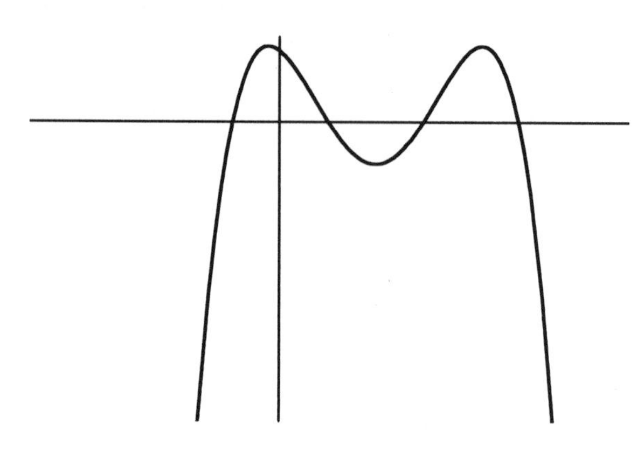

(b) Since f must have a bump between each pair of zeros, f could have at most four zeros.

(c) f could well have no zeros at all. To see this, consider the graph of the above function shifted vertically downwards.

(d) f must have at least two inflection points. Since f has 3 maxima or minima, it has 3 critical points. Consequently f' will have 3 corresponding zeros. Between each pair of these zeroes is a "bump," that is, a maximum or minimum. Thus f' will have at at least two maxima or minima, which implies that f'' will have two zeros. These values, where the second derivative is zero, correspond to points of inflection on the graph of f.

(e) f is of even degree since there are an odd number of critical points. A critical point is a zero of f', so f' has odd degree, implying f has even degree.

(f) The smallest degree f could have is four, since f' has degree at least 3.

(g) For example:

$$f(x) = k(x - a)(x - b)(x - c)(x - d)$$

for some real numbers a, b, c, d, and k. Note that a, b, c, and d are zeros and that k is a stretch factor.

21. (a) When a number grows larger, its reciprocal grows smaller. Therefore, since f is increasing near x_0, we know that g (its reciprocal) must be decreasing. Another, "more mathematical," argument can be made using derivatives. We know that (since f is increasing) $f'(x) > 0$ near x_0. We also know (by the chain rule) that $g'(x) = (f(x)^{-1})' = -\frac{f'(x)}{f(x)^2}$. Since both $f'(x)$ and $f(x)^2$ are positive, this means $g'(x)$ is negative, which in turn means $g(x)$ is decreasing near $x = x_0$.

(b) Since f has a local maximum near x_1, $f(x)$ increases as x nears x_1, and then $f(x)$ decreases as x exceeds x_1. Thus the reciprocal of f, g, decreases as x nears x_1 and then increases as x exceeds x_1. Thus g has a local minimum at $x = x_1$. More formally, since f has a local maximum at $x = x_1$, we know $f'(x_1) = 0$. Since $g'(x) = -\frac{f'(x)}{f(x)^2}$, $g'(x_1) = 0$. To the left of x_1, $f'(x_1)$ is positive, so $g'(x)$ is negative. To the right of x_1, $f'(x_1)$ is negative, so $g'(x)$ is positive. Therefore, g has a local minimum at x_1.

(c) Since f is concave down at x_2, we know $f''(x_2) < 0$. We also know (from above) that

$$g''(x_2) = \frac{2f'(x_2)^2}{f(x_2)^3} - \frac{f''(x_2)f(x_2)}{f(x_2)^3} = \frac{1}{f(x_2)^2}\left(\frac{2f'(x_2)^2}{f(x_2)} - f''(x_2)\right).$$

Since $\frac{1}{f(x_2)^2} > 0$ and $2f'(x_2)^2 > 0$, and since $f(x_2) > 0$ (since f is assumed to be everywhere positive), we see that $g''(x_2)$ is positive. Thus g is concave up at x_2.

Note that for the first two parts of the problem, we didn't need to require f to be positive (only non-zero). However, it was necessary here.

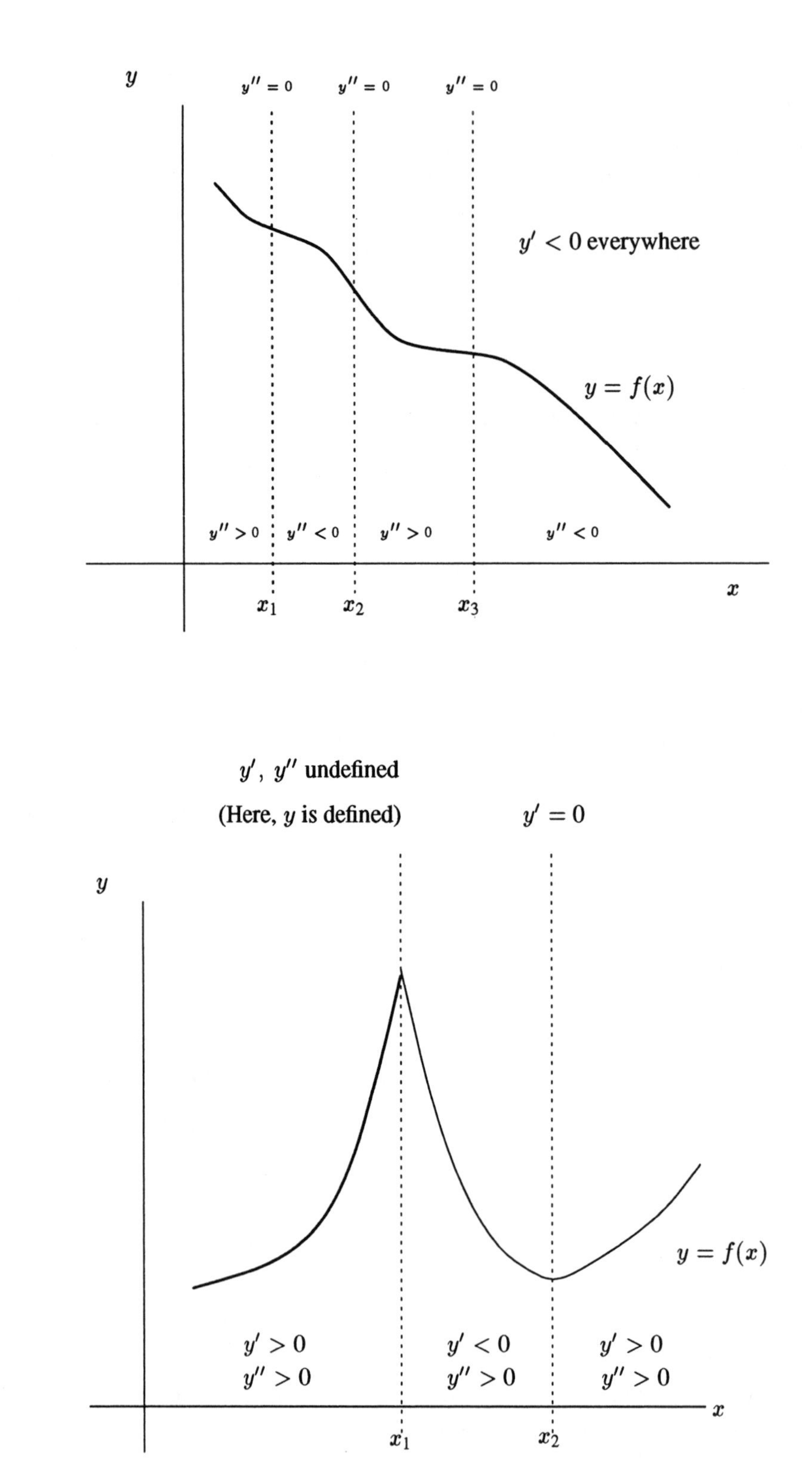

y
y'' = 0
y'' = 0
y'' = 0
y' < 0 everywhere
y = f(x)
y'' > 0
y'' < 0
y'' > 0
y'' < 0
x
x1
x2
x3
y', y'' undefined
(Here, y is defined)
y' = 0
y
y = f(x)
y' > 0
y'' > 0
y' < 0
y'' > 0
y' > 0
y'' > 0
x
x1
x2

25.

27.

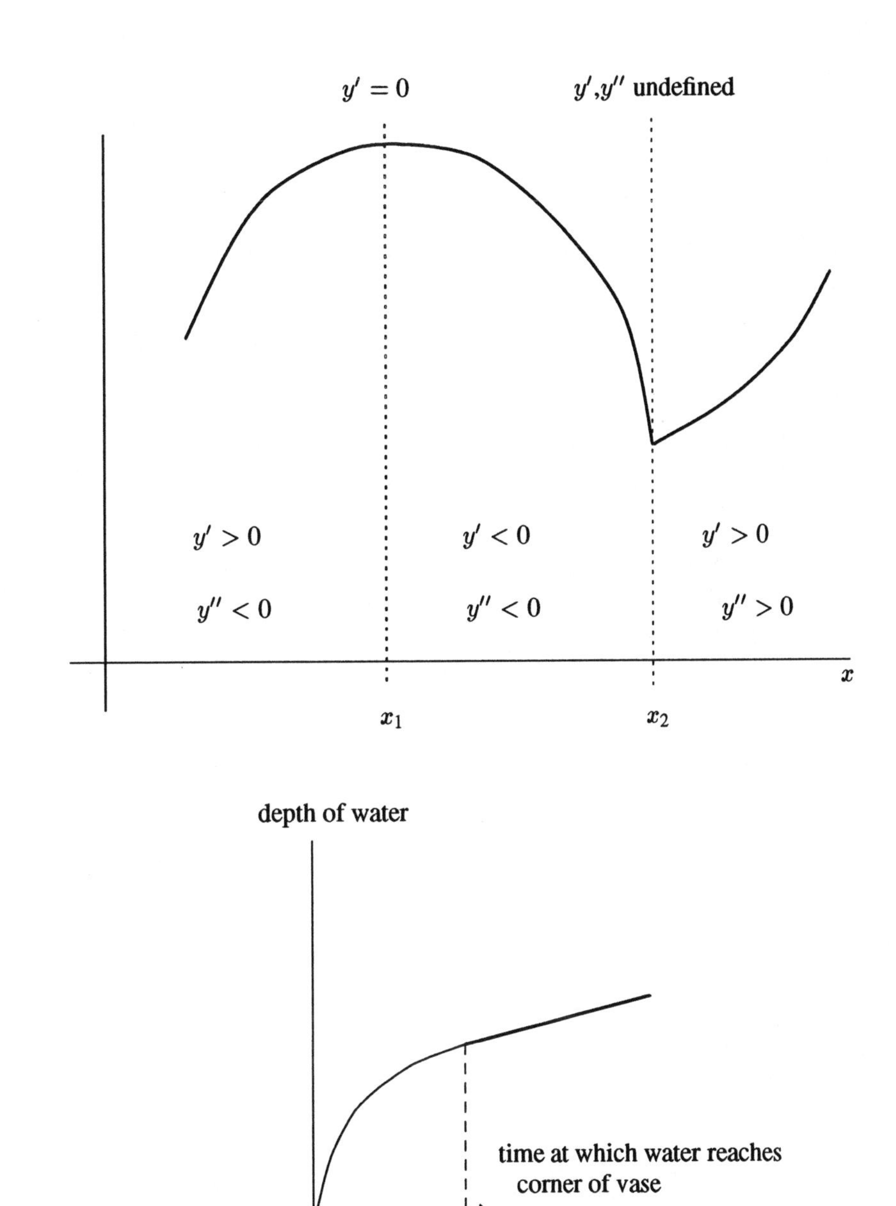

29.

31. (a) We have $g'(t) = \frac{t(1/t) - \ln t}{t^2} = \frac{1-\ln t}{t^2}$, which is zero if $t = e$, negative if $t > e$, and positive if $t < e$, since $\ln t$ is increasing.
Thus $g(e) = \frac{1}{e}$ is a global maximum. Since $t = e$ was the only point at which $g'(t) = 0$, there is no minimum.

(b) Now $\frac{\ln t}{t}$ is increasing for $0 < t < e$, $\frac{\ln 1}{1} = 0$, and $\frac{\ln 5}{5} \approx 0.322 < \frac{\ln(e)}{e}$. Thus, for $0 < t < e$, $\frac{\ln t}{t}$ increases from 0 to above $\frac{\ln 5}{5}$, so there must be a t between 0 and e such that $\frac{\ln t}{t} = \frac{\ln 5}{5}$. For $t > e$, there is only one solution to $\frac{\ln t}{t} = \frac{\ln 5}{5}$, namely $t = 5$, since $\frac{\ln t}{t}$ is decreasing for $t > e$. Thus $\frac{\ln x}{x} = \frac{\ln 5}{5}$ has exactly two solutions.

(c) The graph of $\frac{\ln t}{t}$, intersects the horizontal line $y = \frac{\ln 5}{5}$, at $x = 5$ and $x \approx 1.75$.

5.3 Solutions

1. (a) Let $p(x) = x^3 - ax$, and suppose $a < 0$. Then $p'(x) = 3x^2 - a > 0$ for all x, so $p(x)$ is always increasing.

   (b) Now suppose $a > 0$. We have $p'(x) = 3x^2 - a = 0$ when $x^2 = \frac{a}{3}$, i.e., when $x = \sqrt{\frac{a}{3}}$ and $x = -\sqrt{\frac{a}{3}}$.

   We have $p''(x) = 6x$; since $6\sqrt{a/3} > 0$, $f(\sqrt{a/3})$ is a local minimum, and since $-6\sqrt{a/3} < 0$, $f(-\sqrt{a/3})$ is a local maximum.

   (c) Case 1: $a < 0$
   $p(x)$ is always increasing. $p''(x) = 6x > 0$ if $x > 0$, in which case the graph is concave up; $6x < 0$ if $x < 0$, in which case the graph is concave down. Thus $x = 0$ is an inflection point.

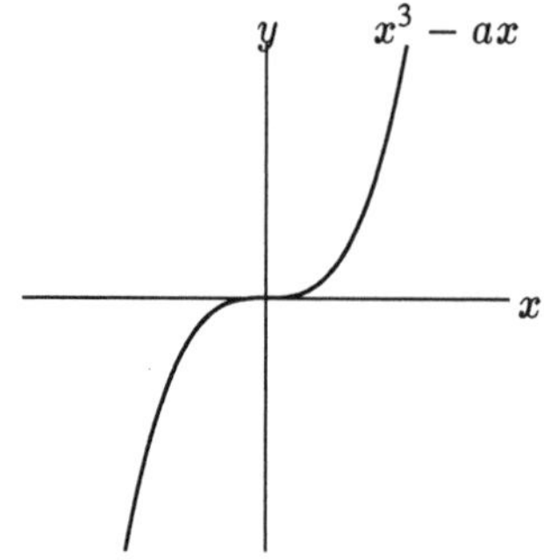

   Case 2: $a > 0$
   We have

$$p\left(\sqrt{\frac{a}{3}}\right) = \left(\sqrt{\frac{a}{3}}\right)^3 - a\sqrt{\frac{a}{3}} = \frac{a\sqrt{a}}{\sqrt{27}} - \frac{a\sqrt{a}}{\sqrt{3}} < 0,$$

$$\text{and} \quad p\left(-\sqrt{\frac{a}{3}}\right) = -\frac{a\sqrt{a}}{\sqrt{27}} + \frac{a\sqrt{a}}{\sqrt{3}} = -p\left(\sqrt{\frac{a}{3}}\right) > 0.$$

$$p'(x) = 3x^2 - a \begin{cases} = 0 & \text{if } |x| = \sqrt{\frac{a}{3}}; \\ > 0 & \text{if } |x| > \sqrt{\frac{a}{3}}; \\ < 0 & \text{if } |x| < \sqrt{\frac{a}{3}}. \end{cases}$$

   So p is increasing for $x < -\sqrt{a/3}$, decreasing for $-\sqrt{a/3} < x < \sqrt{a/3}$, and increasing for $x > \sqrt{a/3}$. Since $p''(x) = 6x$, the graph of $p(x)$ is concave down for values of x less than zero and concave up for values greater than zero. Putting this together:

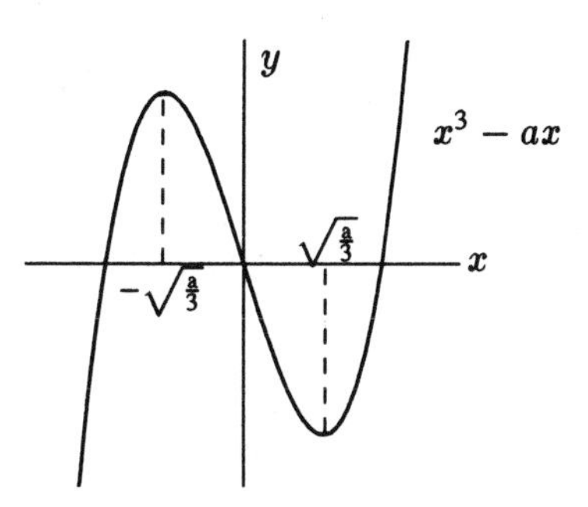

3. We have $f(x) = x^2 + 2ax = x(x + 2a) = 0$ when $x = 0$ or $x = -2a$.

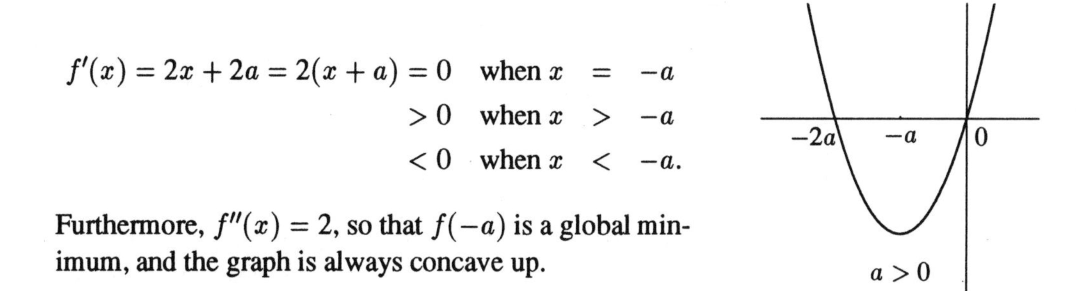

$$\begin{aligned} f'(x) = 2x + 2a = 2(x + a) = 0 \quad &\text{when } x = -a \\ > 0 \quad &\text{when } x > -a \\ < 0 \quad &\text{when } x < -a. \end{aligned}$$

Furthermore, $f''(x) = 2$, so that $f(-a)$ is a global minimum, and the graph is always concave up.

Increasing $|a|$ stretches the graph horizontally. Also, the critical value (the value of f at the critical point) drops further beneath the x-axis.

5. Since $\lim_{t \to \infty} N = a$, we have $a = 200{,}000$. Note that while $N(t)$ will never actually reach 200,000, it will become arbitrarily close to 200,000. Since N represents the number of people, it makes sense to round up long before $t \to \infty \ldots$ When $t = 1$, $N = 0.1(200{,}000) = 20{,}000$ people, so plugging into our formula gives

$$N(1) = 20{,}000 = 200{,}000\left(1 - e^{-k(1)}\right).$$

Solving for k gives

$$\begin{aligned} 2 &= 20\left(1 - e^{-k}\right) \\ 0.1 &= 1 - e^{-k} \\ e^{-k} &= 0.9 \\ k &= -\ln 0.9 \approx 0.105. \end{aligned}$$

7. $T(t) =$ the temperature at time $t = a(1 - e^{-kt}) + b$.

   (a) Since at time $t = 0$ the yam is 20°C, we have

$$T(0) = 20^\circ = a\left(1 - e^{0}\right) + b = a(1 - 1) + b = b.$$

Thus $b = 20°\text{C}$. Now, common sense tells us that after a period of time, the yam will heat up to about 200°, or oven temperature. Thus the temperature T should approach 200° as the time t grows large:

$$\lim_{t \to \infty} T(t) = 200°\text{C} = a(1 - 0) + b = a + b.$$

Since $a + b = 200°$, and $b = 20°$, this means $a = 180°\text{C}$.

(b) Since we're talking about how quickly the yam is heating up, we need to look at the derivative, $T'(t) = ake^{-kt}$:

$$T'(t) = (180°)ke^{-kt}.$$

We know $T'(0) = \frac{2°}{\text{min}}$, so

$$\frac{2°}{\text{min}} = (180°)ke^{-kt} = (180°)ke^{-k(0)} = (180°)(k).$$

So $k = \frac{2°/\text{min}}{180°} = \frac{1}{90}\text{min}^{-1}$.

9. (a) If $B = 0$, $y = \frac{A}{x}$. The larger the value of $|A|$, the steeper the graph. If $A > 0$, then y is positive when $x > 0$, negative when $x < 0$. If $A < 0$, the situation is reversed.

(b) The graph is shifted horizontally by B. The shift is to the left for positive B, to the right for negative B. There is a vertical asymptote at $x = -B$.

(c)

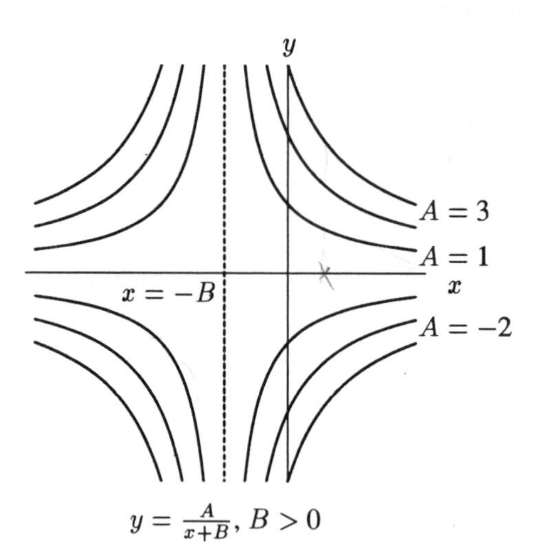

$y = \frac{A}{x+B}$, $B > 0$

11. (a) $f(x) = axe^{-bx}$ (Assume a and $b > 0$) To find the maxima and minima, we solve

$$f'(x) = ae^{-bx} - abxe^{-bx} = ae^{-bx}(1 - bx) \begin{cases} = 0 & \text{if } x = \frac{1}{b}; \\ < 0 & \text{if } x > \frac{1}{b}; \\ > 0 & \text{if } x < \frac{1}{b}. \end{cases}$$

Therefore, f is increasing ($f' > 0$) for $x < \frac{1}{b}$ and decreasing ($f' > 0$) for $x > \frac{1}{b}$. $x = \frac{1}{b}$ is the local maximum. There are no local minima. To find the points of inflection, we solve

$$\begin{aligned} f''(x) &= -abe^{-bx} + ab^2xe^{-bx} - abe^{-bx} \\ &= -2abe^{-bx} + ab^2xe^{-bx} \\ &= ab(bx-2)e^{-bx} = 0 \quad \text{if } x = \frac{2}{b} \end{aligned}$$

Therefore, f is concave up for $x < \frac{2}{b}$ and concave down for $x > \frac{2}{b}$, and the inflection point is $x = \frac{2}{b}$.

(b) Varying a 'stretches' or 'flattens' the graph but does not affect the critical point $x = \frac{1}{b}$ and the inflection point $x = \frac{2}{b}$. Since the critical and inflection points are inversely proportional to b, varying b will change these points, as well as the maximum $f(\frac{1}{b}) = \frac{a}{be}$. For, example, an increase in b will shift the critical and inflection points to the left, and also lower the maximum value of f.

(c) Varying a

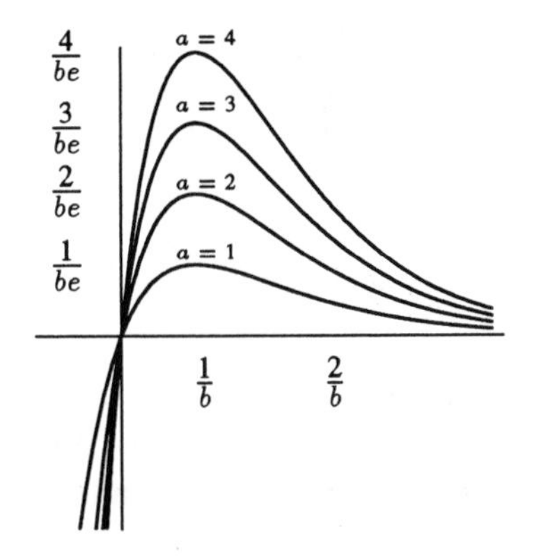

Varying b

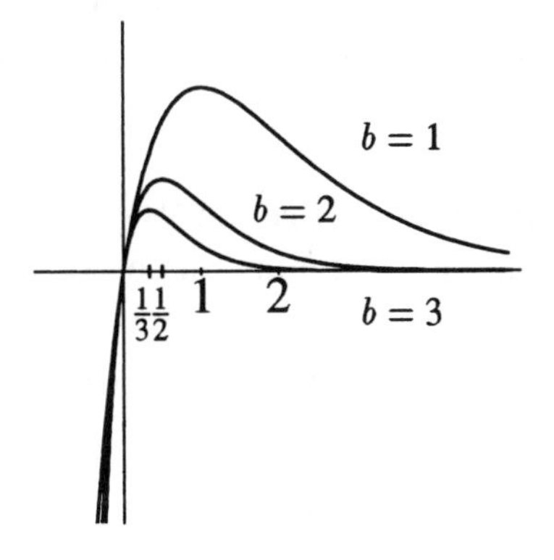

13. (a) $y = e^{-ax}\sin x$, $a > 0$, $x \geq 0$. To find the critical points, we'll solve $y' = 0$.
$y' = e^{-ax}\cos x - ae^{-ax}\sin x = -e^{-ax}(a\sin x - \cos x)$.
Since the $-e^{-ax}$ factor is never zero, $y' = 0$ when $a\sin x - \cos x = 0$, or

$$\begin{aligned} a\frac{\sin x}{\cos x} - 1 &= 0 \\ a\tan x &= 1 \\ \tan x &= \frac{1}{a} \end{aligned}$$

The graph of $e^{-ax}\sin x$ looks like:

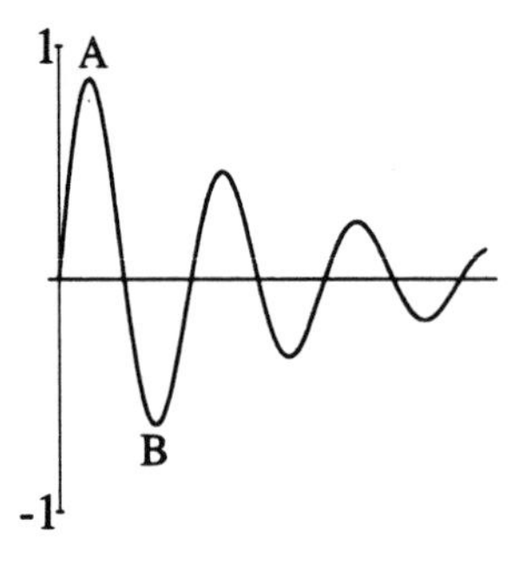

For $x \geq 0$, there are infinitely many local maxima and minima. Since $e^{-ax} \to 0$ as $x \to \infty$, A and B are the global maximum and minimum respectively. That is, A will be the highest peak, and B will be the deepest valley because of the continually decreasing behavior of e^{-ax}.

Since the critical points are where $\tan x = \frac{1}{a}$, the maximum occurs at $x_{max} = \tan^{-1}(\frac{1}{a})$. Since $\frac{1}{a} > 0$, $0 < x_{max} < \frac{\pi}{2}$ by definition of the inverse tangent function. Since tangent has a period of π, the next critical point, which is the global minimum, will be at $x_{min} = x_{max} + \pi$.

(b) As a increases, $\frac{1}{a}$ decreases, so that $\tan^{-1}(\frac{1}{a})$ gets closer and closer to 0. Thus, the x values giving the maximum and minimum become close to 0 and π respectively. Since the values of $\sin x$ lie between -1 and 1, and $\lim_{a\to\infty} e^{-ax} = 0$, the absolute values of the maximum and minimum descend to 0 as $a \to \infty$.

5.4 Solutions

1.

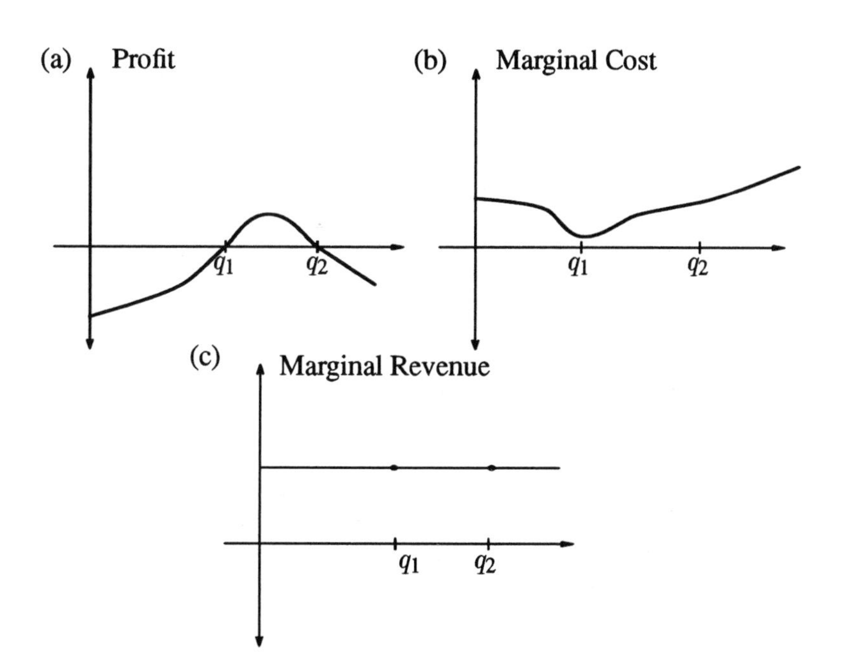

3. It is interesting to note that to draw a graph of $C'(q)$ for this problem, you never have to know what $C(q)$ looks like, although you *could* draw a graph of $C(q)$ if you wanted to. By the formula given in the problem, we know that $C(q) = q \cdot a(q)$. Using the product rule we get that $C'(q) = a(q) + q \cdot a'(q)$.

   We are given a graph of $a(q)$ which is linear, so $a(q) = b + mq$, where $b = a(0)$ is the y-intercept and m is the slope. Therefore

$$\begin{aligned} C'(q) &= a(q) + q \cdot a'(q) = b + mq + q \cdot m \\ &= b + 2mq. \end{aligned}$$

   In other words, $C'(q)$ is also linear, and it has twice the slope and the same y–intercept as $a(q)$.

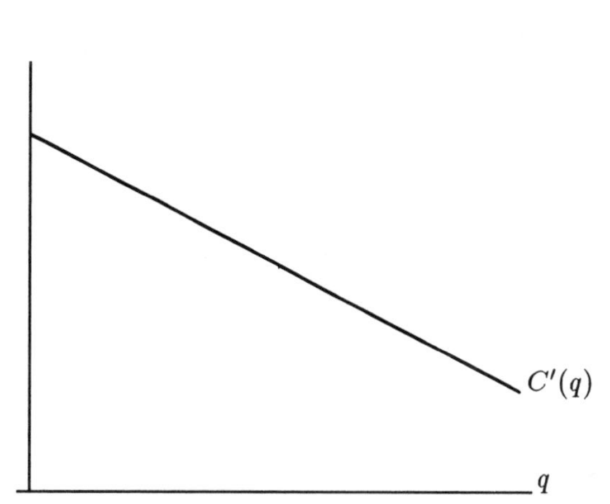

5. (a) $\pi(q)$ is maximized when $R(q) > C(q)$ and they are as far apart as possible:

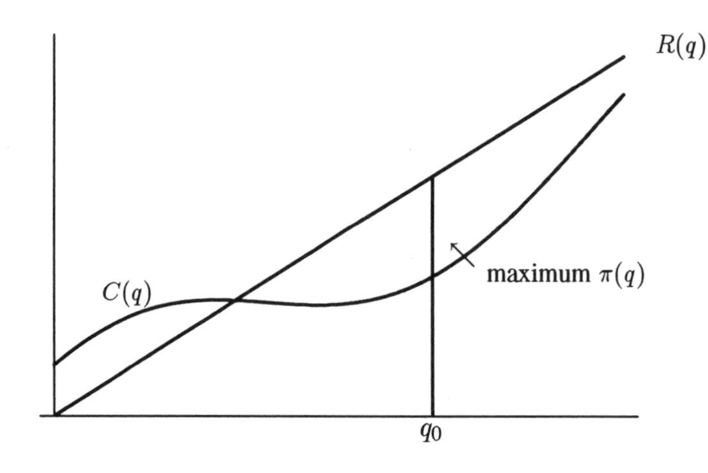

(b) $\pi'(q_0) = R'(q_0) - C'(q_0) = 0 \Rightarrow C'(q_0) = R'(q_0) = p.$

Graphically, the slopes of the two curves at q_0 are equal. This is plausible because if $C'(q_0)$ were greater than p or less than p, the maximum of $\pi(q)$ would be to the left or right of q_0, respectively. In economic terms, if the cost were rising more quickly than revenues, the profit would be maximized at a lower quantity (and if the cost were rising more slowly, at a higher quantity).

(c)

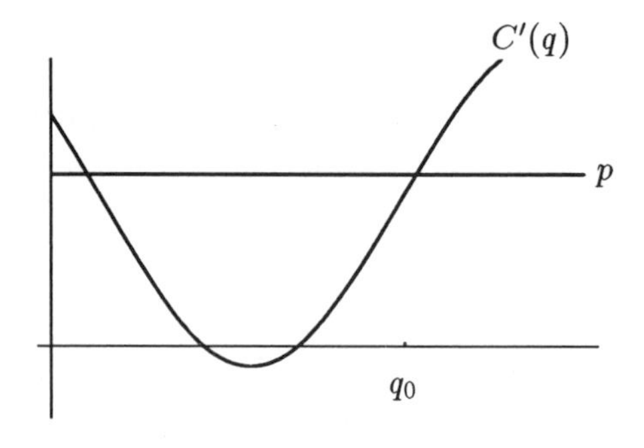

5.5 Solutions

1. We have that $v(r) = a(R - r)r^2 = aRr^2 - ar^3$, and $v'(r) = 2aRr - 3ar^2 = 2ar(R - \frac{3}{2}r)$, which is zero if $r = \frac{2}{3}R$, or if $r = 0$, and so $v(r)$ has critical points here.
$v''(r) = 2aR - 6ar$, and thus $v''(0) = 2aR > 0$, which by the second derivative test implies that v has a minimum at $r = 0$. $v''(\frac{2}{3}R) = 2aR - 4aR = -2aR < 0$, and so by the second derivative test v has a maximum at $r = \frac{2}{3}R$.

3. Call the stacks A and B. (See below.)

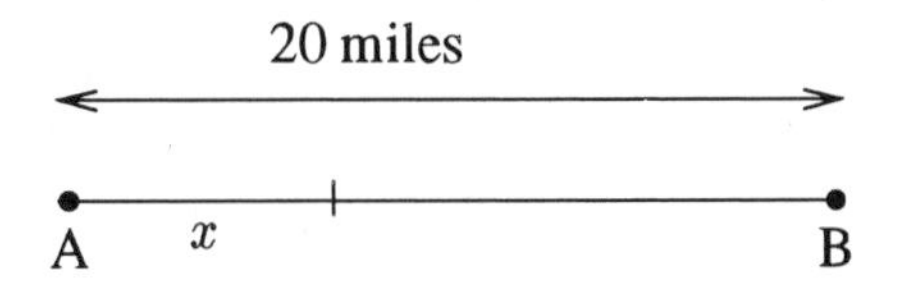

Suppose the point where the concentration of deposit is a minimum occurs at a distance of x miles from stack A. We want to find x such that

$$S = \frac{k_1}{x^2} + \frac{k_2}{(20-x)^2} = k_2\left(\frac{7}{x^2} + \frac{1}{(20-x)^2}\right)$$

is a minimum, which is the same thing as minimizing $f(x) = 7x^{-2} + (20-x)^{-2}$ since k_2 is nonnegative.

We have

$$f'(x) = -14x^{-3} - 2(20-x)^{-3}(-1) = \frac{-14}{x^3} + \frac{2}{(20-x)^3} = \frac{-14(20-x)^3 + 2x^3}{x^3(20-x)^3}.$$

Thus we want to find x such that $-14(20-x)^3 + 2x^3 = 0$, which implies $2x^3 = 14(20-x)^3$. That's equivalent to $x^3 = 7(20-x)^3$, or $\frac{20-x}{x} = (\frac{1}{7})^{1/3} \approx 0.523$. Solving for x, we have $20 - x = 0.523x$, whence $x = \frac{20}{1.523} \approx 13.13$.

To verify that this minimizes f, we take the second derivative:

$$f''(x) = 42x^{-4} + 6(20-x)^{-4} = \frac{42}{x^4} + \frac{6}{(20-x)^4} > 0$$

for any $0 < x < 20$, so by the second derivative test the concentration is minimized 13.13 miles from A.

5. Let $y = \ln(1+x)$. Since $y' = \frac{1}{1+x}$, y is increasing for all $x \geq 0$. The lower bound is at $x = 0$, so, $\ln(1) = 0 \leq y$. There is no upper bound.

7. Let $y = x^3 - 4x^2 + 4x$. To locate the critical points, we solve $y' = 0$. Since $y' = 3x^2 - 8x + 4 = (3x-2)(x-2)$, the critical points are $x = 2/3$ and $x = 2$. To find the global minimum and maximum on $0 \leq x \leq 4$, we check the critical points and the endpoints: $y(0) = 0$; $y(\frac{2}{3}) = \frac{32}{27}$; $y(2) = 0$; $y(4) = 16$. Thus, the global minimum is at $x = 2$, the global maximum is at $x = 4$, and $0 \leq y \leq 16$.

9. (a) If, following the hint, we set $f(x) = \frac{a+x}{2} - \sqrt{ax}$, then $f(x)$ represents the difference between the arithmetic and geometric means for some fixed a and any x. We can find where this difference is minimized by solving $f'(x) = 0$. Since $f'(x) = \frac{1}{2} - \frac{1}{2}\sqrt{a}x^{-\frac{1}{2}}$, $f'(x) = 0 \Rightarrow \frac{1}{2}\sqrt{a}x^{-\frac{1}{2}} = \frac{1}{2}$, or $x = a$. Since $f''(x) = \frac{1}{4}\sqrt{a}x^{-\frac{3}{2}}$ is positive for all

positive x, by the second derivative test $f(x)$ has a minimum at $x = a$ of $f(a) = 0$. Thus $f(x) = \frac{a+x}{2} - \sqrt{ax} \geq 0$ for all $x > 0$, which means $\frac{a+x}{2} \geq \sqrt{ax}$. Taking $x = b$, we have $\frac{a+b}{2} \geq \sqrt{ab}$. This means that the arithmetic average is greater than the geometric average unless $a = b$, in which case the two averages are equal. (Note: a simpler solution exists not using calculus.)

(b) Following the hint, set $f(x) = \frac{a+b+x}{3} - \sqrt[3]{abx}$. Then $f(x)$ represents the difference between the arithmetic and geometric means for some fixed a, b and any x. We can find where this difference is minimized by solving $f'(x) = 0$. Since $f'(x) = \frac{1}{3} - \frac{1}{3}\sqrt[3]{ab}x^{-2/3}$, $f'(x) = 0 \Rightarrow \frac{1}{3}\sqrt[3]{ab}x^{-2/3} = \frac{1}{3}$, or $x = \sqrt{ab}$. Since $f''(x) = \frac{2}{9}\sqrt[3]{ab}x^{-5/3}$ is positive for all positive x, by the second derivative test $f(x)$ has a minimum at $x = \sqrt{ab}$. But

$$f(\sqrt{ab}) = \frac{a+b+\sqrt{ab}}{3} - \sqrt[3]{ab\sqrt{ab}} = \frac{a+b+\sqrt{ab}}{3} - \sqrt{ab} = \frac{a+b-2\sqrt{ab}}{3}.$$

By the first part of this problem, we know that $\frac{a+b}{2} - \sqrt{ab} \geq 0$, which implies that $a+b-2\sqrt{ab} \geq 0$. Thus $f(\sqrt{ab}) = \frac{a+b-2\sqrt{ab}}{3} \geq 0$. Since f has a maximum at $x = \sqrt{ab}$, $f(x)$ is always nonnegative. Thus $f(x) = \frac{a+b+x}{3} - \sqrt[3]{abx} \geq 0 \Rightarrow \frac{a+b+c}{3} \geq \sqrt[3]{abc}$. Note that equality holds only when $a = b = c$. (Note: may also be done without calculus.)

11. (a) Note that

$$f'(x) = 2(x - a_1) + 2(x - a_2) + 2(x - a_3) + 2(x - a_4).$$

Setting $f'(x) = 0$ yields

$$x = \frac{2(a_1 + a_2 + a_3 + a_4)}{8} = \frac{a_1 + a_2 + a_3 + a_4}{4}.$$

Since $f''(x) = 8 > 0$, this is a local minimum.

(b) The mean, $\frac{a_1+a_2+\cdots+a_n}{n}$, minimizes this sum of squares. We check just as before:

$$f'(x) = 2(x - a_1) + 2(x - a_2) + \cdots + 2(x - a_n),$$

so if $f'(x) = 0$, $2nx = 2(a_1 + a_2 + \cdots + a_n)$, so $x = \frac{a_1+a_2+\cdots+a_n}{n}$. This is a minimum because $f''(x) = 2n > 0$.

13. (a) At higher speeds, more energy is used so the graph rises to the right. The initial drop is explained by the fact that the energy it takes a bird to lift off is greater than that needed to maintain the low speed following lift off.

(b) $f(v)$ measures energy per second; $a(v)$ measures energy per meter. A bird traveling at rate v will in 1 second travel v meters, and thus will consume $v \cdot a(v)$ joules of energy in that 1 second period. Thus $v \cdot a(v)$ represents the energy consumption per second, and so $f(v) = v \cdot a(v)$.

(c) Since $v \cdot a(v) = f(v)$, $a(v) = \frac{f(v)}{v}$. But this ratio has the same value as the slope of a line passing from the origin through the point $(v, f(v))$ on the curve (see figure). Thus $a(v)$ is minimal when the slope of this line is minimal. To find the value of v minimizing $a(v)$, we solve $a'(v) = 0$. By the Quotient rule,

$$a'(v) = \frac{vf'(v) - f(v)}{v^2}.$$

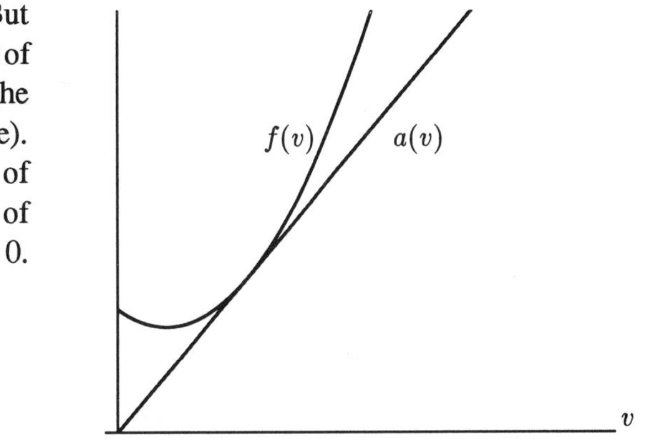

Thus $a'(v) = 0$ when $vf'(v) = f(v)$, or when $f'(v) = \frac{f(v)}{v} = a(v)$. Thus since $a(v)$ is represented by the slope of a line through the origin and a point on the curve, $a(v)$ is minimized when this line is tangent to $f(v)$, since then the slope $a(v)$ equals $f'(v)$.

(d) The bird should minimize $a(v)$ assuming it wants to go from one particular point to another, i.e. where the distance is set. Then minimizing $a(v)$ minimizes the total energy used for the flight.

15.

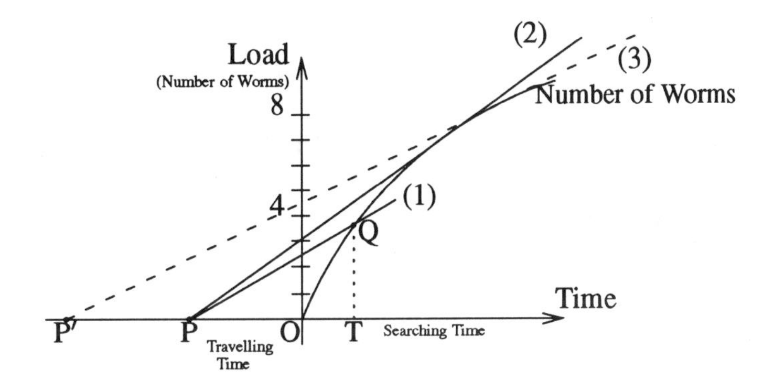

(a) See line (1). For any point Q on the loading curve, the line joining PQ has slope

$$\frac{QT}{PT} = \frac{QT}{PO + OT} = \frac{\text{load}}{\text{traveling time} + \text{searching time}}.$$

(b) The slope of the line PQ is maximized when the line is tangent to the loading curve, which happens with line (2). The load is then approximately 7 worms.

(c) If the traveling time is increased, the point P moves to the left, to point P', say. If line (3) is tangent to the curve, it will be tangent to the curve further to the right than line (2), so the optimal load is larger. This makes sense: if the bird has to fly further, you'd expect it to bring back more worms each time.

5.6 Solutions

1. Let w and l be the width and length, respectively, of the rectangular area you wish to enclose. Then

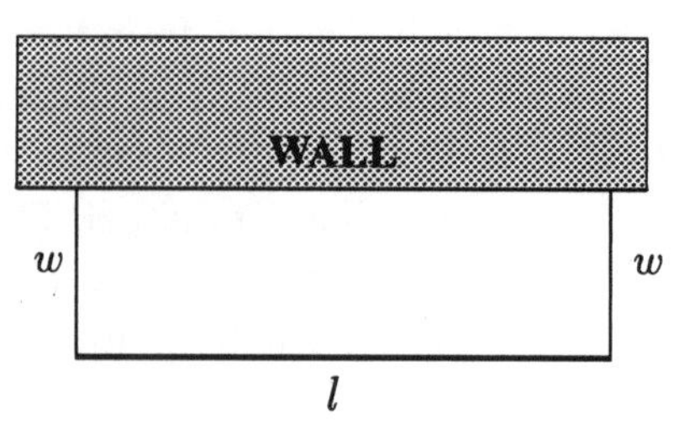

$$\begin{aligned} w + w + l &= 100 \text{ feet} \\ l &= 100 - 2w \\ \text{Area} = w \cdot l &= w(100 - 2w) = 100w - 2w^2 \end{aligned}$$

To maximize area, we solve $A' = 0$ to find critical points. This gives $A' = 100 - 4w = 0 \Rightarrow w = 25,\ l = 50$. So the area is $25 \cdot 50 = 1250$ square feet. This is a local maximum by the second derivative test because $A'' = -4 < 0$. Since the graph of A is a parabola, the local maximum is in fact a global maximum.

3. (a) Since $y^2 = 1 - x^2/9$, we wish to minimize the distance

$$D = \sqrt{(x-2)^2 + (y-0)^2} = \sqrt{(x-2)^2 + 1 - \frac{x^2}{9}}.$$

To do so, we find the value of x minimizing D^2. This x also minimizes D. Since $d = (x-2)^2 + 1 - \frac{x^2}{9}$, we have

$$d' = 2(x-2) - \frac{2x}{9} = \frac{16x}{9} - 4,$$

which is 0 when $x = 9/4$. Since $d'' = 16/9 > 0$, d is at a local minimum when $x = 9/4$. Since the graph of d is a parabola, the local minimum is in fact a global minimum. Solving for y, we have

$$y^2 = 1 - \frac{x^2}{9} = 1 - \left(\frac{9}{4}\right)^2 \cdot \frac{1}{9} = \frac{7}{16},$$

so $y = \pm\frac{\sqrt{7}}{4}$. Therefore, the closest points are $(\frac{9}{4}, \pm\frac{\sqrt{7}}{4})$.

(b) This time, we wish to minimize

$$D = \sqrt{(x-\sqrt{8})^2 + 1 - \frac{x^2}{9}}.$$

Again, let $d = D^2$ and minimize d. Since $d = (x-\sqrt{8})^2 + 1 - \frac{x^2}{9}$,

$$d' = 2(x - 2\sqrt{2}) - \frac{2x}{9} = \frac{16x}{9} - 4\sqrt{2}.$$

Therefore, $d' = 0$ when $x = 9\sqrt{2}/4$. But $9\sqrt{2}/4 > 3$, an impossibility if (x, y) is to lie on the ellipse! The major axis is only 6 units long. Therefore, there aren't any critical points on the interval from -3 to 3, so the minimum distance must be attained at an endpoint. Since $d' < 0$ for all x between -3 and 3, the minimum must be at $x = 3$. So $(3, 0)$ is the point on the ellipse closest to $(\sqrt{8}, 0)$.

5. Let $(x, 0)$ be the coordinates of the bottom left corner of the rectangle. Then the width of the rectangle is $(9 - x)$, and the height is $\sqrt{x}$. Thus the area $A = (9 - x)(\sqrt{x}) = 9x^{\frac{1}{2}} - x^{\frac{3}{2}}$, and $A' = \frac{9}{2}x^{-\frac{1}{2}} - \frac{3}{2}x^{\frac{1}{2}}$. To maximize area, we solve $A' = 0 \Rightarrow \frac{9}{2}x^{-\frac{1}{2}} = \frac{3}{2}x^{\frac{1}{2}} \Rightarrow 3 = x$. Evaluating A at $x = 3$ and at the endpoints, $x = 0$ and $x = 9$, shows the maximum occurs at $x = 3$. The dimensions for the maximal area are $6 \times \sqrt{3}$. Similarly, we let the perimeter be $P = 2(9-x) + 2\sqrt{x} = 18 - 2x + 2x^{\frac{1}{2}}$. So $P' = -2 + x^{-\frac{1}{2}}$. To maximize perimeter, we solve $P' = 0 \Rightarrow x^{-\frac{1}{2}} = 2 \Rightarrow x = \frac{1}{4}$. Evaluating $P(2)$, $P(0)$ and $P(9)$ shows the maximum is $P(2)$. The dimensions for maximal perimeter are $\frac{35}{4} \times \frac{1}{2}$.

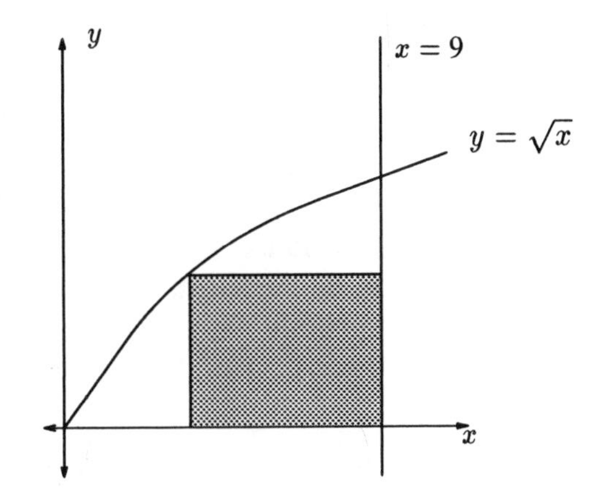

6. Let $(x, 0)$ be the coordinates of the bottom left corner of the rectangle. Then the width of the rectangle is $c - x$ and the height is $\sqrt{x}$. Thus the area is $A = (c - x)(\sqrt{x}) = cx^{\frac{1}{2}} - x^{\frac{3}{2}}$. $A' = \frac{c}{2}x^{-\frac{1}{2}} - \frac{3}{2}x^{\frac{1}{2}}$. To maximize area, we solve $A' = 0 \Rightarrow x = \frac{c}{3}$, so the height $\sqrt{x}$ depends on c.
Similarly, the perimeter is $P = 2(c - x) + 2\sqrt{x}$, so $P' = -2 + x^{-\frac{1}{2}}$. To maximize perimeter, we solve $P' = 0 \Rightarrow x^{-\frac{1}{2}} = 2 \Rightarrow x = \frac{1}{4}$, so the height 0.5 is independent of c.

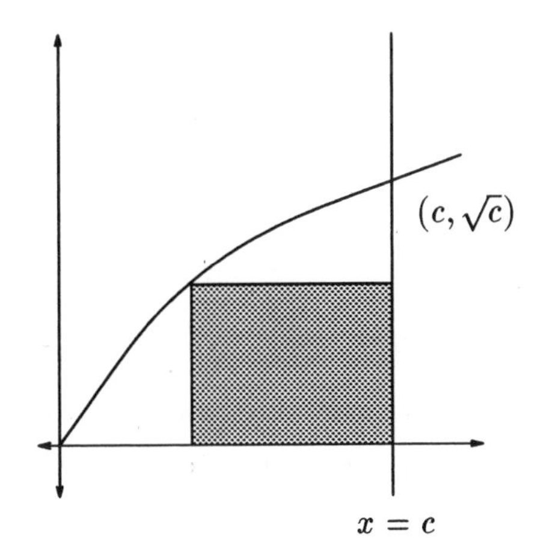

7. Notice that this is an extension of Problem 6. Let $(x, 0)$ be the coordinates of the bottom left corner of the rectangle. Then the width of the rectangle is $(c - x)$ and the height is x^n. Thus the area is $A = (c - x)(x^n)$. $A' = cnx^{n-1} - (n+1)x^n$. To maximize area, we solve $A' = 0 \Rightarrow x = \frac{cn}{n+1}$. So the height is $(\frac{cn}{n+1})^n$, which depends on c. Similarly, the perimeter is $P = 2(c-x) + 2x^n$, so $P' = -2 + 2nx^{n-1}$. To maximize the perimeter, we solve $P' = 0 \Rightarrow 2nx^{n-1} = 2 \Rightarrow x = (\frac{1}{n})^{\frac{1}{n-1}}$. So the height is $(\frac{1}{n})^{\frac{n}{n-1}}$, which is independent of c.

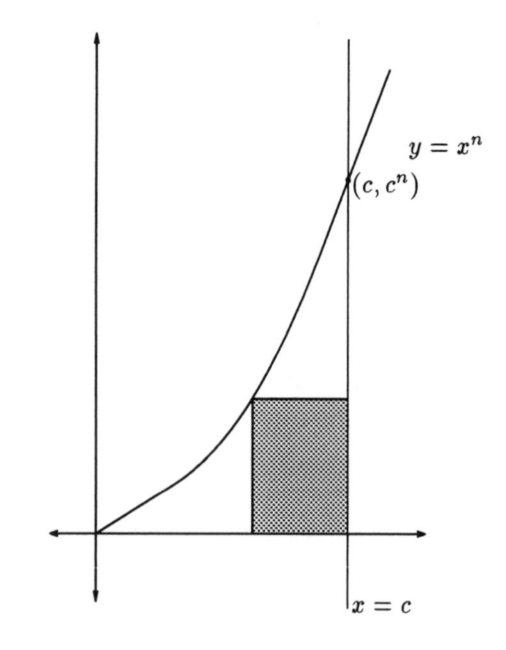

9.

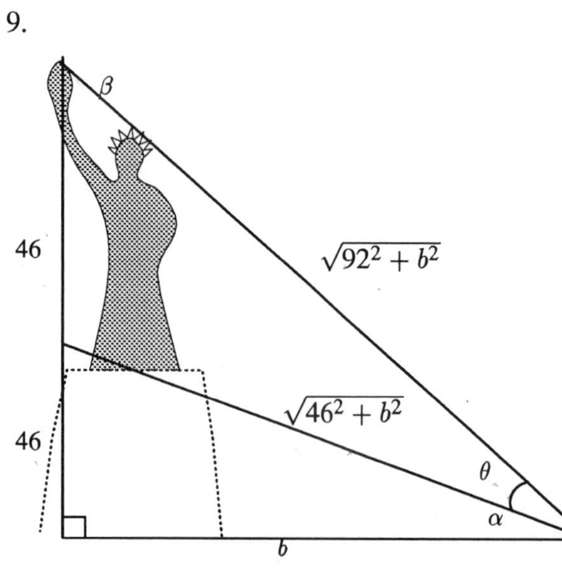

We wish to find b such that angle θ is maximized. By the Law of Sines,

$$\frac{\sin\theta}{46} = \frac{\sin\beta}{\sqrt{46^2 + b^2}},$$

but, by the big triangle,

$$\sin\beta = \frac{b}{\sqrt{92^2 + b^2}}.$$

So $\sin\theta = \frac{46}{\sqrt{46^2+b^2}}(\frac{b}{\sqrt{92^2+b^2}})$. Since θ is less than 90°, maximizing θ is the same as maximizing $\sin\theta$, which is in turn equivalent to maximizing $\sin^2\theta = \frac{46^2 b^2}{(46^2+b^2)(92^2+b^2)}$. This is a function of b^2, so we set $c = b^2$. We will now try to maximize $\sin^2\theta = \frac{46^2 c}{(46^2+c)(92^2+c)}$ as a function of c. This is the same as minimizing

$$1/\sin^2\theta = \frac{(46^2 + c)(92^2 + c)}{46^2 c} = \frac{c}{46^2} + 5 + \frac{92^2}{c}.$$

But the minimum of $f(c) = \frac{c}{46^2} + 5 + \frac{92^2}{c}$ occurs when $f'(c) = 0$, that is, when $\frac{1}{46^2} - \frac{92^2}{c^2} = 0$, which occurs when $c = 46 \cdot 92$. So $b = 46\sqrt{2}$, and $\theta = \arcsin(\frac{46b}{\sqrt{(46^2+b^2)(92^2+b^2)}}) =$

$\arcsin(1/3) \approx 19.47°$.

11. Let x be as indicated on the graph. Then the distance from S to Town 1 is $\sqrt{1+x^2}$ and the distance from S to Town 2 is $\sqrt{(4-x)^2+4^2} = \sqrt{x^2-8x+32}$.

$$\text{Total length of pipe } = f(x) = \sqrt{1+x^2} + \sqrt{x^2-8x+32}.$$

We want to look for critical points of f. The easiest way is to graph f and see that it has a local minimum at about $x = 0.8$ miles. Alternatively, we can use the formula:

$$\begin{aligned} f'(x) &= \frac{2x}{2\sqrt{1+x^2}} + \frac{2x-8}{2\sqrt{x^2-8x+32}} \\ &= \frac{x}{\sqrt{1+x^2}} + \frac{x-4}{\sqrt{x^2-8x+32}} \\ &= \frac{x\sqrt{x^2-8x+32}+(x-4)\sqrt{1+x^2}}{\sqrt{1+x^2}\sqrt{x^2-8x+32}} = 0. \end{aligned}$$

$f'(x)$ is equal to zero when the numerator is equal to zero.

$$\begin{aligned} (*) \qquad \sqrt{x^2-8x+32}+(x-4)\sqrt{1+x^2} &= 0 \\ x\sqrt{x^2-8x+32} &= (4-x)\sqrt{1+x^2}. \end{aligned}$$

Squaring both sides and simplifying, we get

$$\begin{aligned} 15x^2+8x-16 &= 0, \\ (3x+4)(5x-4) &= 0. \end{aligned}$$

So $x = 4/5$. (Discard $x = -4/3$ since this is an extraneous root of the equation (*).) Using the second derivative test, we can verify that $x = 4/5$ is a local minimum.

5.7 Solutions

1. (a) $f'(x) = 3x^2+6x+3 = 3(x+1)^2$. Thus $f'(x) > 0$ everywhere except $x = -1$, so it is increasing everywhere except perhaps at $x = -1$. The function is in fact increasing at $x = -1$ since $f(x) > f(-1)$ for $x > -1$.

   (b) The original equation can have at most one root, since it can only pass through the x-axis once if it never decreases. It must have one root, since $f(0) = -6$ and $f(1) = 1$.

   (c) The root is in the interval $[0, 1]$, since $f(0) < 0 < f(1)$.

   (d) Let $x_0 = 1$.

$$\begin{aligned} x_0 &= 1 \\ x_1 &= 1 - \frac{f(1)}{f'(1)} \end{aligned}$$

$$\begin{aligned}
&= 1 - \frac{1}{12} = \frac{11}{12} \approx 0.917 \\
x_2 &= \frac{11}{12} - \frac{f\left(\frac{11}{12}\right)}{f'\left(\frac{11}{12}\right)} \\
&\approx 0.913 \\
x_3 &= 0.913 - \frac{f(0.913)}{f'(0.913)} \approx 0.913.
\end{aligned}$$

Since the digits repeat, they should be accurate. Thus $x \approx 0.913$.

3. Let $f(x) = x^4 - 100$. Then $f(\sqrt[4]{100}) = 0$, so we can use Newton's method to solve $f(x) = 0$ to obtain $x = \sqrt[4]{100}$. $f'(x) = 4x^3$. Since $3^4 < 100 < 4^4$, try 3.1 as an initial guess.

$$\begin{aligned}
x_0 &= 3.1 \\
x_1 &= 3.1 - \frac{f(3.1)}{f'(3.1)} \\
&\approx 3.164 \\
x_2 &= 3.164 - \frac{f(3.164)}{f'(3.164)} \approx 3.162 \\
x_3 &= 3.162 - \frac{f(3.162)}{f'(3.162)} \approx 3.162
\end{aligned}$$

Thus $\sqrt[4]{100} \approx 3.162$.

5. Let $f(x) = \sin x - 1 + x$; we want to find all zeros of f, because $f(x) = 0$ implies $\sin x = 1 - x$.

Graphing $\sin x$ and $1 - x$, we see that $f(x)$ has one solution at $x \approx \frac{1}{2}$.

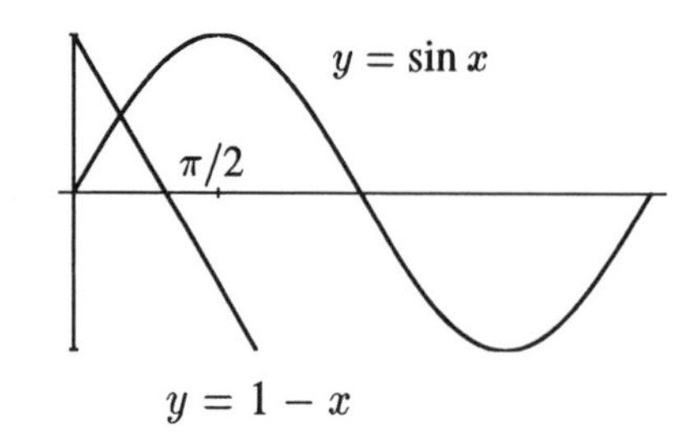

Letting $x_0 = 0.5$, and using Newton's method, we have $f'(x) = \cos x + 1$, so that

$$x_1 = 0.5 - \frac{\sin(0.5) - 1 + 0.5}{\cos(0.5) + 1} \approx 0.511,$$

$$x_2 = 0.511 - \frac{\sin(0.511) - 1 + 0.511}{\cos(0.511) + 1} \approx 0.511.$$

Thus $\sin x = 1 - x$ has one solution at $x \approx 0.511$.

7. Let $f(x) = e^{-x} - \ln x$. Then $f'(x) = -e^{-x} - \frac{1}{x}$. We want to find all zeros of f, because $f(x) = 0$ implies that $e^{-x} = \ln x$. Since e^{-x} is always decreasing and $\ln x$ is always increasing, there must be only 1 solution. Since $e^{-1} > \ln 1 = 0$, and $e^{-e} < \ln e = 1$, then $e^{-x} = \ln x$ for some x, $1 < x < e$. Try $x_0 = 1$. We now use Newton's method.

$$\begin{aligned} x_0 &= 1 \\ x_1 &= 1 - \frac{e^{-1} - 0}{-e^{-1} - 1} \\ &\approx 1.2689 \\ x_2 &\approx 1.309 \\ x_3 &\approx 1.310. \end{aligned}$$

Thus $x \approx 1.310$ is the solution.

9. Let $f(x) = e^x \cos x - 1$. Then $f'(x) = -e^x \sin x + e^x \cos x$
Now we use Newton's method, guessing 1 initially.

$$\begin{aligned} x_0 &= 1 \\ x_1 &= 1 - \frac{f(1)}{f'(1)} \approx 1.5725 \\ \text{Continuing, } x_2 &\approx 1.364 \\ x_3 &\approx 1.299 \\ x_4 &\approx 1.293 \\ x_5 &\approx 1.293 \end{aligned}$$

Thus $x \approx 1.293$ is a solution. Looking at a graph of $f(x)$ suffices to convince us that there is only one.

11. (a) One zero in the interval $0.6 < x < 0.7$.
    (b) Three zeros in the intervals $-1.55 < x < -1.45$, $-0.05 < x < 0.05$, $1.45 < x < 1.55$.
    (c) Two zeros in the intervals $0.1 < x < 0.2$, $3.5 < x < 3.6$.

13. Let $f(x) = x^2 - a$, so $f'(x) = 2x$.
Then by Newton's method, $x_{n+1} = x_n - \frac{x_n^2 - a}{2x_n}$
For $a = 2$:
$x_0 = 1 \Rightarrow x_1 = 1.5 \Rightarrow x_2 = 1.41\overline{6} \Rightarrow x_3 = 1.414215\ldots \Rightarrow x_4 = 1.414213\ldots$ so $\sqrt{2} \approx 1.4142$.
For $a = 10$:
$x_0 = 5 \Rightarrow x_1 = 3.5 \Rightarrow x_2 = 3.17857\ldots \Rightarrow x_3 = 3.162319\ldots \Rightarrow x_4 = 3.162277\ldots$ so

$\sqrt{10} \approx 3.1623$.
For $a = 1000$:
$x_0 = 500 \Rightarrow x_1 = 251 \Rightarrow x_2 = 127.49203\ldots \Rightarrow x_3 = 67.6678\ldots \Rightarrow x_4 = 41.2229\ldots \Rightarrow x_5 = 32.7406\ldots \Rightarrow x_6 = 31.6418 \Rightarrow x_7 = 31.62278\ldots \Rightarrow x_8 = 31.62277\ldots$ so $\sqrt{1000} \approx 31.6228$.
For $a = \pi$:
$x = \frac{\pi}{2} \Rightarrow x_1 = 1.7853\ldots \Rightarrow x_2 = 1.7725\ldots \Rightarrow x_3 = 1.77245\ldots \Rightarrow x_4 = 1.77245\ldots$ so $\sqrt{\pi} \approx 1.77245$.

15. (a) We have

$$f(x) = \sin x - \frac{2}{3}x$$

and

$$f'(x) = \cos x - \frac{2}{3}.$$

Using $x_0 = 0.904$,

$$x_1 = 0.904 - \frac{\sin(0.904) - \frac{2}{3}(0.904)}{\cos(0.904) - \frac{2}{3}} \approx 4.704,$$

$$x_2 = 4.704 - \frac{\sin(4.704) - \frac{2}{3}(4.704)}{\cos(4.704) - \frac{2}{3}} \approx -1.423,$$

$$x_3 = -1.433 - \frac{\sin(-1.423) - \frac{2}{3}(-1.423)}{\cos(-1.423) - \frac{2}{3}} \approx -1.501,$$

$$x_4 = -1.499 - \frac{\sin(-1.501) - \frac{2}{3}(-1.501)}{\cos(-1.501) - \frac{2}{3}} \approx -1.496,$$

$$x_5 = -1.496 - \frac{\sin(-1.496) - \frac{2}{3}(-1.496)}{\cos(-1.496) - \frac{2}{3}} \approx -1.496.$$

Using $x_0 = 0.905$,

$$x_1 = 0.905 - \frac{\sin(0.905) - \frac{2}{3}(0.905)}{\cos(0.905) - \frac{2}{3}} \approx 4.643,$$

$$x_2 = 4.643 - \frac{\sin(4.643) - \frac{2}{3}(4.643)}{\cos(4.643) - \frac{2}{3}} \approx -0.918,$$

$$x_3 = -0.918 - \frac{\sin(-0.918) - \frac{2}{3}(-0.918)}{\cos(-0.918) - \frac{2}{3}} \approx -3.996,$$

$$x_4 = -3.996 - \frac{\sin(-3.996) - \frac{2}{3}(-3.996)}{\cos(-3.996) - \frac{2}{3}} \approx -1.413,$$

$$x_5 = -1.413 - \frac{\sin(-1.413) - \frac{2}{3}(-1.413)}{\cos(-1.413) - \frac{2}{3}} \approx -1.502,$$

$$x_6 = -1.502 - \frac{\sin(-1.502) - \frac{2}{3}(-1.502)}{\cos(-1.502) - \frac{2}{3}} \approx -1.496.$$

Now using $x_0 = 0.906$,

$$x_1 = 0.906 - \frac{\sin(0.906) - \frac{2}{3}(0.906)}{\cos(0.906) - \frac{2}{3}} \approx 4.584,$$

$$x_2 = 4.584 - \frac{\sin(4.584) - \frac{2}{3}(4.584)}{\cos(4.584) - \frac{2}{3}} \approx -0.509,$$

$$x_3 = -0.510 - \frac{\sin(-0.509) - \frac{2}{3}(-0.509)}{\cos(-0.509) - \frac{2}{3}} \approx .207,$$

$$x_4 = -1.300 - \frac{\sin(.207) - \frac{2}{3}(.207)}{\cos(.207) - \frac{2}{3}} \approx -0.009,$$

$$x_5 = -1.543 - \frac{\sin(-0.009) - \frac{2}{3}(-0.009)}{\cos(-0.009) - \frac{2}{3}} \approx 0,$$

(b) Starting with 0.904 and 0.905 yields the same value, but the two paths to get to the root are very different. Starting with 0.906 leads to a different root.

5.8 Solutions

1. $5x$

3. $\frac{1}{3}x^3$

5. $\sin t$

7. $\ln z$

9. $-\dfrac{1}{2z^2}$

11. $-\cos t$

13. $\frac{t^4}{4} - \frac{t^3}{6} - \frac{t^2}{2}$

15. $4x^{\frac{3}{2}} + \dfrac{1}{x} + 10 \ln x$

17. $e^t + 5\dfrac{1}{5}e^{5t} = e^t + e^{5t}$

19. $\dfrac{5^x}{\ln 5}$

21. $\dfrac{1}{2}\sin(t^2)$

23. The general antiderivative of $f(x)$ is

$$\begin{aligned} F(x) &= 3x + C \\ \text{Since } F(0) = 2, \text{ we have } F(0) &= 3(0) + C = C = 2 \\ \text{Thus } C = 2, \text{ and } F(x) &= 3x + 2 \end{aligned}$$

25. The general antiderivative of $f(x)$ is

$$\begin{aligned} F(x) &= \frac{1}{3}x^3 + C \\ \text{Since } F(0) = 2, \text{ we have } F(0) &= \frac{1}{3}(0)^3 + C = C = 2 \\ \text{Thus } C = 2, \text{ and } F(x) &= \frac{1}{3}x^3 + 2 \end{aligned}$$

27. The general antiderivative of $f(x)$ is

$$\begin{aligned} F(x) &= -\cos x + C \\ \text{Since } F(0) = 2, \text{ we have } F(0) &= -\cos 0 + C = -1 + C = 2 \\ \text{Thus } C = 3, \text{ and } F(x) &= -\cos x + 3 \end{aligned}$$

29.

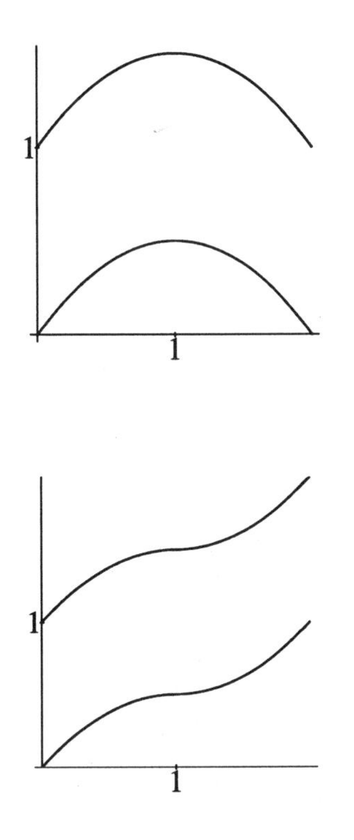

31.

33. Between time $t = 0$ and time $t = B$, the velocity of the cork is always positive, which means the cork is moving upwards. At time $t = B$, the velocity is zero, and so the cork has stopped moving altogether. Since shortly thereafter the velocity of the cork becomes negative, the cork will next begin to move downwards. Thus when $t = B$ the cork has risen as far as it ever will, and is riding on top of the crest of the wave.

    From time $t = B$ to time $t = D$, the velocity of the cork is negative, which means it is falling. When $t = D$, the velocity is again zero, and the cork has ceased to fall. Thus when $t = D$ the cork is riding on the bottom of the trough of the wave.

    Since the cork is on the crest at time B and in the trough at time D, it must be midway between crest and trough when the time is midway between B and D. Thus at time $t = C$ the cork is moving through the equilibrium position on its way down. (The equilibrium position is where the cork would be if the water were absolutely calm.) By symmetry, $t = A$ is the time when the cork is moving through the equilibrium position on the way up.

    Since acceleration is the derivative of velocity, points where the acceleration is zero would be critical points of the velocity function. Since point A (a maximum) and point B (a minimum) are critical points, the acceleration is zero there.

    A possible graph of the height of the cork is shown below. Note that the horizontal axis represents a height equal to the average depth of the ocean at that point (the equilibrium position of the cork).

35. (a)

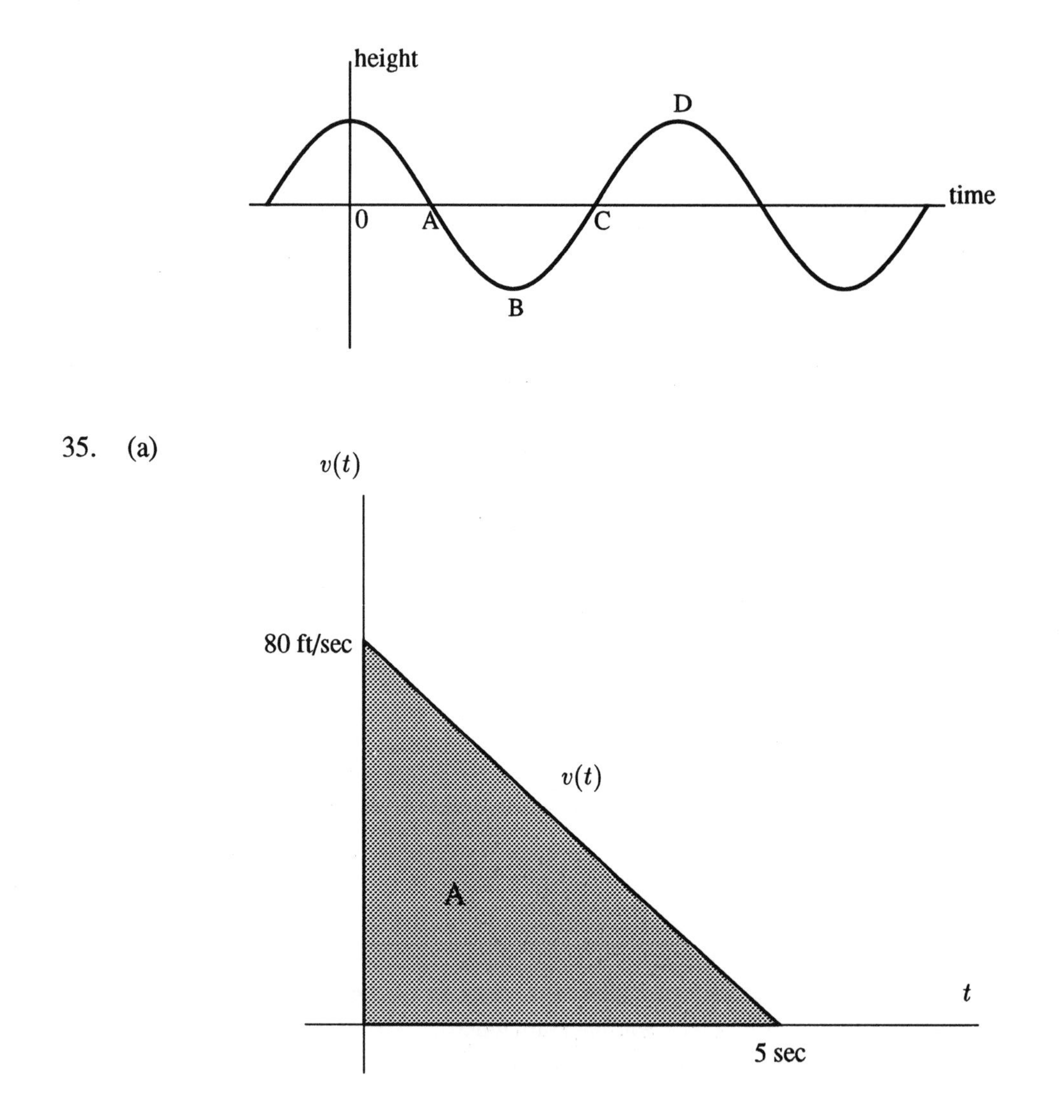

(b) The total distance is represented by the shaded region A, the area under the graph of $v(t)$.

(c) The area A, a triangle, is given by

$$A = \frac{1}{2}(\text{base})(\text{height}) = \frac{1}{2}(5\text{ sec})(80\text{ ft/sec}) = 200\text{ ft}.$$

(d) Using integration and the fundamental theorem of calculus, we have $A = \int_0^5 v(t)\,dt$ or $A = s(5) - s(0)$, where $s(t)$ is an antiderivative of $v(t)$.

We have that $a(t)$, the acceleration, is constant: $a(t) = k$ for some constant k. Therefore $v(t) = kt + C$ for some constant C. We have $80 = v(0) = k(0) + C = C$, so that $v(t) = kt + 80$. Putting in $t = 5$, $0 = v(5) = (k)(5) + 80$, or $k = -80/5 = -16$.

Thus $v(t) = -16t + 80$, and an antiderivative for $v(t)$ is $s(t) = -8t^2 + 80t + C$. Since the total distance traveled at $t = 0$ is 0, we have $s(0) = 0$ which means $C = 0$. Finally,

$A = \int_0^5 v(t)\,dt = s(5) - s(0) = (-8(5)^2 + (80)(5)) - (-8(0)^2 + (80)(0)) = 200$ ft, which agrees with the previous part.

37. (a)

t (sec)	0	0.5	1	1.5	2	2.5	3	3.5	4	4.5	5	5.5	6
$v(t)$ (ft/sec)	30	27.5	25	22.5	20	17.5	15	12.5	10	7.5	5	2.5	0

Since the velocity is constantly decreasing, and $v(6) = 0$, the car comes to rest after 6 seconds.

(b) Over the interval $a \le t \le a + \frac{1}{2}$, the left-hand velocity is $v(a)$, and the right-hand velocity is $v(a + \frac{1}{2})$. Since we are considering half-second intervals, $\Delta t = \frac{1}{2}$, and $n = 12$.

Thus the left-hand sum is given by

$$\sum_{i=0}^{11} v(t_i)\,\Delta t$$

$= 30(\frac{1}{2}) + 27.5(\frac{1}{2}) + 25(\frac{1}{2}) + 22.5(\frac{1}{2}) + 20(\frac{1}{2}) + 17.5(\frac{1}{2}) + 15(\frac{1}{2}) + 12.5(\frac{1}{2}) + 10(\frac{1}{2}) + 7.5(\frac{1}{2}) + 5(\frac{1}{2}) + 2.5(\frac{1}{2}) = 97.5$ ft, and the right-hand sum is given by

$$\sum_{i=1}^{12} v(t_i)\,\Delta t$$

$= 27.5(\frac{1}{2}) + 25(\frac{1}{2}) + 22.5(\frac{1}{2}) + 20(\frac{1}{2}) + 17.5(\frac{1}{2}) + 15(\frac{1}{2}) + 12.5(\frac{1}{2}) + 10(\frac{1}{2}) + 7.5(\frac{1}{2}) + 5(\frac{1}{2}) + 2.5(\frac{1}{2}) + 0(\frac{1}{2}) = 82.5$ ft.

A graph of velocity versus time looks like:

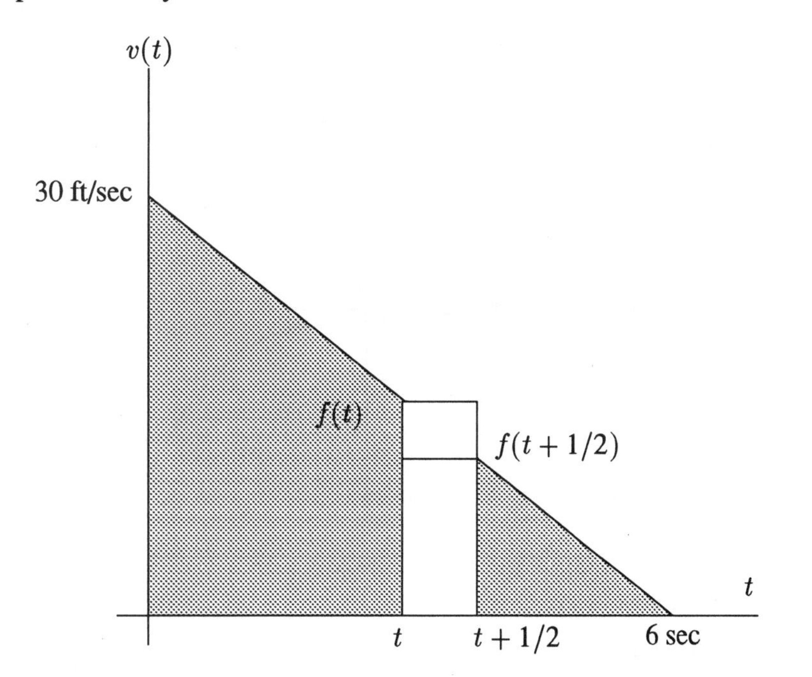

Since $v(t)$ is decreasing, the left-hand sum overestimates the area A, while the right-hand sum underestimates it.

(c) The velocity is constantly decreasing at a rate of 5 ft/sec per second, i.e. after each second the velocity has dropped by 5 units. Therefore $v(t) = 30 - 5t$.

An antiderivative for $v(t)$ is $s(t)$, where $s(t) = 30t - \frac{5}{2}t^2$. Thus by the fundamental theorem of calculus, the area $A = s(6) - s(0) = (30(6) - \frac{5}{2}(6)^2) - (30(0) - \frac{5}{2}(0)^2) =$ 90 ft.

In fact, from the graph, it's clear that the area A should equal the average of the left-hand and right-hand sums: $90\text{ ft} = \frac{97.5\text{ ft}+82.5\text{ ft}}{2}$.

39. (a) $a(t) = 1.6$, so $v(t) = 1.6t + v_0 = 1.6t$, since the initial velocity is 0.

(b) $s(t) = 0.8t^2 + s_0$

41.

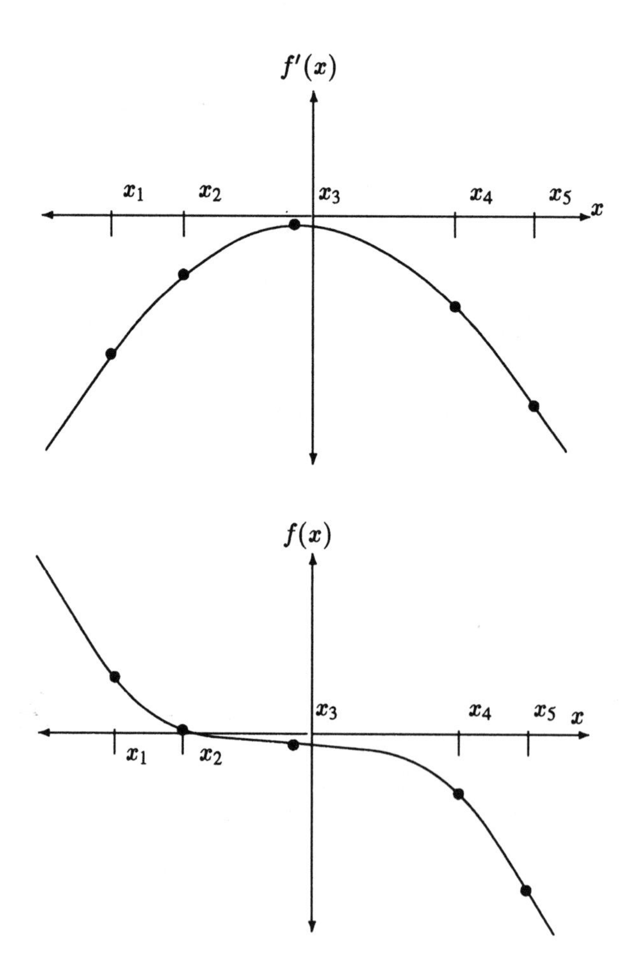

(a) $f(x)$ is greatest at x_1.

(b) $f(x)$ is least at x_5.

(c) $f'(x)$ is greatest at x_3..

(d) $f'(x)$ is least at x_5.

(e) $f''(x)$ is greatest at x_1.

(f) $f''(x)$ is least at x_5.

5.9 Solutions

1. Let the acceleration due to gravity equal $-k$ meters/sec^2, for some positive constant k, and suppose the object falls from an initial height of $s(0)$ meters.

   We have $a(t) = \frac{dv}{dt} = -k$, so that $v(t) = -kt + v_0$. Since the initial velocity is zero, we have $v(0) = -k(0) + v_0 = 0$, which means $v_0 = 0$. Our formula becomes $v(t) = \frac{ds}{dt} = -kt$. This means $s(t) = \frac{-kt^2}{2} + s_0$. Since $s(0) = \frac{-k(0)^2}{2} + s_0$, we have $s_0 = s(0)$, and our formula becomes $s(t) = \frac{-kt^2}{2} + s(0)$. Suppose that the object falls for t seconds. Assuming it hasn't hit the ground, its height is $\frac{-kt^2}{2} + s(0)$, so that the distance traveled is $s(0) - (\frac{-kt^2}{2} + s(0)) = \frac{kt^2}{2}$ meters, which is proportional to t^2.

3. (a) Since $S(t) = gt^2$, the distance a body falls in the first second is

$$S(1) = g \cdot 1^2 = g.$$

In the second second, the body travels

$$S(2) - S(1) = g \cdot 2^2 - g \cdot 1^2 = 4g - g = 3g.$$

In the third second, the body travels

$$S(3) - S(2) = g \cdot 3^2 - g \cdot 2^2 = 9g - 4g = 5g,$$

and in the fourth second, the body travels

$$S(4) - S(3) = g \cdot 4^2 - g \cdot 3^2 = 16g - 9g = 7g.$$

(b) Galileo seems to have been correct. His observation follows from the fact that the differences between consecutive squares are consecutive odd numbers. For, if n is any number, then $n^2 - (n-1)^2 = 2n - 1$, which is the nth odd number (where 1 is the first).

5.10 Answers to Miscellaneous Exercises for Chapter 5

1.

$$\begin{aligned} f(x) &= x^3 - 3x^2 \quad -1 \leq x \leq 3 \\ f'(x) &= 3x^2 - 6x \\ &= 3x(x-2) \\ f''(x) &= 6x - 6 \\ &= 6(x-1) \end{aligned}$$

Critical points: $x = 2, 0$, since $f'(x) = 0$ here. Using the second derivative test, we find that $f(0) = 0$ is a local max since $f'(0) = 0$ and $f''(0) = -6 < 0$, that $f(2) = -4$ is a local min since $f'(2) = 0$ and $f''(2) = 6 > 0$. There is an inflection point at $x = 1$ since f'' changes sign at $x = 1$. At the endpoints: $f(-1) = -4, f(3) = 0$.
So, the global maxima are $f(0) = 0$ and $f(3) = 0$, while the global minima are $f(-1) = -4$ and $f(2) = -4$.

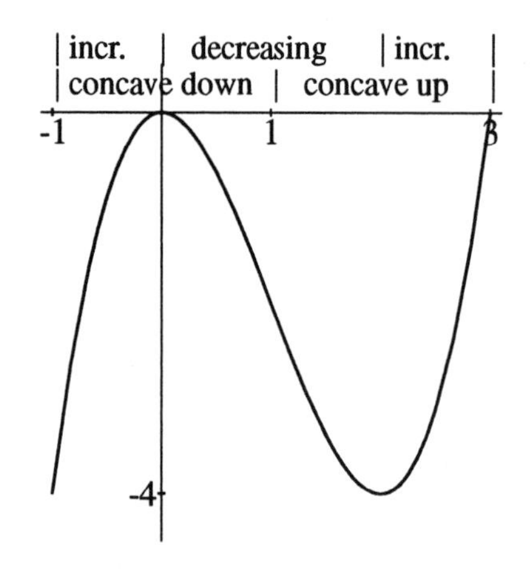

3.

$$\begin{aligned} f(x) &= e^{-x} \sin x \\ f'(x) &= -e^{-x} \sin x + e^{-x} \cos x \\ f''(x) &= e^{-x} \sin x - e^{-x} \cos x - e^{-x} \cos x - e^{-x} \sin x \\ &= -2e^{-x} \cos x \end{aligned}$$

Critical points: $x = \frac{\pi}{4}, \frac{5\pi}{4}$, since $f'(x) = 0$ here. The inflection points are $x = \frac{\pi}{2}, \frac{3\pi}{2}$, since f'' changes sign at these points. At the endpoints, $f(0) = 0,\ f(2\pi) = 0$. So we have $f(\frac{\pi}{4}) = (e^{-\frac{\pi}{4}})(\frac{\sqrt{2}}{2})$ as the local and global max; $f(\frac{5\pi}{4}) = -e^{\frac{-5\pi}{4}}(\frac{\sqrt{2}}{2})$ as the local and global min.

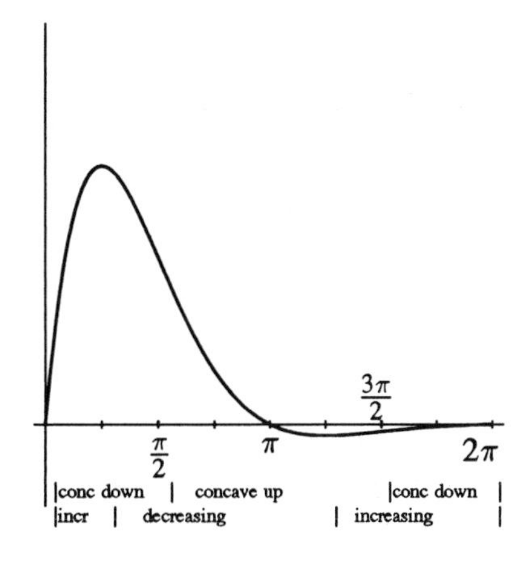

5. The polynomial $f(x)$ behaves like $2x^3$ as $x \to \pm\infty$. Therefore, $\lim_{x\to\infty} f(x) = \infty$ and $\lim_{x\to-\infty} f(x) = -\infty$.

   We have $f'(x) = 6x^2 - 18x + 12 = 6(x-2)(x-1)$, which is zero when $x = 1$ or $x = 2$.

   We have $f''(x) = 12x - 18 = 6(2x-3)$, which is zero when $x = \frac{3}{2}$. We have $f''(x) < 0$ for $x < \frac{3}{2}$, and $f''(x) > 0$ for $x > \frac{3}{2}$. Thus $x = \frac{3}{2}$ is an inflection point.

   The critical points are $x = 1$ and $x = 2$, and $f(1) = 6$, $f(2) = 5$. By the second derivative test, $f''(1) = -6 < 0$, so $x = 1$ is a local max; $f''(2) = 6 > 0$, so $x = 2$ is a local min.

   We have from the above the diagrams

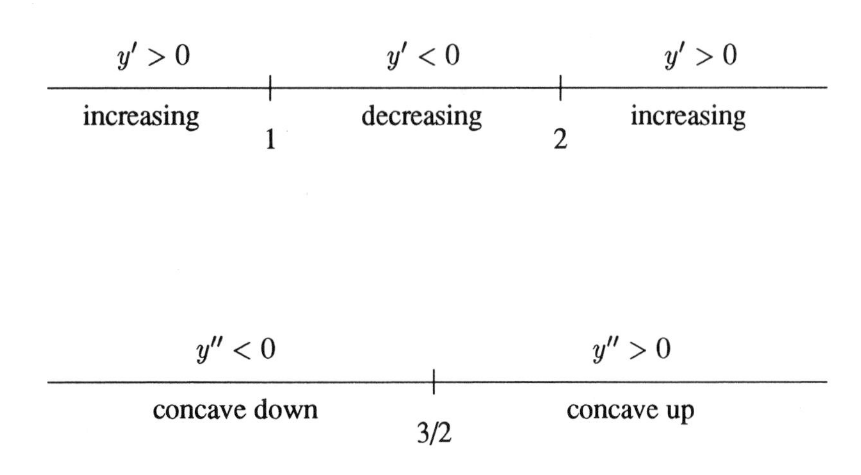

   and the graph of $f(x) = 2x^3 - 9x^2 + 12x + 1$ looks like

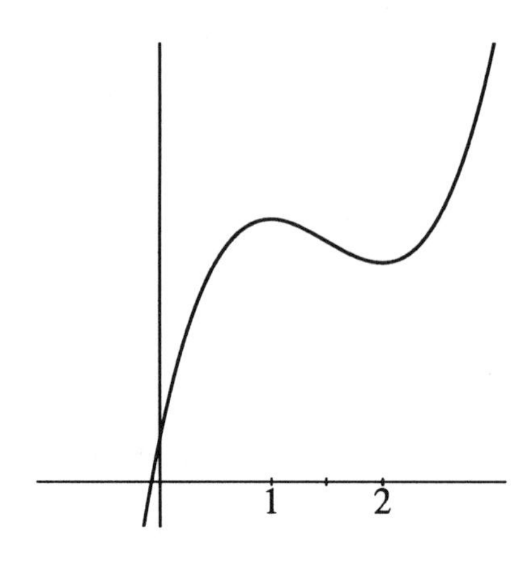

$f(x)$ has no global maximum or minimum.

7. As $x \to -\infty$, $e^{-x} \to \infty$, so $xe^{-x} \to -\infty$. Thus $\lim_{x\to-\infty} xe^{-x} = -\infty$.

   As $x \to \infty$, $\frac{x}{e^x} \to 0$, since e^x grows much more quickly than x. Thus $\lim_{x\to\infty} xe^{-x} = 0$.

   Using the product rule,

$$f'(x) = e^{-x} - xe^{-x} = (1-x)e^{-x},$$

   which is zero when $x = 1$, negative when $x > 1$, and positive when $x < 1$. Thus $f(1) = \frac{1}{e^1} = \frac{1}{e}$ is a local maximum.

   Again, using the product rule,

$$f''(x) = -e^{-x} - e^{-x} + xe^{-x} = xe^{-x} - 2e^{-x} = (x-2)e^{-x},$$

   which is zero when $x = 2$, positive when $x > 2$, and negative when $x < 2$, giving an inflection point at $(2, \frac{2}{e^2})$. With the above, we have the following diagram:

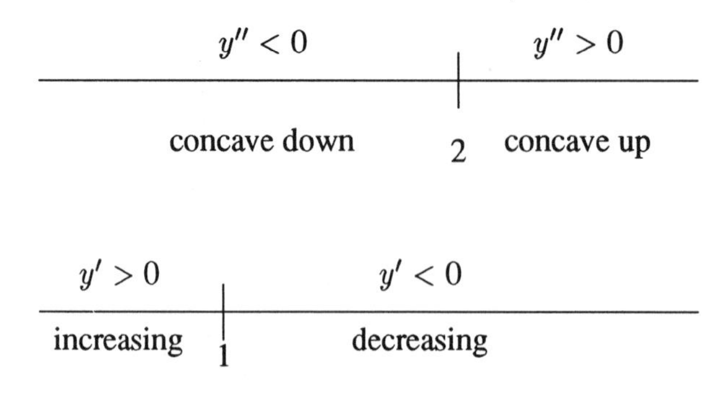

   The graph of f is:

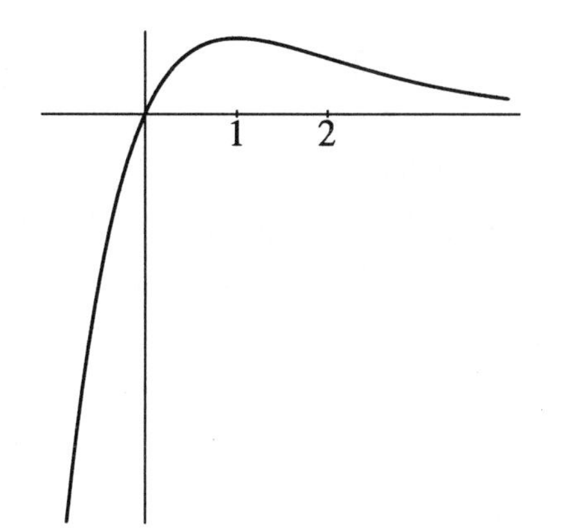

and $f(x)$ has one global maximum and no local or global minima.

9. $\lim_{x\to\infty} f(x) = +\infty$, $\lim_{x\to-\infty} f(x) = -\infty$.
There are no asymptotes.
$f'(x) = 3x^2+6x-9 = 3(x+3)(x-1)$. Critical points are $x=-3$, $x=1$.
$f''(x) = 6(x+1)$.

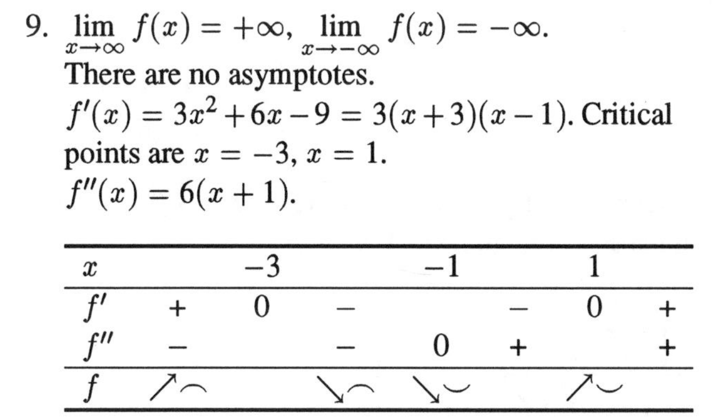

x		-3		-1		1	
f'	$+$	0	$-$		$-$	0	$+$
f''	$-$		$-$	0	$+$		$+$
f	$\nearrow\frown$		$\searrow\frown$	$\searrow\smile$		$\nearrow\smile$	

Thus, $x=-1$ is an inflection point. $f(-3)=12$ is a local maximum; $f(1)=-20$ is a local minimum. There are no global maxima or minima.

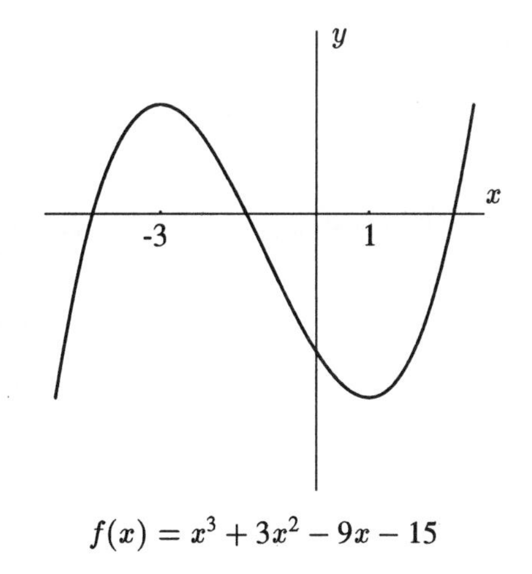

$f(x) = x^3 + 3x^2 - 9x - 15$

11. $\lim_{x\to+\infty} f(x) = +\infty$, $\lim_{x\to0+} f(x) = +\infty$.
Hence, $x = 0$ is a vertical asymptote.

$$f'(x) = 1 - \frac{2}{x} = \frac{x-2}{x}.$$

So, $x = 2$ is the only critical point.
$f''(x) = \frac{2}{x^2}$, which can never be zero. So there are no inflection points.

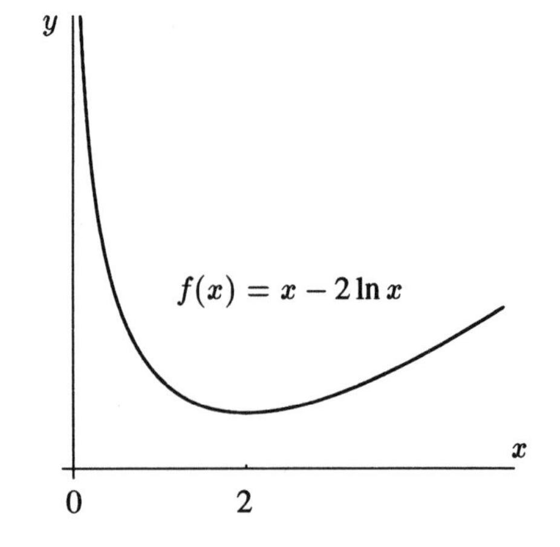

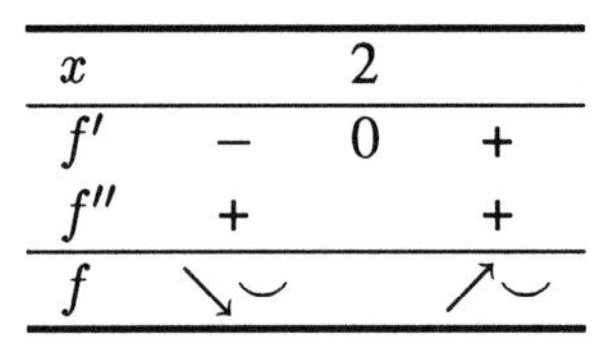

x		2	
f'	$-$	0	$+$
f''	$+$		$+$
f	↘◡		↗◡

Thus, $f(2)$ is a local and global minimum.

13. Since $\lim_{x\to-\infty} f(x) = \lim_{x\to+\infty} f(x) = 0$, $y = 0$ is a horizontal asymptote.
$f'(x) = -2xe^{-x^2}$. So, $x = 0$ is the only critical point.
$f''(x) = -2(e^{-x^2} + x(-2x)e^{-x^2}) = 2e^{-x^2}(2x^2 - 1) = 2e^{-x^2}(\sqrt{2}x - 1)(\sqrt{2}x + 1)$.
Thus, $x = \pm\frac{1}{\sqrt{2}}$ are inflection points.

x		$\frac{-1}{\sqrt{2}}$		0		$\frac{1}{\sqrt{2}}$	
f'	$+$		$+$	0	$-$		$-$
f''	$+$	0	$-$		$-$	0	$+$
f	↗◡		↗◠		↘◠		↘◡

Thus, $f(0) = 1$ is a local and global maximum.

y

$f(x) = e^{-x^2}$

x

15. By Problem 13, Section 5.4, the maximum of $f(x) = e^{-ax}\sin x$, for $a > 0$, $x > 0$, occurs at $x = \arctan(\frac{1}{a})$, and the minimum at $x = \pi + \arctan(\frac{1}{a})$:

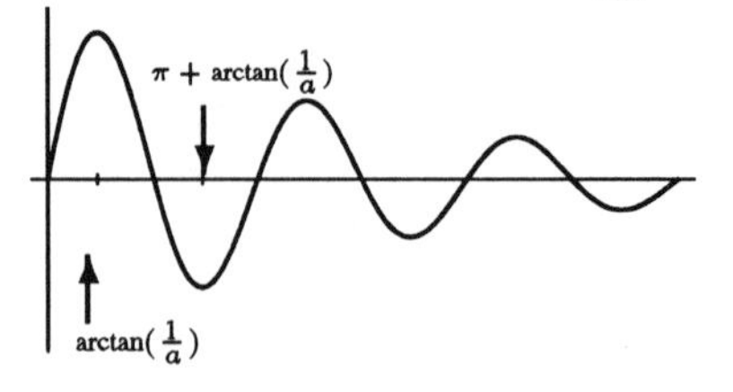

Letting $a = 1$, we have $\arctan(1) = \pi/4$ and $\pi + \arctan(1) = 5\pi/4$.
Thus $e^{-5\pi/4}\sin(5\pi/4) \le e^{-x}\sin x \le e^{-\pi/4}\sin(\pi/4)$, or $-0.014 \le e^{-x}\sin x \le 0.323$.

17. (a) $(-\infty, 0)$ decreasing, $(0, \infty)$ increasing.
    (b) $f(0)$ is a local and global minimum.

19. (a) $(-\infty, 0)$ decreasing, $(0, 4)$ increasing, $(4, \infty)$ decreasing.
    (b) local minimum at $f(0)$, local maximum at $f(4)$.

21. (a) The dots correspond to the lines drawn on the surface of the vase:

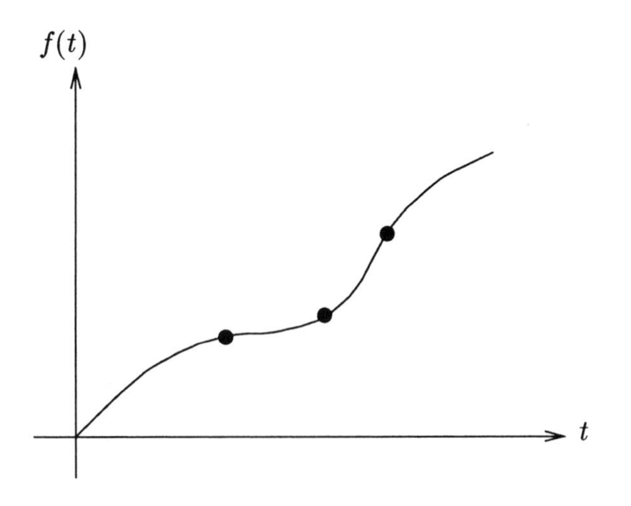

    (b) $f(t)$ grows fastest where the vase is skinniest and slowest where the vase is widest. The diameter of the widest part of the vase looks to be about 4 or 5 times as large as the diameter at the skinniest part. Since the area of a cross section is given by πr^2, where r is the radius, the ratio between areas of cross sections at these two places is 4^2 or 5^2, so the growth rates are in a ratio of about 1 to 20 (the wide part being 20 times slower).

For Problems 23– 30 we use the fact that the antiderivative of x^r is $\frac{x^{r+1}}{r+1}$ for $r \neq -1$. If $r = -1$, the antiderivative of $\frac{1}{x}$ is $\ln|x|$.

23. Antiderivative $F(x) = \frac{x^2}{2} + \frac{x^6}{6} - \frac{x^{-4}}{4} + C$.

25. Antiderivative $G(t) = 5t + \sin t + C$

27. Antiderivative $G(x) = \frac{x^4}{4} + x^3 + \frac{3x^2}{2} + x + C$.

29. $P(y) = \ln|y| + \frac{y^2}{2} + y + C$

30. $F(z) = e^z + 3z + C$

31. $G(\theta) = -\cos\theta - 2\sin\theta + C$

33. From Example 1, we know that the height of an object above the ground which begins at rest and then falls for t seconds is

$$s(t) = -16t^2 + K,$$

where K is the initial height. Here the flower pot falls from 200 ft, so we have

$$s(t) = -16t^2 + 200.$$

To see when the pot hits the ground, solve $s(t) = 0$:

$$-16t^2 + 200 = 0,$$

$$t^2 = \frac{200}{16}.$$

$$t = \sqrt{\frac{200}{16}} = \frac{5\sqrt{2}}{2} \approx 3.54 \text{ seconds.}$$

Now, the velocity of the flower pot is given by $s'(t) = v(t) = -32t$. So, the velocity of the flower pot when it hits the sidewalk is

$$v(3.54) \approx -113.1 \text{ ft/sec.}$$

which is approximately 77 mph downwards.

35. (a) See (b).

(b)

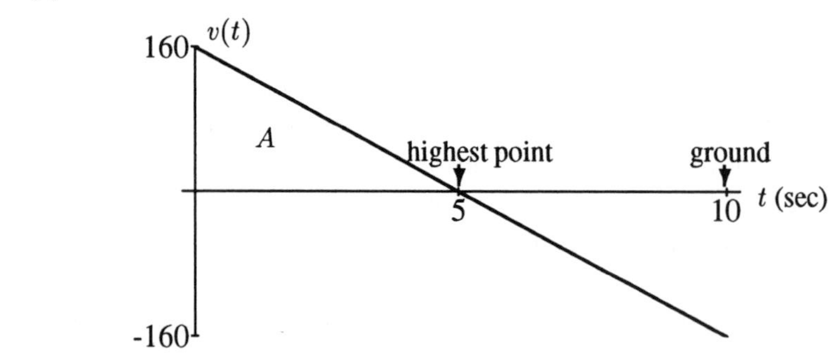

The highest point is at $t = 5$ seconds. The object hits the ground at $t = 10$ seconds, since by symmetry if the object takes 5 seconds to go up, it takes 5 seconds to come back down.

(c) The maximum height is the distance traveled when going up, which is represented by the area A of the triangle above the time axis.

$$\text{Area} = \frac{1}{2}(160 \text{ ft/sec})(5 \text{ sec}) = 400 \text{ feet.}$$

(d) The slope of the line is -32, so $v(t) = -32t + 160$.
Antidifferentiating, we get $s(t) = -16t^2 + 160t + s_0$. $s_0 = 0$, so $s(t) = -16t^2 + 160t$.
At $t = 5$, $s(t) = -400 + 800 = 400$ ft.

37. (a) Since

$$\begin{aligned} F &= -\frac{dV}{dr} \\ F\,dr &= -\,dV \\ \int F\,dr &= -V + C. \end{aligned}$$

Therefore,

$$\begin{aligned} V &= -\int \left(-\frac{A}{r^7} + \frac{B}{r^{13}}\right) dr + C \\ &= -\left(\frac{A}{6r^6} - \frac{B}{12r^{12}}\right) + C. \end{aligned}$$

But as $r \to \infty$, $V \to 0$. Therefore, C must equal 0. So,

$$V = -\frac{A}{6r^6} + \frac{B}{12r^{12}}$$

(b)

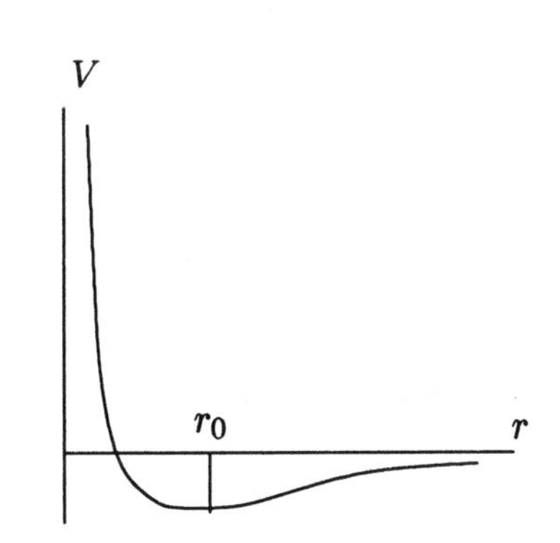

(c) r_0 represents the equilibrium separation distance between the molecules, since $F = 0$ at $r = r_0$. The equilibrium is stable since r_0 is a minimum for the potential energy.

39. (a) $a(q)$ is represented by the slope of the line from the origin to the graph. For example, the slope of line (1) through (0,0) and $(p, C(p))$ is $\frac{C(p)}{p} = a(p)$.

(b) $a(q)$ is minimal where $a(q)$ is tangent to the graph (line (2)).

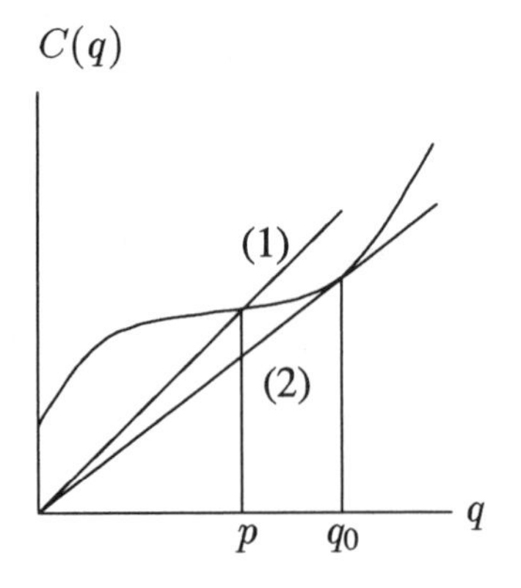

(c) We have $a(q) = \frac{C(q)}{q}$; by the quotient rule

$$a'(q) = \frac{qC'(q) - C(q)}{q^2} = \frac{C'(q) - \frac{C(q)}{q}}{q} = \frac{1}{q}(C'(q) - a(q)).$$

Thus if $q = q_0$, $a'(q_0) = \frac{1}{q}(C'(q_0) - a(q_0)) = 0$, so that $C'(q_0) = a(q_0)$; or, the average cost is minimized when it equals the marginal cost.

(d)

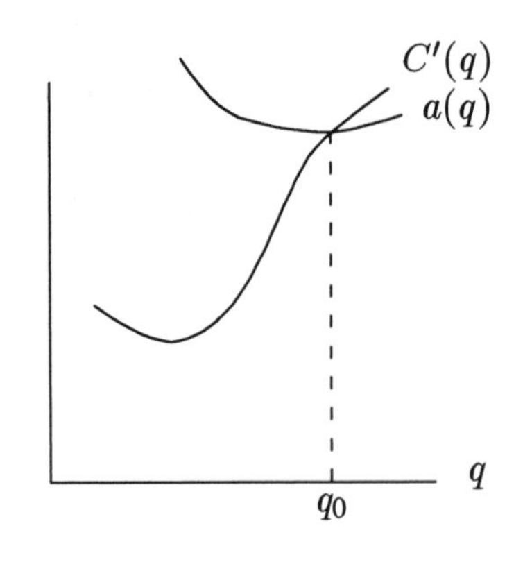

41. $r(\lambda) = a(\lambda)^{-5}(e^{\frac{b}{\lambda}} - 1)^{-1}$
$r'(\lambda) = a(-5\lambda^{-6})(e^{\frac{b}{\lambda}} - 1)^{-1} + a(\lambda^{-5})(\frac{b}{\lambda^2}e^{\frac{b}{\lambda}})(e^{\frac{b}{\lambda}} - 1)^{-2}$
$(0.96, 3.13)$ is a maximum, so:

$$\begin{aligned}
r'(0.96) = 0 \Rightarrow 5\lambda^{-6}(e^{\frac{b}{\lambda}} - 1)^{-1} &= \lambda^{-5}(\frac{b}{\lambda^2}e^{\frac{b}{\lambda}})(e^{\frac{b}{\lambda}} - 1)^{-2} \text{ where } \lambda = 0.96 \\
5\lambda(e^{\frac{b}{\lambda}} - 1) &= be^{\frac{b}{\lambda}} \\
5\lambda e^{\frac{b}{\lambda}} - 5\lambda &= be^{\frac{b}{\lambda}} \\
5\lambda e^{\frac{b}{\lambda}} - be^{\frac{b}{\lambda}} &= 5\lambda
\end{aligned}$$

$$\begin{aligned}\left(\frac{5\lambda - b}{5\lambda}\right) e^{\frac{b}{\lambda}} &= 1 \\ \frac{4.8 - b}{4.8} e^{\frac{b}{0.96}} - 1 &= 0.\end{aligned}$$

Using Newton's method, or some other approximation method, we search for a root. The root should be near 4.8. Using our initial guess, we get $b \approx 4.7665$. At $\lambda = 0.96$, $r = 3.13$, so

$$3.13 = \frac{a}{0.96^5(e^{\frac{b}{0.96}} - 1)} \Rightarrow a = 3.13(0.96)^5(e^{\frac{b}{0.96}} - 1) \approx 363.23.$$

As a check, we try $r(4) \approx 0.155$, which looks about right on the given graph.

Chapter 6

6.1 Solutions

1. Left-hand sum gives: $1^2(1/4) + (1.25)^2(1/4) + (1.5)^2(1/4) + (1.75)^2(1/4) = 1.96875$.

   Right-hand sum gives: $(1.25)^2(1/4) + (1.5)^2(1/4) + (1.75)^2(1/4) + (2)^2(1/4) = 2.71875$.

   We estimate the value of the integral by taking the average, which is 2.34375.

   Since x^2 is monotonic on $1 \le x \le 2$, the true value of the integral lies between 1.96875 and 2.71875. Thus the most our estimate could be off is 0.375. We expect it to be much closer. (And it is — the true value of the integral is $7/3 \approx 2.333$.)

3. Since $2x^2 + 7x$ is an antiderivative of $(4x + 7)$, by the Fundamental Theorem we have

$$\int_1^3 (4x+7)\,dx = 2x^2 + 7x\Big|_1^3 = 30.$$

   Thus the average value is $\dfrac{1}{3-1}\displaystyle\int_1^3 (4x+7)\,dx = \dfrac{30}{2} = 15$. Another way to work this problem is to sketch the graph of f on $1 \le x \le 3$. Then the integral is represented by the area underneath the trapezoid, which can easily be found.

5. Since $\frac{d}{dx}(x^3 + x) = 3x^2 + 1$, by the Fundamental Theorem of Calculus, $\int_0^2 (3x^2+1)\,dx = (x^3 + x)\Big|_0^2 = 10$.

7.

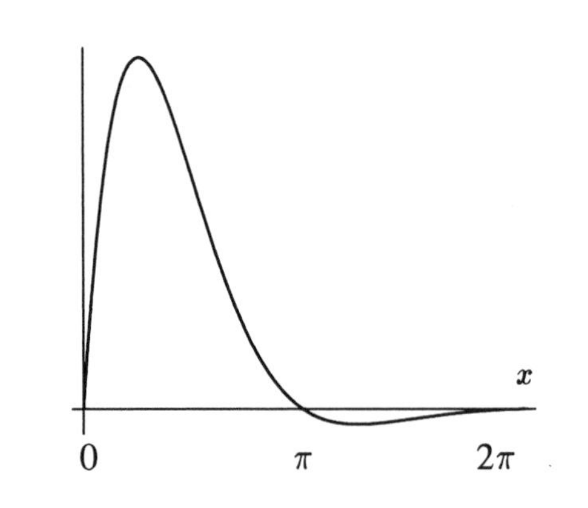

   Looking at the above graph of $e^{-x}\sin x$ for $0 \le x \le 2\pi$, we see that the area above the curve for $0 \le x \le \pi$ is much greater than the area below the curve for $\pi \le x \le 2\pi$. Thus, since the integral is the area above the axis minus the area below the axis,

$$\int_0^{2\pi} e^{-x}\sin x\,dx > 0.$$

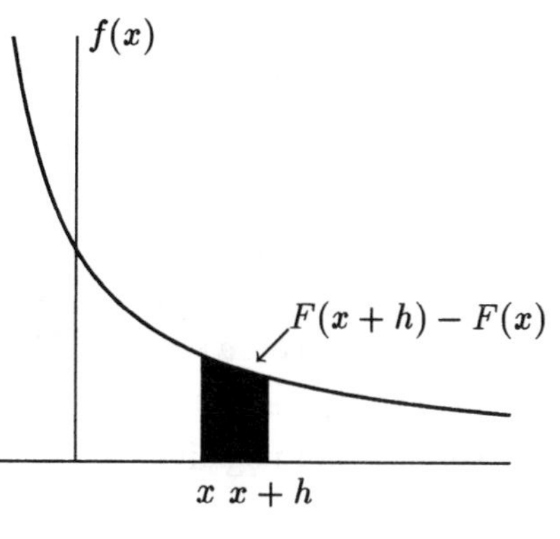

9. (a) $F(x)$ represents the unshaded region under the curve between the y-axis and x; $F(x+h)$ represents this region together with the shaded region. Thus the shaded region alone is given by the difference $F(x+h) - F(x)$.

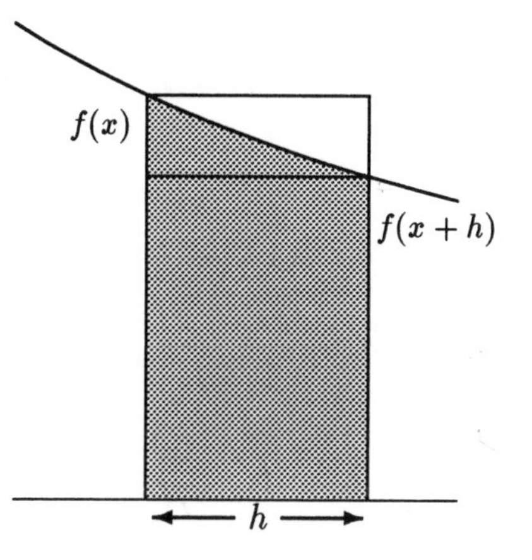

(b) Enlarging the shaded area in part (a), we see that the area of the large box is $h \cdot f(x)$. $F(x+h) - F(x)$ is the area of the shaded region. So $h \cdot f(x+h)$ is the area of the small box. Thus

$$h \cdot f(x) > F(x+h) - F(x) > h \cdot f(x+h).$$

(c) $F'(x) = \lim_{h \to 0} \dfrac{F(x+h) - F(x)}{h}$. From part (b) we have $h \cdot f(x) > F(x+h) - F(x) > h \cdot f(x+h)$. Dividing by h (for $h > 0$) gives $f(x) > \dfrac{F(x+h) - F(x)}{h} > f(x+h)$. Therefore, $\dfrac{F(x+h) - F(x)}{h}$ is trapped between $f(x)$ and $f(x+h)$. Since in the limit as $h \to 0$, $f(x+h) \to f(x)$, we see that $F'(x) = \lim_{h \to 0} \dfrac{F(x+h) - F(x)}{h} = f(x)$.

6.2 Solutions

1. $\int_1^3 (x^2 - x)\, dx = \int_1^3 x^2\, dx - \int_1^3 x\, dx$.
   $\int_1^3 3x^2\, dx = 26$, so $\int_1^3 x^2\, dx = 26/3$, since $\int_1^3 3x^2 dx = 3 \int_1^3 x^2 dx$.

$\int_1^3 2x\,dx = 8$, so $\int_1^3 x\,dx = 4$, since $\int_1^3 2x dx = 2\int_1^3 x dx$.
Thus, $\int_1^3 (x^2 - x)\,dx = \frac{26}{3} - 4 = \frac{14}{3}$.

3. (a) The integral represents the area above the x-axis and below the line $y = x$ between $x = a$ and $x = b$. This area is $A_1 + A_2 = a(b-a)+\frac{1}{2}(b-a)^2 = (a+\frac{b-a}{2})(b-a) = \frac{b+a}{2}(b-a) = \frac{b^2-a^2}{2}$. The formula holds similarly for negative values.

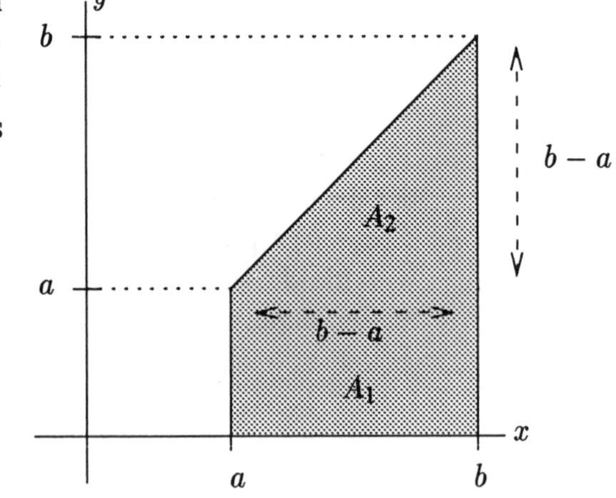

(b) i. $\int_2^5 x\,dx = 21/2$. ii. $\int_{-3}^8 x\,dx = 55/2$. iii. $\int_1^3 5x\,dx = 5\int_1^3 x\,dx = 20$.

5. (a)

$$\begin{aligned}\frac{1}{\sqrt{2\pi}}\int_1^3 e^{-\frac{x^2}{2}}\,dx &= \frac{1}{\sqrt{2\pi}}\int_0^3 e^{-\frac{x^2}{2}}\,dx - \frac{1}{\sqrt{2\pi}}\int_0^1 e^{-\frac{x^2}{2}}\,dx \\ &\approx 0.4987 - 0.3413 = 0.1574.\end{aligned}$$

(b)

$$\begin{aligned}\frac{1}{\sqrt{2\pi}}\int_{-2}^3 e^{-\frac{x^2}{2}}\,dx &= \frac{1}{\sqrt{2\pi}}\int_{-2}^0 e^{-\frac{x^2}{2}}\,dx + \frac{1}{\sqrt{2\pi}}\int_0^3 e^{-\frac{x^2}{2}}\,dx \\ \text{by symmetry of } e^{x^2/2} &= \frac{1}{\sqrt{2\pi}}\int_0^2 e^{-\frac{x^2}{2}}\,dx + \frac{1}{\sqrt{2\pi}}\int_0^3 e^{-\frac{x^2}{2}}\,dx \\ &\approx 0.4772 + 0.4987 = 0.9759.\end{aligned}$$

7. $\int_{-1}^1 e^{x^2}\,dx$ is positive, since e^{x^2} is always positive, and $\int_{-1}^1 e^{x^2}\,dx$ represents the area below the curve $y = e^{x^2}$, as indicated on the graph.

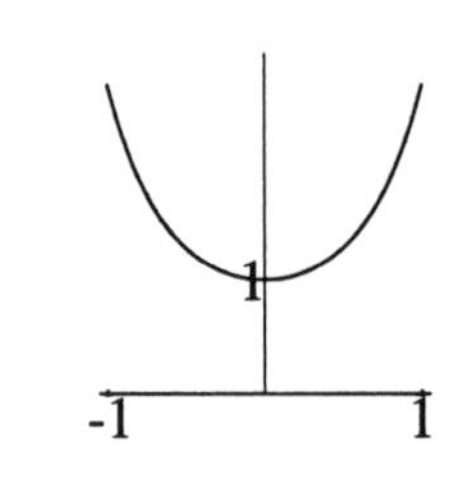

9. The sine function is monotonically increasing on $0 \le x \le 1$, so if we take left and right sums using n intervals, the difference between them will be (by the error formula in Chapter 3)

$$\frac{1}{n}(\sin 1 - \sin 0) \approx \frac{0.84}{n}.$$

Since we want to find the integral to the ten-thousandths place, we let $n = 10000$. We find the left sum ≈ 0.459656 and the right sum ≈ 0.459740. Thus $\int_0^1 \sin x\,dx \approx 0.4597$.

6.3 Solutions

1. $f(x) = 3 \Rightarrow F(x) = 3x + C$. $F(0) = 0 \Rightarrow 3 \cdot 0 + C = 0$, so $C = 0$. Thus $F(x) = 3x$ is the only possibility.

3. $f(x) = -7x \Rightarrow F(x) = \frac{-7x^2}{2} + C$. $F(0) = 0 \Rightarrow -\frac{7}{2} \cdot 0^2 + C = 0$, so $C = 0$. Thus $F(x) = -7x^2/2$ is the only possibility.

5. $f(x) = x^2 \Rightarrow F(x) = \frac{x^3}{3} + C$. $F(0) = 0 \Rightarrow \frac{0^3}{3} + C = 0$, so $C = 0$. Thus $F(x) = \frac{x^3}{3}$ is the only possibility.

7. $f(x) = 2 + 4x + 5x^2 \Rightarrow F(x) = 2x + 2x^2 + \frac{5}{3}x^3 + C$. $F(0) = 0 \Rightarrow C = 0$. Thus $F(x) = 2x + 2x^2 + \frac{5}{3}x^3$ is the only possibility.

9. $f(x) = \sin x \Rightarrow F(x) = -\cos x + C$. $F(0) = 0 \Rightarrow -\cos 0 + C = 0$, so $C = 1$. Thus $F(x) = -\cos x + 1$ is the only possibility.

11. One antiderivative is $\frac{1}{4}x^4 + \frac{10}{3}x^{\frac{3}{2}} + 2x^{-1}$. The general form for all antiderivatives is $\frac{1}{4}x^4 + \frac{10}{3}x^{\frac{3}{2}} + 2x^{-1} + C$.

13. One antiderivative is $7t - \frac{t^9}{72} + \ln|t|$. The general form for all antiderivatives is $7t - \frac{t^9}{72} + \ln|t| + C$.

15. One antiderivative is $\sin\theta - \cos\theta$. The general form for all antiderivatives is $\sin\theta - \cos\theta + C$.

17. Let $F(x) = \frac{x^2 \sin x}{2}$. Then by the product rule $F'(x) = x\sin x + \frac{x^2 \cos x}{2} \neq x\cos x$. Thus $F(x)$ is not an antiderivative of $x\cos x$.

19. Let $F(y) = \arctan 2y$. $F'(y) = \frac{1}{1+(2y)^2} \cdot 2 = \frac{2}{1+4y^2}$, so $\arctan 2y$ is an antiderivative of $\frac{2}{1+4y^2}$.

21. $3\ln|t| + \dfrac{2}{t} + C$

23. $\frac{2}{5}x^{\frac{5}{2}} + \frac{2}{15}x^{\frac{3}{2}} - 2\ln|x| + C$

25. Since $f(x) = x + 1 + \frac{1}{x}$, we find that the indefinite integral is $\frac{1}{2}x^2 + x + \ln|x| + C$

27. $3\sin x + 7\cos x + C$.

29. $2e^x - 8\sin x + C$.

31. $\frac{1}{\ln 2}2^x + C$, since $\frac{d}{dx}(2^x) = (\ln 2)\cdot 2^x$.

33. $\int (x+1)^3\,dx = \frac{(x+1)^4}{4} + C$.
Another way to work the problem is to expand $(x+1)^3$ to $x^3 + 3x^2 + 3x + 1$:

$$\int (x+1)^3\,dx = \int (x^3 + 3x^2 + 3x + 1)\,dx = \frac{x^4}{4} + x^3 + \frac{3}{2}x^2 + x + C.$$

It can be shown that these answers are the same by expanding $\frac{(x+1)^4}{4}$.

35. $\ln|x+1| + C$

37. $e^{5+x} + \frac{1}{5}e^{5x} + C$, since $\frac{d}{dx}(e^{5x}) = 5e^{5x}$.

39. (a) $\frac{1}{2}e^{2t}$ (b) $-\frac{1}{3}e^{-3\theta}$

41.

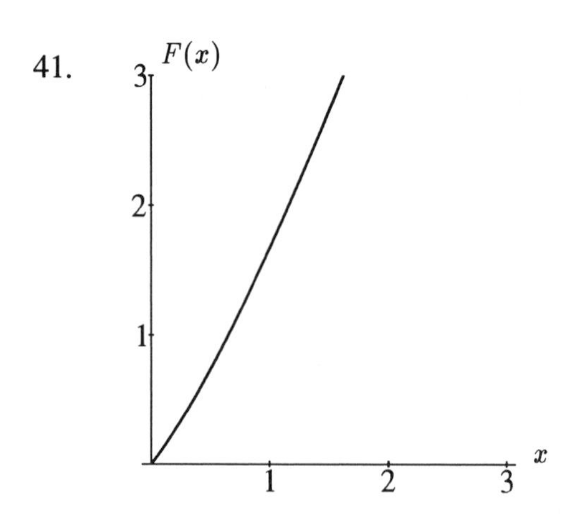

F is increasing because f is positive; F is concave up because f is increasing.

43.

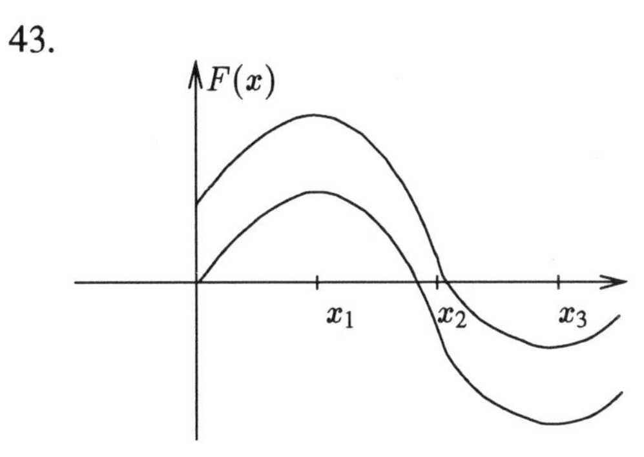

Note that since $f(x_1) = 0$, $f'(x_1) < 0 \Rightarrow F(x_1)$ is a local maximum; since $f(x_3) = 0$, $f'(x_3) > 0 \Rightarrow F(x_3)$ is a local minimum. Also, since $f'(x_2) = 0$ and f changes from decreasing to increasing about $x = x_2$, F has an inflection point at $x = x_2$.

45.

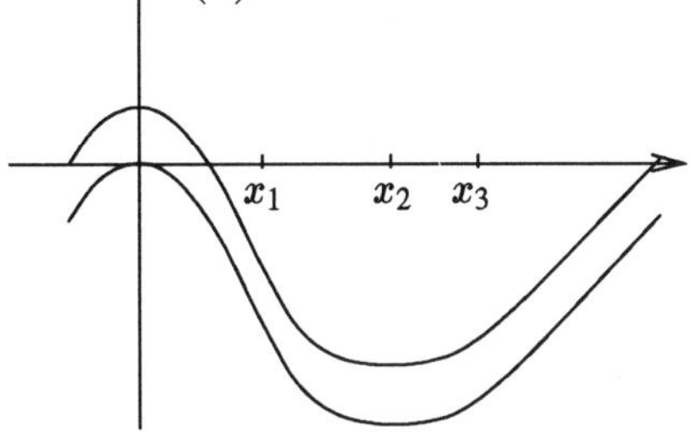

Note that since $f(x_2) = 0$, $f'(x_2) > 0 \Rightarrow F(x_2)$ is a local minimum. Also, since $f'(x_1) = 0$ and f changes from decreasing to increasing about $x = x_1$, F has an inflection point at $x = x_1$.

47. $\displaystyle\int_0^1 \sin\theta \, d\theta = -\cos\theta\Big|_0^1 = 1 - \cos 1 \approx 0.460$

49. $\displaystyle\int_0^2 \left(\frac{x^3}{3} + 2x\right) dx = \left(\frac{x^4}{12} + x^2\right)\Big|_0^2 = \frac{4}{3} + 4 = 16/3 \approx 5.333$

51. $\displaystyle\int_{-3}^{-1} \frac{2}{r^3} \, dr = -r^{-2}\Big|_{-3}^{-1} = -1 + \frac{1}{9} = -8/9 \approx 0.888$

53. Since $(\tan x)' = \dfrac{1}{\cos^2 x}$, $\displaystyle\int_0^{\frac{\pi}{4}} \frac{1}{\cos^2 x} \, dx = \tan x\Big|_0^{\frac{\pi}{4}} = \tan\frac{\pi}{4} - \tan 0 = 1.$

55. $\displaystyle\int 2^x \, dx = \frac{1}{\ln 2} 2^x + C$, since $\dfrac{d}{dx} 2^x = \ln 2 \cdot 2^x$, so
$\displaystyle\int_{-1}^{1} 2^x \, dx = \frac{1}{\ln 2}\left[2^x\Big|_{-1}^{1}\right] = \frac{3}{2\ln 2} \approx 2.164.$

57. (a) i. $\dfrac{d}{dx} \sin 5x = 5\cos 5x$

ii. $\dfrac{d}{dx} \cos x^2 = -2x \sin x^2$

iii. $\dfrac{d}{dx}e^{\sin x} = (\cos x)e^{\sin x}$

iv. $\dfrac{d}{dx}\sin(\cos x) = -\cos(\cos x)(-\sin x) = -\sin x\,\cos(\cos x)$

v. $\dfrac{d}{dx}\ln(\cos x) = \dfrac{1}{\cos x}(-\sin x) = -\tan x$

vi. $\dfrac{d}{dx}\ln(\cos x^4) = \dfrac{1}{\cos x^4}(-\sin x^4)(4x^3) = -4x^3\tan x^4$

(b) i. $\displaystyle\int \cos 5x\,dx = \frac{1}{5}\sin 5x + C$

ii. $\displaystyle\int x\sin x^2 dx = -\frac{1}{2}\cos x^2 + C$

iii. $\displaystyle\int \cos x\, e^{\sin x} dx = e^{\sin x} + C$

iv. $\displaystyle\int \sin x\,\cos(\cos x)dx = -\sin(\cos x) + C$

v. $\displaystyle\int \tan x\,dx = -\ln(|\cos x|) + C$

vi. $\displaystyle\int x^3\tan x^4 dx = -\frac{1}{4}\ln(|\cos x^4|) + C$

(c) We use the chain rule in part (a) and obtain derivatives which we recognize as integrands in part (b), up to a constant multiple. Therefore, we suspect there may be some sort of "chain rule in reverse" rule for integration.

59. The average value of $v(x)$ on the interval $1 \le x \le c]$ is

$$\frac{1}{c-1}\int_1^c \frac{6}{x^2}\,dx = \frac{1}{c-1}\left(-\frac{6}{x}\right)\Big|_1^c = \frac{1}{c-1}\left(\frac{-6}{c}+6\right) = \frac{6}{c}.$$

Since $\dfrac{1}{c-1}\displaystyle\int_1^c \frac{6}{x^2}dx = 1$, $\frac{6}{c} = 1 \Rightarrow c = 6$.

61. The curves intersect at $(0,0)$ and $(\pi,0)$. At any x-coordinate the "height" between the two curves is $\sin x - x(x-\pi)$.

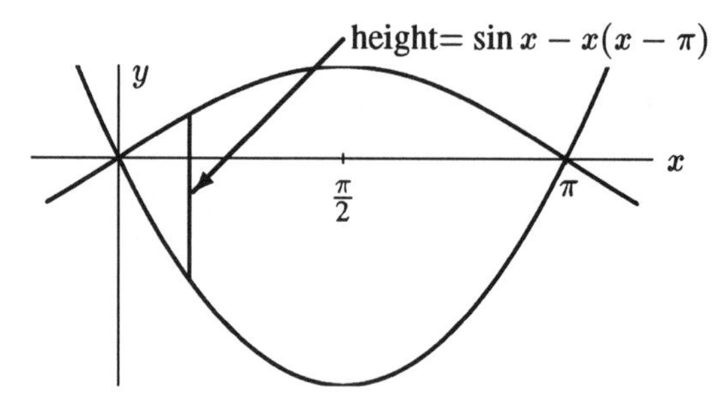

Thus the total area is

$$\int_0^\pi [\sin x - x(x-\pi)]\,dx = \;=\; \int_0^\pi (\sin x - x^2 + \pi x)\,dx$$
$$= \left(-\cos x - \frac{x^3}{3} + \frac{\pi x^2}{2}\right)\Big|_0^\pi$$

$$= (1 - \frac{\pi^3}{3} + \frac{\pi^3}{2}) - (-1)$$
$$= 2 + \frac{\pi^3}{6}.$$

Another approach is to notice that the area between the two curves is (area A) + (area B).

$$\begin{aligned}
\text{area A} &= -\int_0^{\pi} x(x-\pi)\,dx \text{ since the function is negative on } 0 \le x \le \pi \\
&= -\left(\frac{x^3}{3} - \frac{\pi x^2}{2}\right)\Big|_0^{\pi} \\
&= \frac{\pi^3}{2} - \frac{\pi^3}{3} = \frac{\pi^3}{6}; \\
\text{area B} &= \int_0^{\pi} \sin x\,dx = -\cos x\Big|_0^{\pi} = 2.
\end{aligned}$$

Thus the area is $2 + \frac{\pi^3}{6}$.

63. (a)

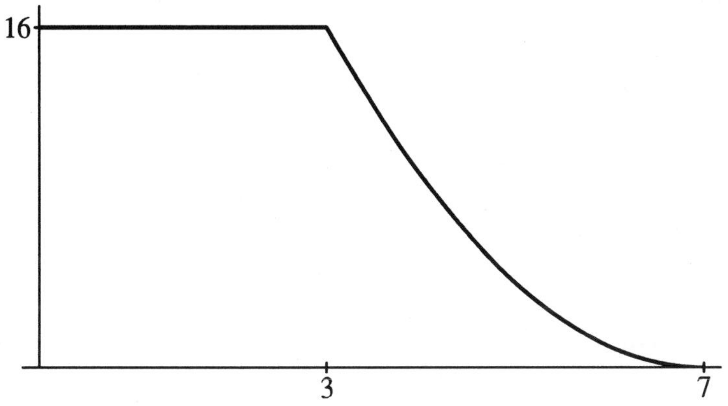

16

3 7

(b) 7 years, because $t^2 - 14t + 49 = (t-7)^2$ indicates that the rate of flow was zero after 7 years.

(c)

$$\begin{aligned}
\text{Area under the curve} &= 3(16) + \int_3^7 (t^2 - 14t + 49)\,dt \\
&= 48 + \left(\frac{1}{3}t^3 - 7t^2 + 49t\right)\Big|_3^7 \\
&= 48 + \frac{343}{3} - 343 + 343 - 9 + 63 - 147 \\
&= \frac{208}{3}
\end{aligned}$$

6.4 Solutions

1. (a) $\frac{d}{dx}\sin(x^2+1) = 2x\cos(x^2+1)$; $\frac{d}{dx}\sin(x^3+1) = 3x^2\cos(x^3+1)$
   (b) i. $\frac{1}{2}\sin(x^2+1)+C$ ii. $\frac{1}{3}\sin(x^3+1)+C$
   (c) i. $-\frac{1}{2}\cos(x^2+1)+C$ ii. $-\frac{1}{3}\cos(x^3+1)+C$

3. Make the substitution $w = 3x$, $dw = 3\,dx$. We have
$$\int \sin 3x\,dx = \frac{1}{3}\int \sin w\,dw = \frac{1}{3}(-\cos w) + C = -\frac{1}{3}\cos 3x + C.$$

5. Make the substitution $w = \sin x$, $dw = \cos x\,dx$. We have
$$\int e^{\sin x}\cos x\,dx = \int e^w\,dw = e^w + C = e^{\sin x} + C$$

7. Make the substitution $w = 2x$, the $dw = 2dx$. We have
$$\begin{aligned}\int \frac{1}{3\cos^2 2x}\,dx &= \frac{1}{3}\int \frac{1}{\cos^2 w}\left(\frac{1}{2}\right)\,dw \\ &= \frac{1}{6}\int \frac{1}{\cos^2 w}\,dw = \frac{1}{6}\tan w + C = \frac{1}{6}\tan 2x + C.\end{aligned}$$

9. For Problem 12 we use the substitution $w = -x^2$, $dw = -2x\,dx$.
   For Problem 13 we use the substitution $w = y^2+5$, $dw = 2y\,dy$.
   For Problem 14 we use the substitution $w = t^3-3$, $dw = 3t^2dt$.
   For Problem 15 we use the substitution $w = x^2-4$, $dw = 2xdx$.
   For Problem 16 we use the substitution $w = y+5$, $dw = dy$.
   For Problem 17 we use the substitution $w = 2t-7$, $dw = 2dt$.
   For Problem 18 we use the substitution $w = x^2+3$, $dw = 2xdx$.
   For Problem 19 it would be easier if we just multiply out $(x^2+3)^2$ and then integrate.
   For Problem 20 use the substitution $w = 4-x$, $dw = -dx$.
   For Problem 21 use the substitution $w = \cos\theta + 5$, $dw = -\sin\theta d\theta$.
   For Problem 22 use the substitution $w = x^3 = 1$, $dw = 3x^2dx$.
   For Problem 23 use the substitution $w = \sin\alpha$, $dw = \cos\alpha d\alpha$.

11. For Problem 34 use $w = \cos 2x$, $dw = -2\sin 2xdx$.
    For Problem 35 use $w = x^2+2x+19$, $dw = 2(x+1)dx$.
    For Problem 36 use $w = \sin(x^2)$, $dw = 2x\cos(x^2)dx$.
    For Problem 37 use $w = 1+2x^3$, $dw = 6x^2dx$.
    For Problem 38 it would be easier not to use any substitution. Just multiply out and then integrate.
    For Problem 39 use $w = 1+3t^2$, $dw = 6tdt$.
    For Problem 40 it would be easier not to substitute.

12. We use the substitution $w = -x^2$, $dw = -2x\,dx$.

$$\begin{aligned}\int xe^{-x^2}\,dx &= -\frac{1}{2}\int e^{-x^2}(-2x\,dx) = -\frac{1}{2}\int e^w\,dw \\ &= -\frac{1}{2}e^w + C = -\frac{1}{2}e^{-x^2} + C.\end{aligned}$$

Check: $\frac{d}{dx}(-\frac{1}{2}e^{-x^2} + C) = (-2x)(-\frac{1}{2}e^{-x^2}) = xe^{-x^2}$.

13. We use the substitution $w = y^2 + 5$, $dw = 2y\,dy$.

$$\begin{aligned}\int y(y^2+5)^8\,dy &= \frac{1}{2}\int (y^2+5)^8(2y\,dy) \\ &= \frac{1}{2}\int w^8\,dw = \frac{1}{2}\frac{w^9}{9} + C \\ &= \frac{1}{18}(y^2+5)^9 + C.\end{aligned}$$

Check: $\frac{d}{dy}(\frac{1}{18}(y^2+5)^9 + C) = \frac{1}{18}[9(y^2+5)^8(2y)] = y(y^2+5)^8$.

14. We use the substitution $w = t^3 - 3$, $dw = 3t^2\,dt$.

$$\begin{aligned}\int t^2(t^3-3)^{10}\,dt &= \frac{1}{3}\int (t^3-3)^{10}(3t^2dt) = \int w^{10}(\frac{1}{3}\,dw) \\ &= \frac{1}{3}\frac{w^{11}}{11} + C = \frac{1}{33}(t^3-3)^{11} + C.\end{aligned}$$

Check: $\frac{d}{dt}[\frac{1}{33}(t^3-3)^{11} + C] = \frac{1}{3}(t^3-3)^{10}(3t^2) = t^2(t^3-3)^{10}$.

15. We use the substitution $w = x^2 - 4$, $dw = 2x\,dx$.

$$\begin{aligned}\int x(x^2-4)^{\frac{7}{2}}\,dx &= \frac{1}{2}\int (x^2-4)^{\frac{7}{2}}(2xdx) = \frac{1}{2}\int w^{\frac{7}{2}}\,dw \\ &= \frac{1}{2}(\frac{2}{9}w^{\frac{9}{2}}) + C = \frac{1}{9}(x^2-4)^{\frac{9}{2}} + C.\end{aligned}$$

Check: $\frac{d}{dx}[\frac{1}{9}(x^2-4)^{\frac{9}{2}} + C] = \frac{1}{9}[\frac{9}{2}(x^2-4)^{\frac{7}{2}}]2x = x(x^2-4)^{\frac{7}{2}}$.

16. We use the substitution $w = y + 5$, $dw = dy$, to get

$$\int \frac{dy}{y+5} = \int \frac{dw}{w} = \ln|w| + C = \ln|y+5| + C.$$

Check: $\frac{d}{dy}(\ln|y+5| + C) = \frac{1}{y+5}$.

17. We use the substitution $w = 2t - 7$, $dw = 2\,dt$.

$$\int (2t-7)^{73}\,dt = \frac{1}{2}\int w^{73}\,dw = \frac{1}{(2)(74)}w^{74} + C = \frac{1}{148}(2t-7)^{74} + C.$$

Check: $\frac{d}{dt}[\frac{1}{148}(2t-7)^{74} + C] = \frac{74}{148}(2t-7)^{73}(2) = (2t-7)^{73}$.

18. We use the substitution $w = x^2 + 3$, $dw = 2x\,dx$.

$$\int x(x^2+3)^2\,dx = \int w^2(\frac{1}{2}\,dw) = \frac{1}{2}\frac{w^3}{3} + C = \frac{1}{6}(x^2+3)^3 + C.$$

Check: $\frac{d}{dx}[\frac{1}{6}(x^2+3)^3 + C] = \frac{1}{6}[3(x^2+3)^2(2x)] = x(x^2+3)^2$.

19. In this case, it seems easier not to substitute.

$$\int (x^2+3)^2\,dx = \int (x^4 + 6x^2 + 9)\,dx = \frac{x^5}{5} + 2x^3 + 9x + C.$$

Check: $\frac{d}{dx}[\frac{x^5}{5} + 2x^3 + 9x + C] = x^4 + 6x^2 + 9 = (x^2+3)^2$.

20. We use the substitution $w = 4 - x$, $dw = -dx$.

$$\int \frac{1}{\sqrt{4-x}}\,dx = -\int \frac{1}{\sqrt{w}}\,dw = -2\sqrt{w} + C = -2\sqrt{4-x} + C.$$

Check: $\frac{d}{dx}(-2\sqrt{4-x} + C) = -2 \cdot \frac{1}{2} \cdot \frac{1}{\sqrt{4-x}} \cdot -1 = \frac{1}{\sqrt{4-x}}$.

21. We use the substitution $w = \cos\theta + 5$, $dw = -\sin\theta\,d\theta$.

$$\begin{aligned}\int \sin\theta(\cos\theta+5)^7\,d\theta &= -\int w^7\,dw = -\frac{1}{8}w^8 + C \\ &= -\frac{1}{8}(\cos\theta+5)^8 + C.\end{aligned}$$

Check:

$$\begin{aligned}\frac{d}{d\theta}[-\frac{1}{8}(\cos\theta+5)^8 + C] &= -\frac{1}{8}\cdot 8(\cos\theta+5)^7 \cdot (-\sin\theta) \\ &= \sin\theta(\cos\theta+5)^7\end{aligned}$$

22. We use the substitution $w = x^3 + 1$, $dw = 3x^2\,dx$, to get

$$\int x^2 e^{x^3+1}\,dx = \frac{1}{3}\int e^w\,dw = \frac{1}{3}e^w + C = \frac{1}{3}e^{x^3+1} + C.$$

Check: $\frac{d}{dx}(\frac{1}{3}e^{x^3+1} + C) = \frac{1}{3}e^{x^3+1} \cdot 3x^2 = x^2 e^{x^3+1}$.

23. We use the substitution $w = \sin\alpha$, $dw = \cos\alpha\, d\alpha$.

$$\int \sin^3\alpha\cos\alpha\, d\alpha = \int w^3\, dw = \frac{w^4}{4} + C = \frac{\sin^4\alpha}{4} + C.$$

Check: $\frac{d}{d\alpha}(\frac{\sin^4\alpha}{4} + C) = \frac{1}{4}\cdot 4\sin^3\alpha\cdot\cos\alpha = \sin^3\alpha\cos\alpha.$

25. We use the substitution $w = \ln z, dw = \frac{1}{z}\, dz$.

$$\int \frac{(\ln z)^2}{z}\, dz = \int w^2\, dw = \frac{w^3}{3} + C = \frac{(\ln z)^3}{3} + C.$$

Check: $\frac{d}{dz}[\frac{(\ln z)^3}{3} + C] = 3\cdot\frac{1}{3}(\ln z)^2\cdot\frac{1}{z} = \frac{(\ln z)^2}{z}.$

26. We use the substitution $w = \sin\theta$, $dw = \cos\theta\, d\theta$.

$$\int \sin^6\theta\cos\theta\, d\theta = \int w^6\, dw = \frac{w^7}{7} + C = \frac{\sin^7\theta}{7} + C.$$

Check: $\frac{d}{d\theta}[\frac{\sin 7\theta}{7} + C] = \sin^6\theta\cos\theta.$

27. We use the substitution $w = \sin 5\theta$, $dw = 5\cos 5\theta\, d\theta$.

$$\int \sin^6 5\theta\cos 5\theta\, d\theta = \frac{1}{5}\int w^6\, dw = \frac{1}{5}(\frac{w^7}{7}) + C = \frac{1}{35}\sin^7 5\theta + C.$$

Check: $\frac{d}{d\theta}(\frac{1}{35}\sin^7 5\theta + C) = \frac{1}{35}[7\sin^6 5\theta](5\cos 5\theta) = \sin^6 5\theta\cos 5\theta.$
Note that we could also use Problem 26 to solve this problem, substituting $w = 5\theta$ and $dw = 5\, d\theta$ to get:

$$\begin{aligned}\int \sin^6 5\theta\cos 5\theta\, d\theta &= \frac{1}{5}\int \sin^6 w\cos w\, dw \\ &= \frac{1}{5}(\frac{\sin^7 w}{7}) + C = \frac{1}{35}\sin^7 5\theta + C.\end{aligned}$$

29. We use the substitution $w = e^t + t$, $dw = (e^t + 1)\, dt$.

$$\int \frac{e^t + 1}{e^t + t}\, dt = \int \frac{1}{w}\, dw = \ln|w| + C = \ln|e^t + t| + C.$$

Check: $\frac{d}{dt}(\ln|e^t + t| + C) = \frac{e^t + 1}{e^t + t}.$

31. We use the substitution $w = y^2 + 4$, $dw = 2y\, dy$.

$$\int \frac{y}{y^2 + 4}\, dy = \frac{1}{2}\int \frac{dw}{w} = \frac{1}{2}\ln|w| + C = \frac{1}{2}\ln(y^2 + 4) + C.$$

(We can drop the absolute value signs since $y^2+4\geq 0$ for all y.)
Check: $\frac{d}{dy}[\frac{1}{2}\ln(y^2+4)+C]=\frac{1}{2}\cdot\frac{1}{y^2+4}\cdot 2y=\frac{y}{y^2+4}$.

33. We use the substitution $w=\sqrt{y}$, $dw=\frac{1}{2\sqrt{y}}\,dy$.

$$\int\frac{e^{\sqrt{y}}}{\sqrt{y}}\,dy=2\int e^w\,dw=2e^w+C=2e^{\sqrt{y}}+C.$$

Check: $\frac{d}{dy}(2e^{\sqrt{y}}+C)=2e^{\sqrt{y}}\cdot\frac{1}{2\sqrt{y}}=\frac{e^{\sqrt{y}}}{\sqrt{y}}$.

34. We use the substitution $w=\cos 2x$, $dw=-2\sin 2x\,dx$.

$$\begin{aligned}\int\tan 2x\,dx &= \int\frac{\sin 2x}{\cos 2x}\,dx=-\frac{1}{2}\int\frac{dw}{w}\\ &= -\frac{1}{2}\ln|w|+C=-\frac{1}{2}\ln|\cos 2x|+C.\end{aligned}$$

Check:

$$\begin{aligned}\frac{d}{dx}[-\frac{1}{2}\ln|\cos 2x|+C] &= -\frac{1}{2}\cdot\frac{1}{\cos 2x}\cdot -2\sin 2x\\ &= \frac{\sin 2x}{\cos 2x}=\tan 2x.\end{aligned}$$

35. We use the substitution $w=x^2+2x+19$, $dw=2(x+1)dx$.

$$\int\frac{(x+1)dx}{x^2+2x+19}=\frac{1}{2}\int\frac{dw}{w}=\frac{1}{2}\ln|w|+C=\frac{1}{2}\ln(x^2+2x+19)+C.$$

(We can drop the absolute value signs, since $x^2+2x+19=(x+1)^2+18\geq 0$ for all x.)
Check: $\frac{1}{dx}[\frac{1}{2}\ln(x^2+2x+19)]=\frac{1}{2}\frac{1}{x^2+2x+19}(2x+2)=\frac{x+1}{x^2+2x+19}$.

36. We use the substitution $w=\sin(x^2)$, $dw=2x\cos(x^2)\,dx$.

$$\int\frac{x\cos(x^2)}{\sqrt{\sin(x^2)}}\,dx=\frac{1}{2}\int w^{-\frac{1}{2}}\,dw=\frac{1}{2}(2w^{\frac{1}{2}})+C=\sqrt{\sin(x^2)}+C.$$

Check: $\frac{d}{dx}(\sqrt{\sin(x^2)}+C)=\frac{1}{2\sqrt{\sin(x^2)}}[\cos(x^2)]2x=\frac{x\cos(x^2)}{\sqrt{\sin(x^2)}}$.

37. We use the substitution $w=1+2x^3$, $dw=6x^2\,dx$.

$$\int x^2(1+2x^3)^2\,dx=\int w^2(\frac{1}{6}\,dw)=\frac{1}{6}(\frac{w^3}{3})+C=\frac{1}{18}(1+2x^3)^3+C.$$

Check: $\frac{d}{dx}[\frac{1}{18}(1+2x^2)^3+C]=\frac{1}{18}[3(1+2x^3)^2(6x^2)]=x^2(1+2x^3)^2$.

38. In this case, it seems easier not to substitute.

$$\begin{aligned}\int y^2(1+y)^2\,dy &= \int y^2(y^2+2y+1)\,dy = \int (y^4+2y^3+y^2)\,dy\\ &= \frac{y^5}{5}+\frac{y^4}{2}+\frac{y^3}{3}+C.\end{aligned}$$

Check: $\frac{d}{dy}(\frac{y^5}{5}+\frac{y^4}{2}+\frac{y^3}{3}+C) = y^4+2y^3+y^2 = y^2(y+1)^2.$

39. We use the substitution $w = 1+3t^2$, $dw = 6t\,dt$.

$$\int \frac{t}{1+3t^2}\,dt = \int \frac{1}{w}(\frac{1}{6}\,dw) = \frac{1}{6}\ln|w| + C = \frac{1}{6}\ln(1+3t^2)+C.$$

(We can drop the absolute value signs since $1+3t^2 > 0$ for all t).
Check: $\frac{d}{dt}[\frac{1}{6}\ln(1+3t^2)+C] = \frac{1}{6}\frac{1}{1+3t^2}(6t) = \frac{t}{1+3t^2}.$

40. It seems easier not to substitute.

$$\begin{aligned}\int \frac{(t+1)^2}{t^2}\,dt &= \int \frac{(t^2+2t+1)}{t^2}\,dt\\ &= \int (1+\frac{2}{t}+\frac{1}{t^2})\,dt = t+2\ln|t|-\frac{1}{t}+C.\end{aligned}$$

Check: $\frac{d}{dt}(t+2\ln|t|-\frac{1}{t}+C) = 1+\frac{2}{t}+\frac{1}{t^2} = \frac{(t+1)^2}{t^2}.$

41. We use the substitution $w = e^x + e^{-x}$, $dw = (e^x - e^{-x})\,dx$.

$$\int \frac{e^x - e^{-x}}{e^x + e^{-x}}\,dx = \int \frac{dw}{w} = \ln|w| + C = \ln(e^x+e^{-x})+C.$$

(We can drop the absolute value signs since $e^x + e^{-x} > 0$ for all x).
Check: $\frac{d}{dx}[\ln(e^x+e^{-x})+C] = \frac{1}{e^x+e^{-x}}(e^x - e^{-x}).$

43. (a) $\int 4x(x^2+1)\,dx = \int (4x^3+4x)\,dx = x^4+2x^2+C.$

(b) If $w = x^2+1$, then $dw = 2x\,dx$.

$$\int 4x(x^2+1)\,dx = \int 2w\,dw = w^2+C = (x^2+1)^2+C.$$

(c) The expressions from parts (a) and (b) look different, but they are both correct. Note that $(x^2+1)^2+C = x^4+2x^2+1+C$. In other words, the expressions from parts (a) and (b) differ only by a constant, so they are both correct antiderivatives.

45. Since $v = \frac{dh}{dt}$, it follows that $h(t) = \int v(t)\,dt$ and $h(0) = h_0$. Since

$$v(t) = \frac{mg}{k}\left(1 - e^{-\frac{k}{m}t}\right) = \frac{mg}{k} - \frac{mg}{k}e^{-\frac{k}{m}t},$$

we have

$$h(t) = \int v(t)\,dt = \frac{mg}{k}\int dt - \frac{mg}{k}\int e^{-\frac{k}{m}t}\,dt.$$

The first integral is simply $\frac{mg}{k}t + C$. To evaluate the second integral, make the substitution $w = -\frac{k}{m}t$. Then

$$dw = -\frac{k}{m}\,dt,$$

so

$$\int e^{-\frac{k}{m}t}\,dt = \int e^{w}\left(-\frac{m}{k}\right)\,dw = -\frac{m}{k}e^{w} + C = -\frac{m}{k}e^{-\frac{k}{m}t} + C.$$

Thus

$$\begin{aligned} h(t) &= \int v\,dt = \frac{mg}{k}t - \frac{mg}{k}\left(-\frac{m}{k}e^{-\frac{k}{m}t}\right) + C \\ &= \frac{mg}{k}t + \frac{m^2 g}{k^2}e^{-\frac{k}{m}t} + C. \end{aligned}$$

Since $h(0) = h_0$,

$$h_0 = \frac{mg}{k}\cdot 0 + \frac{m^2 g}{k^2}e^{0} + C;$$

$$C = h_0 - \frac{m^2 g}{k^2}.$$

Thus

$$h(t) = \frac{mg}{k}t + \frac{m^2 g}{k^2}e^{-\frac{k}{m}t} - \frac{m^2 g}{k^2} + h_0$$

$$h(t) = \frac{mg}{k}t - \frac{m^2 g}{k^2}\left(1 - e^{-\frac{k}{m}t}\right) + h_0.$$

6.5 Solutions

1. (a) We substitute $w = 1 + x^2$, $dw = 2x\,dx$.

$$\int_{x=0}^{x=1} \frac{x}{1+x^2}\,dx = \frac{1}{2}\int_{w=1}^{w=2} \frac{1}{w}\,dw = \frac{1}{2}\ln|w|\Big|_1^2 = \frac{1}{2}\ln 2.$$

(b) We substitute $w = \cos x$, $dw = -\sin x\,dx$.

$$\int_{x=0}^{x=\frac{\pi}{4}} \frac{\sin x}{\cos x}\,dx = -\int_{w=1}^{w=\sqrt{2}/2} \frac{1}{w}\,dw$$

$$= -\ln|w|\Big|_1^{\sqrt{2}/2} = -\ln\frac{\sqrt{2}}{2} = \frac{1}{2}\ln 2.$$

3. We substitute $w = \sqrt[3]{x} = x^{\frac{1}{3}}$. Then $dw = \frac{1}{3}x^{-\frac{2}{3}}\,dx = \frac{1}{3\sqrt[3]{x^2}}\,dx$.

$$\int_1^8 \frac{e^{\sqrt[3]{x}}}{\sqrt[3]{x^2}}dx = \int_{x=1}^{x=8} e^w(3\,dw) = 3e^w\Big|_{x=1}^{x=8} = 3e^{\sqrt[3]{x}}\Big|_1^8 = 3(e^2 - e) \approx 14.01.$$

5. We substitute $w = t + 2$, so $dw = dt$.

$$\int_{t=-1}^{t=e-2} \frac{1}{t+2}dt = \int_{w=1}^{w=e} \frac{dw}{w} = \ln|w|\Big|_1^e = \ln e - \ln 1 = 1.$$

7. No immediate substitution is apparent, so we must try to modify the integral in some way. First, we put the integral into a more convenient form by using the fact that $\sin^2\theta = 1 - \cos^2\theta$.
Thus: $\int_{-\frac{\pi}{4}}^{\frac{\pi}{4}} \cos^2\theta\sin^5\theta\,d\theta = \int_{-\frac{\pi}{4}}^{\frac{\pi}{4}} \cos^2\theta(1-\cos^2\theta)^2\sin\theta\,d\theta.$
Now, we can make a substitution which helps. We let $w = \cos\theta$, so $dw = -\sin\theta\,d\theta$.
Note that $w = \frac{\sqrt{2}}{2}$ when $\theta = -\frac{\pi}{4}$ and when $\theta = \frac{\pi}{4}$. Thus after our substitution, we get $-\int_{w=\frac{\pi}{4}}^{w=\frac{\pi}{4}} w^2(1-w^2)\,dw$. Since the upper and lower limits of integration are the same, this definite integral must equal 0. Notice that we could have deduced this fact immediately, since $\cos^2\theta$ is even and $\sin^5\theta$ is odd, so $\cos^2\theta\sin^5\theta$ is odd.
Thus $\int_{-\frac{\pi}{4}}^{0} \cos^2\theta\sin^5\theta\,d\theta = -\int_0^{\frac{\pi}{4}} \cos^2\theta\sin^5\theta\,d\theta$, and the given integral must evaluate to 0.

9.

$$\int_{-1}^{3}(x^3+5x)\,dx = \left.\frac{x^4}{4}\right|_{-1}^{3} + \left.\frac{5x^2}{2}\right|_{-1}^{3} = 40.$$

11. We substitute $w = \cos\theta + 5$, $dw = -\sin\theta d\theta$. Then

$$\int_{\theta=0}^{\theta=\pi} \sin\theta d\theta(\cos\theta+5)^7 = -\int_{w=6}^{w=4} w^7 dw = \int_{w=4}^{w=6} w^7 dw = \left.\frac{w^8}{8}\right|_4^6 = 201,760.$$

13. Substitute $w = 1 + x^2$, $dw = 2x\,dx$. Then $x\,dx = \frac{1}{2}\,dw$, and

$$\int_{x=0}^{x=1} x(1+x^2)^{20}\,dx = \frac{1}{2}\int_{w=1}^{w=2} w^{20}\,dw = \left.\frac{w^{21}}{42}\right|_1^2 = \frac{299593}{6}.$$

15. Substitute $w = 3\alpha$, $dw = 3\,d\alpha$. Then $d\alpha = \frac{1}{3}\,dw$. We have

$$\begin{aligned}\int_{\alpha=0}^{\alpha=\frac{\pi}{12}} \sin 3\alpha\,d\alpha &= \frac{1}{3}\int_{w=0}^{w=\frac{\pi}{4}} \sin w\,dw \\ &= -\frac{1}{3}\cos w\Big|_0^{\frac{\pi}{4}} \\ &= -\frac{1}{3}\left(\frac{\sqrt{2}}{2}-1\right) = \frac{1}{3}\left(1-\frac{\sqrt{2}}{2}\right).\end{aligned}$$

17. Substitute $w = x^2 + 4$, $dw = 2x\,dx$. Then,

$$\begin{aligned}\int_{x=4}^{x=1} x\sqrt{x^2+4}dx &= \frac{1}{2}\int_{w=20}^{w=5} w^{\frac{1}{2}}\,dw = \frac{1}{3}w^{\frac{3}{2}}\Big|_{20}^{5} \\ &= \frac{1}{3}\left(5^{\frac{3}{2}} - 8\cdot 5^{\frac{3}{2}}\right) = -\frac{7}{3}\cdot 5^{\frac{3}{2}} = -\frac{7}{3}\sqrt{125}.\end{aligned}$$

19. Let $w = x^2$, $dw = 2x\,dx$. When $x = 0$, $w = 0$, and when $x = \frac{1}{\sqrt{2}}$, $w = \frac{1}{2}$. Then

$$\int_0^{\frac{1}{\sqrt{2}}} \frac{x\,dx}{\sqrt{1-x^4}} = \int_0^{\frac{1}{2}} \frac{\frac{1}{2}\,dw}{\sqrt{1-w^2}} = \frac{1}{2}\arcsin w\Big|_0^{\frac{1}{2}} = \frac{1}{2}(\arcsin\frac{1}{2} - \arcsin 0) = \frac{\pi}{12}.$$

21. $f(t) = \sin\frac{1}{t}$ has no elementary antiderivative, so we will have to use left and right sums. With $n = 100$, left sum $= 0.5462$ and right sum $= 0.5582$. However, since f is not monotonic on $[1/4, 1]$ (see figure), we cannot be sure that the integral is between these values.

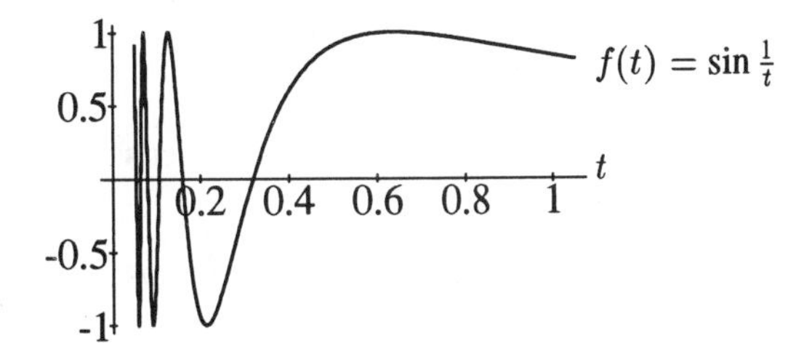

To get an upper and lower bound for this integral, divide the interval $[1/4, 1]$ into subintervals in such a way that f is monotonic on each one. Since $f'(t) = -\frac{1}{t^2}\cos\frac{1}{t} = 0$ when $\frac{1}{t} = \frac{\pi}{2}$ or $t = \frac{2}{\pi}$ (and this is the only point in $[1/4, 1]$ where $f'(t) = 0$), we will write

$$\int_{1/4}^{1} \sin\frac{1}{t}\,dt = \int_{1/4}^{2/\pi} \sin\frac{1}{t}\,dt + \int_{2/\pi}^{1} \sin\frac{1}{t}\,dt$$

Now, using $n = 100$, we find $0.209 < \int_{1/4}^{2/\pi} \sin\frac{1}{t}\,dt < 0.216$ (left sum $= 0.209$, right sum $= 0.216$, f is increasing), and $0.339 < \int_{2/\pi}^{1} \sin\frac{1}{t}\,dt < 0.340$ (here left sum $= 0.340$, right sum $= 0.339$ because f is decreasing). Thus

$$0.548 < \int_{1/4}^{1} \sin\frac{1}{t}\,dt < 0.556$$

23. The substitution $w = \ln x$, $dw = \frac{1}{x}\,dx$ transforms the first integral into $\int w\,dw$, which is just a respelling of the integral $\int x\,dx$.

25. For the first integral, let $w = x + 1$, $dw = dx$. Then

$$\int \sqrt{x+1}\,dx = \int \sqrt{w}\,dw.$$

For the second integral, let $w = 1 + \sqrt{x}$, $dw = \frac{1}{2}x^{-\frac{1}{2}}\,dx = \frac{1}{2\sqrt{x}}dx$. Then, $\frac{dx}{\sqrt{x}} = 2dw$, and

$$\int \frac{\sqrt{1+\sqrt{x}}}{\sqrt{x}}\,dx = \int \sqrt{1+\sqrt{x}}\left(\frac{dx}{\sqrt{x}}\right) = 2\int \sqrt{w}dw.$$

27.

$$\int \frac{dx}{x^2+4x+5} = \int \frac{dx}{(x+2)^2+1}.$$

We make the substitution $\tan\theta = x+2$. Then $dx = \frac{1}{\cos^2\theta}\,d\theta$.

$$\begin{aligned}\int \frac{dx}{(x+2)^2+1} &= \int \frac{d\theta}{\cos^2\theta(\tan^2\theta+1)} \\ &= \int \frac{d\theta}{\cos^2\theta(\frac{\sin^2\theta}{\cos^2\theta}+1)} \\ &= \int \frac{d\theta}{\sin^2\theta+\cos^2\theta} \\ &= \int d\theta = \theta + C\end{aligned}$$

But since $\tan\theta = x+2$, $\theta = \arctan(x+2)$, and so $\theta + C = \arctan(x+2) + C$.

29. To find the area under $f(x) = xe^{x^2}$, we need to evaluate the definite integral

$$\int_0^2 xe^{x^2}\,dx.$$

This is done in Example 1, Section 6.5, using the substitution $w = x^2$, the result being

$$\int_0^2 xe^{x^2}\,dx = \frac{1}{2}(e^4-1) \approx 26.7991.$$

31. (a) We first try the substitution $w = \sin\theta$, $dw = \cos\theta\,d\theta$. Then

$$\int \sin\theta\cos\theta\,d\theta = \int w\,dw = \frac{w^2}{2} + C = \frac{\sin^2\theta}{2} + C.$$

(b) If we instead try the substitution $w = \cos\theta$, $dw = -\sin\theta\,d\theta$, we get

$$\int \sin\theta\cos\theta\,d\theta = -\int w\,dw = -\frac{w^2}{2} + C = -\frac{\cos^2\theta}{2} + C.$$

(c) Once we note that $\sin 2\theta = 2\sin\theta\cos\theta$, we can also say

$$\int \sin\theta\cos\theta\,d\theta = \frac{1}{2}\int \sin 2\theta\,d\theta.$$

Substituting $w = 2\theta$, $dw = 2\,d\theta$, the above equals

$$\frac{1}{4}\int \sin w\,dw = -\frac{\cos w}{4} + C = -\frac{\cos 2\theta}{4} + C.$$

(d) All these answers are correct. Although they have different forms, they differ from each other only in terms of a constant, and thus they are all acceptable antiderivatives. For example, $1 - \cos^2\theta = \sin^2\theta$, so $\frac{\sin^2\theta}{2} = -\frac{\cos^2\theta}{2} + \frac{1}{2}$. Thus the first two expressions differ only in terms of the constant C.

Similarly, $\cos 2\theta = \cos^2\theta - \sin^2\theta = 2\cos^2\theta - 1$, so $-\frac{\cos 2\theta}{4} = -\frac{\cos^2\theta}{2} + \frac{1}{4}$, and thus the second and third expressions differ only by a constant. Of course, if the first two expressions and the last two expressions differ only in the constant C, then the first and last only differ in the constant as well.

6.6 Solutions

1. (a) i.) $-x\sin x + \cos x$

   ii.) $-2\sin 2x$

   iii.) $-x^2\sin x + 2x\cos x$

   iv.) $1 + \ln x$

   (b) i.) Since $(x\ln x)' = 1 + \ln x$, it follows that $(x\ln x - x)' = 1 + \ln x - 1 = \ln x$. Therefore, $\int \ln x\, dx = x\ln x - x + C$.

   ii.) Since $(\cos 2x)' = -2\sin 2x$, it follows that $(-\frac{1}{2}\cos 2x)' = \sin 2x$. Therefore, $\int \sin 2x\, dx = -\frac{1}{2}\cos 2x + C$.

   iii.) Since $(x\cos x)' = -x\sin x + \cos x$, it follows that $(-x\cos x + \sin x)' = x\sin x - \cos x + \cos x = x\sin x$. Therefore, $\int x\sin x\, dx = -x\cos x + \sin x + C$.

   iv.) Since $(x^2\cos x)' = -x^2\sin x + 2x\cos x$, and since $(2x\sin x)' = 2x\cos x + 2\sin x$, it follows that

$$\begin{aligned}(-x^2\cos x + 2x\sin x + 2\cos x)' &= x^2\sin x - 2x\cos x + 2x\cos x + 2\sin x - 2\sin x \\ &= x^2\sin x.\end{aligned}$$

   Therefore, $\int x^2\sin x\, dx = -x^2\cos x + 2x\sin x + 2\cos x + C$.

3. Let $u = t$ and $v' = e^{5t}$, so $u' = 1$ and $v = \frac{1}{5}e^{5t}$.

   Then $\int te^{5t}\, dt = \frac{1}{5}te^{5t} - \int \frac{1}{5}e^{5t}\, dt = \frac{1}{5}te^{5t} - \frac{1}{25}e^{5t} + C$.

5. Let $u = p$ and $v' = e^{(-0.1)p}$, $u' = 1$. Thus, $v = \int e^{(-0.1)p}\, dp = -10e^{(-0.1)p}$. With this choice of u and v, integration by parts gives:

$$\begin{aligned}\int pe^{(-0.1)p}\, dp &= p(-10e^{(-0.1)p}) - \int (-10e^{(-0.1)p})\, dp \\ &= -10pe^{(-0.1)p} + 10\int e^{(-0.1)p}\, dp \\ &= -10pe^{(-0.1)p} - 100e^{(-0.1)p} + C.\end{aligned}$$

7. Let $u = \ln x$ and $v' = x^3$, so $u' = \frac{1}{x}$ and $v = \frac{x^4}{4}$.

   Then $\int x^3\ln x\, dx = \frac{x^4}{4}\ln x - \int \frac{x^3}{4}\, dx = \frac{x^4}{4}\ln x - \frac{x^4}{16} + C$.

9. Let $u = t^2$, $v' = \sin t$ implying $v = -\cos t$ and $u' = 2t$. Integrating by parts, we get:

$$\int t^2\sin t\, dt = -t^2\cos t - \int 2t(-\cos t)\, dt.$$

Again, applying integration by parts with $u = t$, $v' = \cos t$, we have:

$$\int t \cos t\, dt = t \sin t + \cos t + C.$$

Thus

$$\int t^2 \sin t\, dt = -t^2 \cos t + 2t \sin t + 2 \cos t + C.$$

11. Let $u = z + 1$, $v' = e^{2z}$. Thus, $v = \frac{1}{2}e^{2z}$ and $u' = 1$. Integrating by parts, we get:

$$\begin{aligned}
\int (z+1)e^{2z}\, dz &= (z+1) \cdot \frac{1}{2}e^{2z} - \int \frac{1}{2}e^{2z}\, dz \\
&= \frac{1}{2}(z+1)e^{2z} - \frac{1}{4}e^{2z} + C \\
&= \frac{1}{4}(2z+1)e^{2z} + C.
\end{aligned}$$

13. Let $u = \theta + 1$ and $v' = \sin(\theta + 1)$, so $u' = 1$ and $v = -\cos(\theta + 1)$.
$\int (\theta+1)\sin(\theta+1)\, d\theta = -(\theta+1)\cos(\theta+1) + \int \cos(\theta+1)\, d\theta = -(\theta+1)\cos(\theta+1) + \sin(\theta+1) + C.$

14. Let $u = \sin\theta$ and $v' = \sin\theta$, so $u' = \cos\theta$ and $v = -\cos\theta$. Then

$$\begin{aligned}
\int \sin^2\theta\, d\theta &= -\sin\theta\cos\theta + \int \cos^2\theta\, d\theta \\
&= -\sin\theta\cos\theta + \int (1 - \sin^2\theta)\, d\theta \\
&= -\sin\theta\cos\theta + \int 1\, d\theta - \int \sin^2\theta\, d\theta.
\end{aligned}$$

By adding $\int \sin^2\theta\, d\theta$ to both sides of the above equation, we find that $2\int \sin^2\theta\, d\theta = -\sin\theta\cos\theta + \theta + C$, so $\int \sin^2\theta\, d\theta = -\frac{1}{2}\sin\theta\cos\theta + \frac{\theta}{2} + C'$.

15. Let $u = \cos(3\alpha + 1)$ and $v' = \cos(3\alpha + 1)$, so $u' = -3\sin(3\alpha + 1)$, and $v = \frac{1}{3}\sin(3\alpha + 1)$. Then

$$\begin{aligned}
\int \cos^2(3\alpha+1)\, d\alpha &= \int (\cos(3\alpha+1))\cos(3\alpha+1)\, d\alpha \\
&= \frac{1}{3}\cos(3\alpha+1)\sin(3\alpha+1) + \int \sin^2(3\alpha+1)\, d\alpha \\
&= \frac{1}{3}\cos(3\alpha+1)\sin(3\alpha+1) + \int \left(1 - \cos^2(3\alpha+1)\right)\, d\alpha \\
&= \frac{1}{3}\cos(3\alpha+1)\sin(3\alpha+1) + \alpha - \int \cos^2(3\alpha+1)\, d\alpha
\end{aligned}$$

By adding $\int \cos^2(3\alpha+1)\,d\alpha$ to both sides of the above equation, we find that

$$2\int \cos^2(3\alpha+1)\,d\alpha = \frac{1}{3}\cos(3\alpha+1)\sin(3\alpha+1)+\alpha+C,$$

which gives

$$\int \cos^2(3\alpha+1)\,d\alpha = \frac{1}{6}\cos(3\alpha+1)\sin(3\alpha+1)+\frac{\alpha}{2}+C.$$

17. Let $u=\ln x$, $v'=x^{-2}$. Then $v=-x^{-1}$ and $u'=x^{-1}$. Integrating by parts, we get:

$$\begin{aligned}\int x^{-2}\ln x\,dx &= -x^{-1}\ln x-\int(-x^{-1})\cdot x^{-1}\,dx\\ &= -x^{-1}\ln x - x^{-1}+C.\end{aligned}$$

19. Let $u=y$ and $v'=\frac{1}{\sqrt{5-y}}$, so $u'=1$ and $v=-2(5-y)^{1/2}$.

$$\int \frac{y}{\sqrt{5-y}}\,dy = -2y(5-y)^{1/2}+2\int(5-y)^{1/2}\,dy = -2y(5-y)^{1/2}-\frac{4}{3}(5-y)^{3/2}+C.$$

21. $\displaystyle\int \frac{t+7}{\sqrt{5-t}}\,dt = \int\frac{t}{\sqrt{5-t}}\,dt+7\int(5-t)^{-1/2}\,dt.$

To calculate the first integral, we use integration by parts. Let $u=t$ and $v'=\frac{1}{\sqrt{5-t}}$, so $u'=1$ and $v=-2(5-t)^{1/2}$. Then

$$\int\frac{t}{\sqrt{5-t}}\,dt = -2t(5-t)^{1/2}+2\int(5-t)^{1/2}\,dt = -2t(5-t)^{1/2}-\frac{4}{3}(5-t)^{3/2}+C.$$

We can calculate the second integral directly: $7\int(5-t)^{-1/2} = -14(5-t)^{1/2}+C_1$. Thus

$$\int\frac{t+7}{\sqrt{5-t}}\,dt = -2t(5-t)^{1/2}-\frac{4}{3}(5-t)^{3/2}-14(5-t)^{1/2}+C_2.$$

23. Let $u=(\ln x)^4$ and $v'=x$, so $u'=\frac{4(\ln x)^3}{x}$ and $v=\frac{x^2}{2}$. Then

$$\int x(\ln x)^4\,dx = \frac{x^2(\ln x)^4}{2}-2\int x(\ln x)^3\,dx.$$

$\int x(\ln x)^3\,dx$ is somewhat less complicated than $\int x(\ln x)^4\,dx$. To calculate it, we again try integration by parts, this time letting $u=(\ln x)^3$ (instead of $(\ln x)^4$) and $v'=x$. We find

$$\int x(\ln x)^3\,dx = \frac{x^2}{2}(\ln x)^3-\frac{3}{2}\int x(\ln x)^2\,dx.$$

Once again, we are led to an expression in terms of a less-complicated integral than before. Using integration by parts two more times, we find that

$$\int x(\ln x)^2\,dx = \frac{x^2}{2}(\ln x)^2 - \int x(\ln x)\,dx$$

and that

$$\int x\ln x\,dx = \frac{x^2}{2}\ln x - \frac{x^2}{4} + C.$$

Putting this all together, we have

$$\int x(\ln x)^4\,dx = \frac{x^2}{2}(\ln x)^4 - x^2(\ln x)^3 + \frac{3}{2}x^2(\ln x)^2 - \frac{3}{2}x^2\ln x + \frac{3}{4}x^2 + C.$$

25. This integral can first be simplified by making the substitution $w = x^2$, $dw = 2x\,dx$. Then

$$\int x\arctan x^2\,dx = \frac{1}{2}\int \arctan w\,dw.$$

To evaluate $\int \arctan w\,dw$, we'll use integration by parts. Let $u = \arctan w$ and $v' = 1$, so $u' = \frac{1}{1+w^2}$ and $v = w$. Then

$$\int \arctan w\,dw = w\arctan w - \int \frac{w}{1+w^2}\,dw = w\arctan w - \frac{1}{2}\ln|1+w^2| + C.$$

Since $1 + w^2$ is never negative, we can drop the absolute value signs. Thus, we have

$$\begin{aligned}\int x\arctan x^2\,dx &= \frac{1}{2}\left(x^2\arctan x^2 - \frac{1}{2}\ln(1+(x^2)^2) + C\right)\\ &= \frac{1}{2}x^2\arctan x^2 - \frac{1}{4}\ln(1+x^4) + C.\end{aligned}$$

27. Let $u = x^2$ and $v' = xe^{x^2}$, so $u' = 2x$ and $v = \frac{1}{2}e^{x^2}$. Then

$$\int x^3e^{x^2}\,dx = \frac{1}{2}x^2e^{x^2} - \int xe^{x^2}\,dx = \frac{1}{2}x^2e^{x^2} - \frac{1}{2}e^{x^2} + C.$$

Note that we can also do this problem by substitution and integration by parts. If we let $w = x^2$, so $dw = 2x\,dx$, then $\int x^3e^{x^2}\,dx = \frac{1}{2}\int we^w\,dw$. We could then perform integration by parts on this integral to get the same result.

29. From integration by parts in Problem 14, we obtain

$$\int \sin^2\theta\,d\theta = -\frac{1}{2}\sin\theta\cos\theta + \frac{1}{2}\theta + C.$$

Using the identity given in the book, we have

$$\int \sin^2\theta\,d\theta = \int \frac{1-\cos 2\theta}{2}\,d\theta = \frac{1}{2}\theta - \frac{1}{4}\sin 2\theta + C.$$

Although the answers differ in form, they are really the same, since (by one of the standard double angle formulas) $-\frac{1}{4}\sin 2\theta = -\frac{1}{4}(2\sin\theta\cos\theta) = -\frac{1}{2}\sin\theta\cos\theta$.

31. First, let $u = e^x$ and $v' = \sin x$, so $u' = e^x$ and $v = -\cos x$.
Thus $\int e^x \sin x\,dx = -e^x\cos x + \int e^x \cos x\,dx$. To calculate $\int e^x\cos x\,dx$, we again need to use integration by parts. Let $u = e^x$ and $v' = \cos x$, so $u' = e^x$ and $v = \sin x$.
Thus $\int e^x\cos x\,dx = e^x\sin x - \int e^x \sin x\,dx$. This gives

$$\int e^x \sin x\,dx = e^x\sin x - e^x\cos x - \int e^x\sin x\,dx.$$

By adding $\int e^x\sin x\,dx$ to both sides, we obtain

$$\begin{aligned} 2\int e^x\sin x\,dx &= e^x(\sin x - \cos x) + C. \\ \text{Thus } \int e^x\sin x\,dx &= \frac{1}{2}e^x(\sin x - \cos x) + C. \end{aligned}$$

This problem could also be done in other ways; for example, we could have started with $u = \sin x$ and $v' = e^x$ as well.

32. Let $u = e^\theta$ and $v' = \cos\theta$, so $u' = e^\theta$ and $v = \sin\theta$.
Then $\int e^\theta\cos\theta\,d\theta = e^\theta\sin\theta - \int e^\theta\sin\theta\,d\theta$.
In Problem 31 we found that $\int e^x\sin x dx = \frac{1}{2}e^x(\sin x - \cos x) + C$.

$$\begin{aligned} \int e^\theta\cos\theta\,d\theta &= e^\theta\sin\theta - [\tfrac{1}{2}e^\theta(\sin\theta - \cos\theta)] + C \\ &= \frac{1}{2}e^\theta(\sin\theta + \cos\theta) + C. \end{aligned}$$

33. We integrate by parts. Since in Problem 31 we found that $\int e^x\sin x\,dx = \frac{1}{2}e^x(\sin x - \cos x)$, we let $u = x$ and $v' = e^x\sin x$, so $u' = 1$ and $v = \frac{1}{2}e^x(\sin x - \cos x)$.

$$\begin{aligned} \text{Then } \int xe^x\sin x\,dx &= \frac{1}{2}xe^x(\sin x - \cos x) - \frac{1}{2}\int e^x(\sin x - \cos x)\,dx \\ &= \frac{1}{2}xe^x(\sin x - \cos x) - \frac{1}{2}\int e^x\sin x\,dx + \frac{1}{2}\int e^x\cos x\,dx. \end{aligned}$$

Using Problems 31 and 32, we see that this equals

$$\begin{aligned} &= \frac{1}{2}xe^x(\sin x - \cos x) - \frac{1}{4}e^x(\sin x - \cos x) + \frac{1}{4}e^x(\sin x + \cos x) + C \\ &= \frac{1}{2}xe^x(\sin x - \cos x) + \frac{1}{2}e^x\cos x + C. \end{aligned}$$

35. We integrate by parts. Since we know what the answer is supposed to be, it's easier to choose u and v'. Let $u = x^n$ and $v' = e^x$, so $u' = nx^{n-1}$ and $v = e^x$. Then
$$\int x^n e^x\,dx = x^n e^x - n\int x^{n-1}e^x\,dx.$$

37. We integrate by parts. Let $u = x^n$ and $v' = \cos ax$, so $u' = nx^{n-1}$ and $v = \frac{1}{a}\sin ax$.
$$\text{Then } \int x^n \cos ax\,dx = \frac{1}{a}x^n \sin ax - \int (nx^{n-1})(\frac{1}{a}\sin ax)\,dx = \frac{1}{a}x^n \sin ax - \frac{n}{a}\int x^{n-1}\sin ax\,dx.$$

39. $\displaystyle\int_1^5 \ln t\,dt = (t\ln t - t)\Big|_1^5 = 5\ln 5 - 4 \approx 4.047$

41. $\displaystyle\int_3^5 x\cos x\,dx = (\cos x + x\sin x)\Big|_3^5 = \cos 5 + 5\sin 5 - \cos 3 - 3\sin 3 \approx -3.944.$

43. $\displaystyle\int_0^5 \ln(1+t)\,dt = ((1+t)\ln(1+t) - (1+t))\Big|_0^5 = 6\ln 6 - 5 \approx 5.751.$

44. We use integration by parts. Let $u = \arctan y$ and $v' = 1$, so $u' = \frac{1}{1+y^2}$ and $v = y$. Thus
$$\begin{aligned}\int_0^1 \arctan y\,dy &= (\arctan y)y\Big|_0^1 - \int_0^1 \frac{y}{1+y^2}\,dy\\ &= \frac{\pi}{4} - \frac{1}{2}\ln|1+y^2|\Big|_0^1\\ &= \frac{\pi}{4} - \frac{1}{2}\ln 2\\ &\approx 0.439.\end{aligned}$$

45. First we make the substitution $y = x^2$, so $dy = 2x\,dx$.
Thus
$$\int_{x=0}^{x=1} x\arctan x^2\,dx = \frac{1}{2}\int_{y=0}^{y=1}\arctan y\,dy.$$

From Problem 44, we know that

$$\int_0^1 \arctan y\, dy = \frac{\pi}{4} - \frac{\ln 2}{2}.$$

Thus

$$\int_0^1 x \arctan x^2\, dx = \frac{1}{2}\left(\frac{\pi}{4} - \frac{1}{2}\ln 2\right) \approx 0.219.$$

46. We use integration by parts. Let $u = \arcsin z$ and $v' = 1$, so $u' = \dfrac{1}{\sqrt{1-z^2}}$ and $v = z$. Then

$$\int_0^1 \arcsin z\, dz = z \arcsin z\Big|_0^1 - \int_0^1 \frac{z}{\sqrt{1-z^2}}\, dz = \frac{\pi}{2} - \int_0^1 \frac{z}{\sqrt{1-z^2}}\, dz.$$

To find $\displaystyle\int_0^1 \frac{z}{\sqrt{1-z^2}}\, dz$, we substitute $w = 1 - z^2$, so $dw = -2z\, dz$.
Then

$$\int_{z=0}^{z=1} \frac{z}{\sqrt{1-z^2}}\, dz = -\frac{1}{2}\int_{w=1}^{w=0} w^{-\frac{1}{2}}\, dw = \frac{1}{2}\int_{w=0}^{w=1} w^{-\frac{1}{2}}\, dw = w^{\frac{1}{2}}\Big|_0^1 = 1.$$

Thus our final answer is $\frac{\pi}{2} - 1 \approx 0.571$.

47. To simplify the integral, we first make the substitution $z = u^2$, so $dz = 2u\, du$. Then

$$\int_{u=0}^{u=1} u \arcsin u^2\, du = \frac{1}{2}\int_{z=0}^{z=1} \arcsin z\, dz.$$

From Problem 46, we know that $\int_0^1 \arcsin z\, dz = \frac{\pi}{2} - 1$. Thus,

$$\int_0^1 u \arcsin u^2\, du = \frac{1}{2}\left(\frac{\pi}{2} - 1\right) \approx 0.285.$$

49. (a) One way to avoid integrating by parts is to take the derivative of the right hand side instead. Since $\int e^{ax}\sin bx\, dx$ is the antiderivative of $e^{ax}\sin bx$,

$$\begin{aligned} e^{ax}\sin bx &= \frac{d}{dx}[e^{ax}(A\sin bx + B\cos bx) + C] \\ &= ae^{ax}(A\sin bx + B\cos bx) + e^{ax}(Ab\cos bx - Bb\sin bx) \\ &= e^{ax}[(aA - bB)\sin bx + (aB + bA)\cos bx]. \end{aligned}$$

Thus $aA - bB = 1$ and $aB + bA = 0$. Solving for A and B in terms of a and b, we get

$$A = \frac{a}{a^2+b^2}, \quad B = -\frac{b}{a^2+b^2}.$$

Thus

$$\int e^{ax}\sin bx = e^{ax}\left(\frac{a}{a^2+b^2}\sin bx - \frac{b}{a^2+b^2}\cos bx\right) + C.$$

(b) If we go through the same process, we find

$$ae^{ax}[(aA - bB)\sin bx + (aB + bA)\cos bx] = e^{ax}\cos bx.$$

Thus $aA - bB = 0$, and $aB + bA = 1$.
In this case, solving for A and B yields

$$A = \frac{b}{a^2+b^2},\ B = \frac{a}{a^2+b^2}.$$

Thus $\int e^{ax}\cos bx = e^{ax}(\frac{b}{a^2+b^2}\sin bx + \frac{a}{a^2+b^2}\cos bx) + C.$

6.7 Solutions

1. See the solutions to Problems 5– 13. Answers may vary, as there may be more than one way to approach a problem.

3. See the solutions to Problems 26– 38.

5. $(\frac{1}{2}x^3 - \frac{3}{4}x^2 + \frac{3}{4}x - \frac{3}{8})e^{2x} + C.$
(Let $a = 2, p(x) = x^3$ in III- 14.)

7. $\frac{3}{16}\cos 3\theta \sin 5\theta - \frac{5}{16}\sin 3\theta\cos 5\theta + C.$
(Let $a = 3, b = 5$ in II- 10.)

9. $\frac{1}{6}x^6\ln x - \frac{1}{36}x^6 + C.$ (Let $n = 5$ in III- 13.)

11. Let $m = 3$ in IV- 21.

$$\begin{aligned}\int \frac{1}{\cos^3 x}\,dx &= \frac{1}{2}\frac{\sin x}{\cos^2 x} + \frac{1}{2}\int\frac{1}{\cos x}\,dx \\ &= \frac{1}{2}\frac{\sin x}{\cos^2 x} + \frac{1}{4}\ln\left|\frac{\sin x + 1}{\sin x - 1}\right| + C \text{ by IV- 22.}\end{aligned}$$

13. $\frac{1}{\sqrt{3}}\arctan\frac{y}{\sqrt{3}} + C.$
(Let $a = \sqrt{3}$ in V- 24).

15. $(\frac{1}{3}x^4 - \frac{4}{9}x^3 + \frac{4}{9}x^2 - \frac{8}{27}x + \frac{8}{81})e^{3x} + C.$
(Let $a = 3, p(x) = x^4$ in III- 14.)

17.

$$\begin{aligned}\int y^2 \sin 2y\, dy &= -\frac{1}{2}y^2 \cos 2y + \frac{1}{4}(2y)\sin 2y + \frac{1}{8}(2)\cos 2y + C \\ &= -\frac{1}{2}y^2 \cos 2y + \frac{1}{2}y \sin 2y + \frac{1}{4}\cos 2y + C.\end{aligned}$$

(Use $a = 2, p(y) = y^2$ in III- 15.)

19. $\frac{1}{34}e^{5x}(5\sin 3x - 3\cos 3x) + C.$
(Let $a = 5, b = 3$ in II- 8.)

21. Since $\ln 5u = \ln 5 + \ln u$,

$$\begin{aligned}\int u^5 \ln 5u\, du &= \ln 5 \int u^5\, du + \int u^5 \ln u\, du \\ &= \ln 5 \left(\frac{u^6}{6}\right) + \int u^5 \ln u\, du.\end{aligned}$$

By III- 13, with $n = 5$, we find

$$\begin{aligned}\int u^5 \ln 5u\, du &= \ln 5 \int u^5\, du + \int u^5 \ln u\, du \\ &= \frac{u^6}{6}\ln 5 + \frac{1}{6}u^6 \ln u - \frac{1}{36}u^6 + C \\ &= \frac{u^6}{6}\ln 5u - \frac{1}{36}u^6 + C.\end{aligned}$$

23. Since $\ln y^3 = 3\ln y$,

$$\begin{aligned}\int y^7 \ln(y^3)\, dy &= 3\int y^7 \ln y\, dy \\ &= 3\left(\frac{1}{8}y^8 \ln y - \frac{1}{64}y^8\right) + C \quad \text{(using III- 13)} \\ &= \frac{3}{8}y^8 \ln y - \frac{3}{64}y^8 + C.\end{aligned}$$

25. Substitute $w = z^2$, $dw = 2z\,dz$. Using IV- 17,

$$\begin{aligned}\int z\sin^3(z^2)\,dz &= \frac{1}{2}\int \sin^3 w\,dw = \frac{1}{2}[-\frac{1}{3}\sin^2 w\cos w + \frac{2}{3}\int \sin w\,dw] \\ &= -\frac{1}{6}\sin^2 w\cos w - \frac{1}{3}\cos w + C \\ &= -\frac{1}{6}\sin^2(z^2)\cos(z^2) - \frac{1}{3}\cos(z^2) + C.\end{aligned}$$

26. $\arcsin\dfrac{x}{\sqrt{2}} + C$.
(Let $a = \sqrt{2}$ in VI- 28.)

27. $\displaystyle\int \frac{1}{\sqrt{1-9x^2}}\,dx = \frac{1}{3}\int \frac{1}{\sqrt{\frac{1}{9}-x^2}}\,dx = \frac{1}{3}\arcsin 3x + C.$
(Let $a = \frac{1}{3}$ in VI- 28.)

29. Substitute $w = 2\theta$, $dw = 2\,d\theta$. Then use IV- 19, letting $m = 2$.
$$\int \frac{1}{\sin^2 2\theta}\,d\theta = \frac{1}{2}\int \frac{1}{\sin^2 w}\,dw = \frac{1}{2}(-\frac{\cos w}{\sin w}) + C = -\frac{1}{2\tan w} + C = -\frac{1}{2\tan 2\theta} + C.$$

31. Substitute $w = 7x$, $dw = 7\,dx$. Then use IV- 21.

$$\begin{aligned}\int \frac{1}{\cos^4 7x}\,dx = \frac{1}{7}\int \frac{1}{\cos^4 w}\,dw &= \frac{1}{7}\left[\frac{1}{3}\frac{\sin w}{\cos^3 w} + \frac{2}{3}\int \frac{1}{\cos^2 w}\,dw\right] \\ &= \frac{1}{21}\frac{\sin w}{\cos^3 w} + \frac{2}{21}\left[\frac{\sin w}{\cos w} + C\right] \\ &= \frac{1}{21}\frac{\tan w}{\cos^2 w} + \frac{2}{21}\tan w + C \\ &= \frac{1}{21}\frac{\tan 7x}{\cos^2 7x} + \frac{2}{21}\tan 7x + C.\end{aligned}$$

33. $-\dfrac{1}{4}(9-4x^2)^{\frac{1}{2}} + C$.
(Substitute $w = 9 - 4x^2$, $dw = -8x\,dx$). You need not use the table.)

35.

$$\begin{aligned}\int \sin^3 3\theta\cos^2 3\theta\,d\theta &= \int (\sin 3\theta)(\cos^2 3\theta)(1-\cos^2 3\theta)\,d\theta \\ &= \int \sin 3\theta(\cos^2 3\theta - \cos^4 3\theta)\,d\theta.\end{aligned}$$

Using an extension of the tip given in rule IV- 23, we let $w = \cos 3\theta$, $dw = -3\sin 3\theta\, d\theta$.

$$\begin{aligned}
\int \sin 3\theta(\cos^2 3\theta - \cos^4 3\theta)\, d\theta &= -\frac{1}{3}\int (w^2 - w^4)\, dw \\
&= -\frac{1}{3}(\frac{w^3}{3} - \frac{w^5}{5}) + C \\
&= -\frac{1}{9}(\cos^3 3\theta) + \frac{1}{15}(\cos^5 3\theta) + C.
\end{aligned}$$

37. $\int \frac{dz}{z(z-3)} = -\frac{1}{3}(\ln|z| - \ln|z-3|) + C.$
(Let $a = 0, b = 3$ in V- 26.)

38. $\int \frac{dy}{4-y^2} = -\int \frac{dy}{(y+2)(y-2)} = -\frac{1}{4}(\ln|y-2| - \ln|y+2|) + C.$
(Let $a = 2, b = -2$ in V- 26.)

39. $\arctan(z+2) + C.$
(Substitute $w = z + 2$ and use V- 24, letting $a = 1$.)

41. $\arcsin \frac{x+1}{\sqrt{2}} + C.$
(Substitute $w = x + 1$, and then apply VI- 28 with $a = \sqrt{2}$).

43. $\int \frac{1}{x^2+4x+4}\, dx = \int \frac{1}{(x+2)^2}\, dx = -\frac{1}{x+2} + C.$
You need not use the table.

45. Using long division, we find that

$$\frac{x^3+3}{x^2-3x+2} = x + 3 + \frac{7x-3}{x^2-3x+2}.$$

Thus

$$\begin{aligned}
\int \frac{x^3+3}{x^2-3x+2}\, dx &= \int (x + 3 + \frac{7x-3}{x^2-3x+2})\, dx \\
&= \int (x+3)\, dx + \int \frac{7x-3}{(x-1)(x-2)}\, dx.
\end{aligned}$$

Using V- 27 (with $a = 1, b = 2, c = 7$, and $d = -3$) we have

$$\int \frac{7x-3}{(x-1)(x-2)}\, dx = -4\ln|x-1| + 11\ln|x-2| + C.$$

Thus

$$\int \frac{x^3+3}{x^2-3x+2}\,dx = \frac{x^2}{2} + 3x - 4\ln|x-1| + 11\ln|x-2| + C.$$

47. Completing the square, we find that $x^2+8x+7=(x+4)^2-9$. Substitute $w=x+4$ and $dw=dx$.
$\int \sqrt{x^2+8x+7}\,dx = \int \sqrt{(x+4)^2-9}\,dx = \int \sqrt{w^2-9}\,dw.$
Now using VI- 31 and VI- 29 with $a=3$:

$$\begin{aligned}\int \sqrt{w^2-9}\,dw &= \frac{1}{2}\left(w\sqrt{w^2-9} - 9\int \frac{1}{\sqrt{w^2-9}}\,dw\right) + C\\ &= \frac{1}{2}w\sqrt{w^2-9} - \frac{9}{2}\ln|w+\sqrt{w^2-9}| + C\\ &= \frac{1}{2}(x+4)\sqrt{x^2+8x+7} - \frac{9}{2}\ln|x+4+\sqrt{x^2+8x+7}| + C.\end{aligned}$$

49.

$$\int \frac{5z-13}{z^2-5z+6}\,dz = \int \frac{5z-13}{(z-3)(z-2)}\,dz.$$

Let $a=3, b=2, c=5$, and $d=-13$ in V- 27.

$$\int \frac{5z-13}{(z-3)(z-2)}\,dz = 2\ln|z-3| + 3\ln|z-2| + C.$$

51.

$$\begin{aligned}\int_0^2 \frac{1}{4+x^2}\,dx &= \frac{1}{2}\arctan\frac{x}{2}\Big|_0^2 \quad \text{using V- 24}\\ &= \frac{1}{2}\arctan 1 - \frac{1}{2}\arctan 0 = \frac{\pi}{8} \approx 0.3927.\end{aligned}$$

Since $\frac{1}{4+x^2}$ is monotonically decreasing on $0 \le x \le 2$, we expect the integral to be between the left- and right-hand sums. Using 100 subintervals, we find that

$$0.3939 > \int_0^2 \frac{1}{4+x^2}\,dx > 0.3914$$

so our answer checks.

53. $\int_0^1 \frac{dx}{x^2+2x+5} = \int_0^1 \frac{dx}{(x+1)^2+4} = \frac{1}{2}\arctan\frac{x+1}{2}\Big|_0^1 = \frac{1}{2}\arctan 1 - \frac{1}{2}\arctan\frac{1}{2} \approx 0.1609.$
(Substitute $w = x+1$ and use V- 24).
$\frac{1}{x^2+2x+5}$ is monotonic over $0 \le x \le 1$, so we expect the value of the integral to be between the left- and right-hand sums. Using 100 subintervals, we find $0.1605 < \int_0^1 \frac{dx}{x^2+2x+5} < 0.1613$, which matches our result.

55.

$$\int_{-3}^{-1} \frac{dx}{\sqrt{x^2+6x+10}} = \int_{-3}^{-1} \frac{dx}{\sqrt{(x+3)^2+1}} = \ln|x+3+\sqrt{x^2+6x+10}|\Big|_{-3}^{-1}.$$

(Substitute $w = x+3$ and then use VI- 29)

$$= \ln|2+\sqrt{5}| \approx 1.4436.$$

$\frac{1}{\sqrt{x^2+6x+10}}$ is monotonic over $-3 \le x \le -1$, so we expect the value of the integral to be between the left- and right-hand sums. Using 100 subintervals, we find out $1.4381 < \int_{-3}^{-1} \frac{dx}{\sqrt{x^2+6x+10}} < 1.4492$, which matches our result.

57. Use $\sin 2\theta = 2\sin\theta\cos\theta$. Then
$\int_0^2 \cos\theta\sin 2\theta\, d\theta = 2\int_0^2 \cos^2\theta\sin\theta\, d\theta = (-2)\frac{\cos^3\theta}{3}\Big|_0^2 \approx 0.7147.$
$\cos\theta\sin 2\theta$ is not monotonic over the interval $0 \le \theta \le 2$, but we find using 100 subintervals that the left-hand sum ≈ 0.7115, the right-hand sum ≈ 0.7178, and the average ≈ 0.7147, so these results match our answer very well.

59. We use VI- 30 and VI- 28:

$$\begin{aligned}\int_0^1 \sqrt{4-x^2}\,dx &= \frac{1}{2}(x\sqrt{4-x^2})\Big|_0^1 + \frac{1}{2}(2^2)\int_0^1 \frac{1}{\sqrt{4-x^2}}\,dx \\ &= \frac{1}{2}\sqrt{3} + 2\arcsin\frac{x}{2}\Big|_0^1 \\ &\approx 1.913.\end{aligned}$$

Since $\sqrt{4-x^2}$ is monotonically decreasing, we expect the value of the integral to fall between the left- and right-hand sums. Using 100 intervals, we find $1.912 < \int_0^1 \sqrt{4-x^2}\,dx < 1.915$, which matches our result.

61.

$$\int_1^2 (x-2x^3)\ln x\,dx = \int_1^2 x\ln x\,dx - 2\int_1^2 x^3\ln x\,dx$$

$$\begin{aligned}
&= \left(\frac{1}{2}x^2\ln x - \frac{1}{4}x^2\right)\Big|_1^2 - \left(\frac{1}{2}x^4\ln x - \frac{1}{8}x^4\right)\Big|_1^2 \quad \text{using III- 13} \\
&= 2\ln 2 - \frac{3}{4} - \left(8\ln 2 - \frac{15}{8}\right) \\
&= \frac{9}{8} - 6\ln 2 \approx -3.034.
\end{aligned}$$

By looking at the graph or taking the derivative, we see that the function is monotonically decreasing. Thus, using 100 subintervals, we expect that $-3.083 < \int_1^2 (x - 2x^3)\ln x\, dx < -2.986$, which matches our result.

63. Use $\sin 2x = 2\sin x \cos x$. Then

$$\int_{-\pi}^{\pi} \sin 5x \cos 5x\, dx = \frac{1}{2}\int_{-\pi}^{\pi} \sin 10x\, dx = -\frac{1}{20}\cos 10x\Big|_{-\pi}^{\pi} = 0.$$

This makes sense because $\sin 5x \cos 5x$ is odd. This can also be checked numerically.

65. (a) Since $R(T)$ is the rate or production, we find the total production by integrating:

$$\begin{aligned}
\int_0^N R(t)\,dt &= \int_0^N (A + Be^{-t}\sin(2\pi t))\,dt \\
&= NA + B\int_0^N e^{-t}\sin(2\pi t)\,dt. \\
&\quad \text{Let } a = -1 \text{ and } b = 2\pi \text{ in II- 8.} \\
&= NA + \frac{B}{1+4\pi^2}e^{-t}(-\sin(2\pi t) - 2\pi\cos(2\pi t))\Big|_0^N.
\end{aligned}$$

Since N is an integer (so $\sin 2\pi N = 0$ and $\cos 2\pi N = 1$),

$$\int_0^N R(t)\,dt = NA + B\frac{2\pi}{1+4\pi^2}(1 - e^{-N}).$$

Thus the total production is $NA + \frac{2\pi B}{1+4\pi^2}(1 - e^{-N})$ over the first N years.

(b) The average production over the first N years is

$$\int_0^N \frac{R(t)\,dt}{N} = A + \frac{2\pi B}{1+4\pi^2}\left(\frac{1-e^{-N}}{N}\right).$$

(c) As $N \to \infty$, $A + \frac{2\pi B}{1+4\pi^2}\frac{1-e^{-N}}{N} \to A$, since the second term in the sum goes to 0. This is why A is called the average!

(d) When t gets large, the term $Be^{-t}\sin(2\pi t)$ gets very small. Thus, $R(t) \approx A$ for most t, so it makes sense that the average of $\int_0^N R(t)\,dt$ is A as $N \to \infty$.

(e) This model is not reasonable for long periods of time, since an oil well has finite capacity and will eventually "run dry." Thus, we cannot expect average production to be close to constant over a long period of time.

67. (a) $\frac{2}{x} + \frac{1}{x+3} = \frac{2(x+3)}{x(x+3)} + \frac{x}{x(x+3)} = \frac{3x+6}{x^2+3x}$. Thus

$$\int \frac{3x+6}{x^2+3x}\,dx = \int (\frac{2}{x} + \frac{1}{x+3})\,dx = 2\ln|x| + \ln|x+3| + C.$$

(b) Let $a = 0, b = -3, c = 3$, and $d = 6$ in V- 27.

$$\begin{aligned}\int \frac{3x+6}{x^2+3x}\,dx &= \int \frac{3x+6}{x(x+3)}\,dx \\ &= \frac{1}{3}(6\ln|x| + 3\ln|x+3|) + C = 2\ln|x| + \ln|x+3| + C.\end{aligned}$$

69. Using formula II- 11, if $m \neq \pm n$, then

$$\int_{-\pi}^{\pi} \cos m\theta \cos n\theta\,d\theta = \frac{1}{n^2 - m^2}(n\cos m\theta \sin n\theta - m\sin m\theta \cos n\theta)\Big|_{-\pi}^{\pi}.$$

We see that in the evaluation, each term will have a $\sin k\pi$ term, so the expression reduces to 0.

6.8 Solutions

1.

	$n=1$	$n=2$	$n=4$
Left	40	40.7846	41.7116
Right	51.2250	46.3971	44.5179
Trap	45.6125	43.5909	43.1147
Mid	41.5692	42.6386	42.8795

3. $g(x) = \sqrt{x}\, e^x \Longrightarrow g'(x) = e^x\left(\sqrt{x} + \frac{1}{2\sqrt{x}}\right) \Longrightarrow g''(x) = e^x\left(\sqrt{x} + \frac{1}{\sqrt{x}} - \frac{1}{4\sqrt{x^3}}\right)$

   The first and second derivatives are always positive for $1 \le x \le 2$, so g is increasing and concave up on this interval. Hence, the left sum and the mipoint sum underestimate the integral, and the right sum and the trapezoid sum overestimate the integral:

N	Left Sum	Right Sum	Trapezoid	Midpoint
10	5.47115	6.24429	5.85772	5.84649
100	5.81165	5.88896	5.85031	5.85023
1000	5.84637	5.85410	5.85023	5.85023

5.

N	Left Sum	Right Sum	Trapezoid	Midpoint
10	0.14047	0.18842	0.16445	0.16335
100	0.16132	0.16612	0.16372	0.16371
1000	0.16347	0.16395	0.16371	0.16371

   The curve is increasing on the interval $0 \le \theta \le 1$, so left sums give an underestimate and right sums an overestimate. The curve is concave down on all of the interval, so the trapezoid sum gives an underestimate and the midpoint sum gives an overestimate.

7.

N	Left Sum	Right Sum	Trapezoid	Midpoint
10	6.71685	8.00430	7.36058	7.33181
100	7.27723	7.40597	7.34160	7.34131
1000	7.33497	7.34785	7.34141	7.34141

   The curve is increasing on the interval $0 \le x \le 3$, so left sums give an underestimate and right sums an overestimate. The curve is concave up on all of the interval, so the trapezoid sum gives an overestimate and the midpoint sum gives an underestimate.

9. (a) i. LEFT(32) = 13.6961
   RIGHT(32) = 14.3437
   TRAP(32) = 14.0199

   Exact value $= (x \ln x - x)\Big|_1^{10} \approx 14.02585093$

ii. LEFT(32) $= 50.3180$
RIGHT(32) $= 57.0178$
TRAP(32) $= 53.6679$

Exact value $= e^x\Big|_0^4 \approx 53.59815003$

(b) Both $\ln x$ and e^x are increasing, so the left sum underestimates and the right sum overestimates.

i. LEFT(32) $\leq$ TRAP(32) $\leq$ Actual value $\leq$ RIGHT(32)

ii. LEFT(32) $\leq$ Actual value $\leq$ TRAP(32) $\leq$ RIGHT(32)

For the trapezoidal rule, the principle was stated in the chapter: "f concave down: trapezoid underestimates. f concave up: trapezoid overestimates."

Since $\ln x$ is concave down, the trapezoidal estimate is too small. Since e^x is concave up, the trapezoidal estimate is too large. In each case, however, the trapezoidal estimate should be better than the left- or right-hand sums, since it is the average of the two.

11. Let $s(t)$ be the distance traveled at time t and $v(t)$ be the velocity at time t. Then the distance traveled during the interval $0 \leq t \leq 6$ is

$$\begin{aligned} s(6) - s(0) &= s(t)\Big|_0^6 \\ &= \int_0^6 s'(t)\, dt \quad \text{(by the Fundamental Theorem)} \\ &= \int_0^6 v(t)\, dt. \end{aligned}$$

We estimate the distance by estimating this integral.

From the table, we find:
LEFT(6) $= 31$
RIGHT(6) $= 39$
TRAP(6) $= 35$

13. (a) $\displaystyle\int_0^{2\pi} \sin\theta\, d\theta = -\cos\theta\Big|_0^{2\pi} = 0.$

(b) MID(1) is 0 since the midpoint of 0 and 2π is π, and $\sin\pi = 0$. Thus MID(1) $= 2\pi(\sin\pi) = 0$. MID(2) is 0 since the midpoints we use are $\pi/2$ and $3\pi/2$, and $\sin(\pi/2) = -\sin(3\pi/2)$. So MID(2) $= \pi\sin(\pi/2) + \pi\sin(3\pi/2) = 0$.

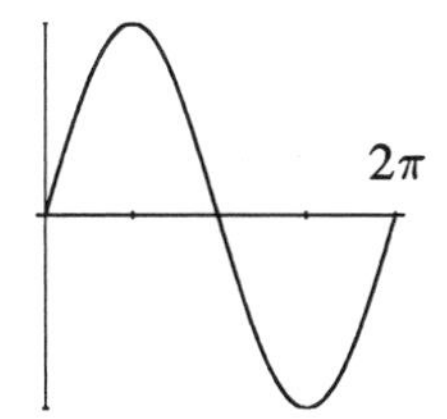

(c) MID(3) does equal 0.

In general, MID$(n) = 0$ for all n, even though your calculator (because of round-off error) might not return it as such. The reason is that $\sin(x) = -\sin(2\pi - x)$. If we use MID($n$), we will always take sums where we are adding pairs of the form $\sin(x)$ and $\sin(2\pi - x)$, so the sum will cancel to 0. (If n is odd, we will get a $\sin\pi$ in the sum which doesn't pair up with anything — but $\sin\pi$ is already 0!)

15.

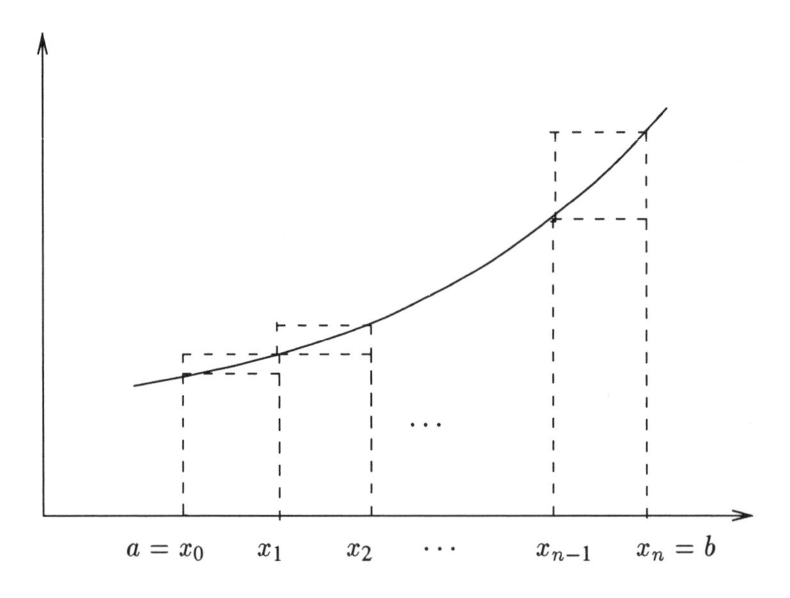

From the diagram, the difference between RIGHT(n) and LEFT(n) is the area of the shaded rectangles.

RIGHT(n) $= f(x_1)\Delta x + f(x_2)\Delta x + \cdots + f(x_n)\Delta x$
LEFT(n) $= f(x_0)\Delta x + f(x_1)\Delta x + \cdots + f(x_{n-1})\Delta x$
Notice that the terms in these two sums are the same, except that RIGHT(n) contains $f(x_n)\Delta x$ $(= f(b)\Delta x)$, and LEFT(n) contains $f(x_0)\Delta x$ $(= f(a)\Delta x)$. Thus

$$\begin{aligned}\text{RIGHT}(n) &= \text{LEFT}(n) + f(x_n)\Delta x - f(x_0)\Delta x \\ &= \text{LEFT}(n) + f(b)\Delta x - f(a)\Delta x\end{aligned}$$

17.

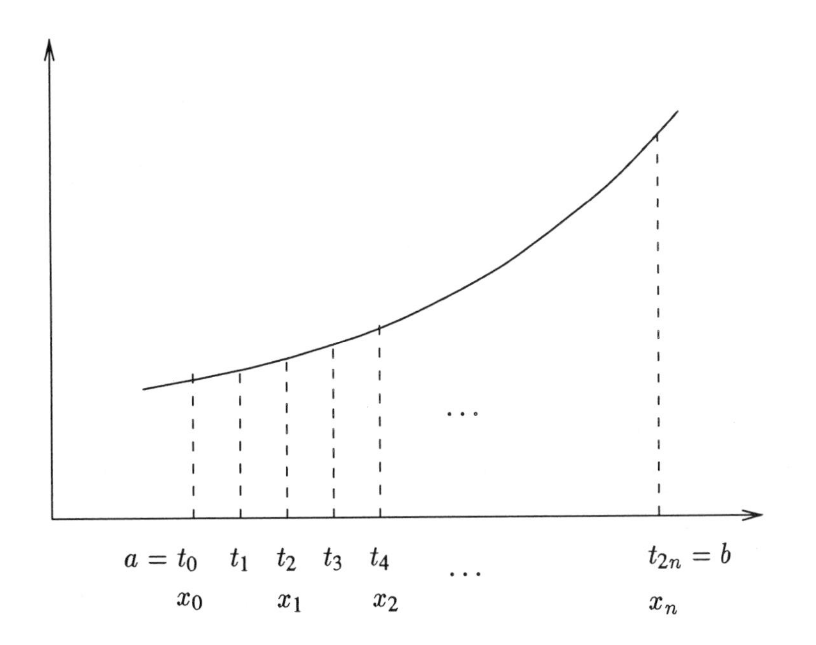

Divide the interval $[a, b]$ into n pieces, by $x_0, x_1, x_2, \ldots, x_n$, and also into $2n$ pieces, by $t_0, t_1, t_2, \ldots, t_{2n}$. Then the x's coincide with the even t's, so $x_0 = t_0, x_1 = t_2, x_2 = t_4, \ldots, x_n = t_{2n}$ and $\Delta t = \frac{1}{2}\Delta x$.

$$\text{LEFT}(n) = f(x_0)\Delta x + f(x_1)\Delta x + \cdots + f(x_{n-1})\Delta x$$

Since MID(n) involves evaluating at the midpoints of the x intervals, which are $t_1, t_3, t_5, \ldots$ etc., we get

$$\text{MID}(n) = f(t_1)\Delta x + f(t_3)\Delta x + \cdots + f(t_{2n-1})\Delta x$$

Now

$$\text{LEFT}(2n) = f(t_0)\Delta t + f(t_1)\Delta t + f(t_2)\Delta t + \cdots + f(t_{2n-1})\Delta t.$$

Regroup terms, putting all the even t's first, the odd t's last:

$$\text{LEFT}(2n) = f(t_0)\Delta t + f(t_2)\Delta t + \cdots + f(t_{2n-2})\Delta t + f(t_1)\Delta t + f(t_3)\Delta t + \cdots + f(t_{2n-1})\Delta t$$

$$= \underbrace{f(x_0)\frac{\Delta x}{2} + f(x_1)\frac{\Delta x}{2} + \cdots + f(x_{n-1})\frac{\Delta x}{2}}_{\frac{\text{LEFT}(n)}{2}} + \underbrace{f(t_1)\frac{\Delta x}{2} + f(t_3)\frac{\Delta x}{2} + \cdots + f(t_{2n-1})\frac{\Delta x}{2}}_{\frac{\text{MID}(n)}{2}}$$

So

$$\text{LEFT}(2n) = \frac{1}{2}(\text{LEFT}(n) + \text{MID}(n))$$

19. First, we compute:

$$\begin{aligned}(f(b)-f(a))\Delta x &= (f(b)-f(a))\left(\frac{b-a}{n}\right)\\ &= (f(5)-f(2))\left(\frac{3}{n}\right)\\ &= (21-13)\left(\frac{3}{n}\right)\\ &= \frac{24}{n}\end{aligned}$$

RIGHT(10) = LEFT(10) + 24 = 3.156 + 2.4 = 5.556
TRAP(10) = LEFT(10) + $\frac{1}{2}$(2.4) = 3.156 + 1.2 = 4.356
LEFT(20) = $\frac{1}{2}$(LEFT(10) + MID(10)) = $\frac{1}{2}$(3.156 + 3.242) = 3.199
RIGHT(20) = LEFT(20) + 2.4 = 3.199 + 1.2 = 4.399
TRAP(20) = LEFT(20) + $\frac{1}{2}$(1.2) = 3.199 + 0.6 = 3.799

21. Suppose you approximate $\int_1^{100,001} \frac{1}{x}\,dx$, by rectangles, with $n = 100,000$ so $\Delta x = 1$.

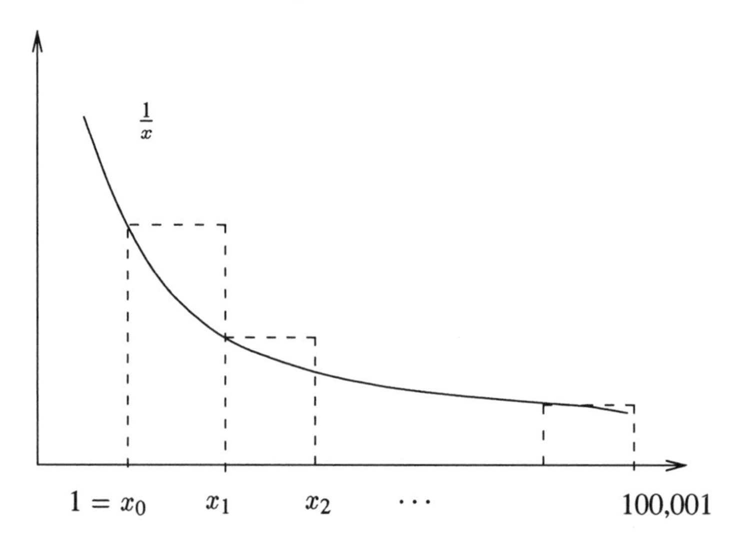

Then

$$\begin{aligned}\text{LEFT(100,000)} &= f(1)\cdot 1 + f(2)\cdot 1 + \cdots + f(100,000)\cdot 1\\ &= \frac{1}{1}+\frac{1}{2}+\cdots+\frac{1}{100,000} = \sum_{k=1}^{100,000}\frac{1}{k}\end{aligned}$$

Since the left sum is an overestimate,

$$\int_1^{100,001}\frac{1}{x}\,dx < \text{LEFT(100,000)},$$

and since

$$\int_1^{100,001} \frac{1}{x}\,dx = \ln(100,001) - \ln 1 = \ln(100,001),$$

so

$$\ln 100,001 < \sum_{k=1}^{100,000} \frac{1}{k}.$$

Now imagine all the rectangles moved one unit to the left; they are the right sum approximation to

$$\int_1^{100,000} \frac{1}{x}\,dx + \text{area of first rectangle}$$

and this time they give an underestimate.

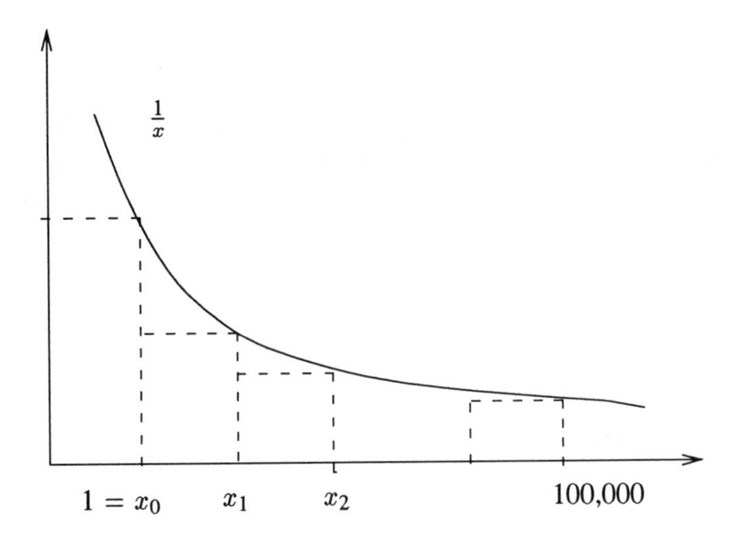

The area of the rectangles is our sum, so

$$\sum_{k=1}^{100,000} \frac{1}{k} < \int_1^{100,000} \frac{1}{x}\,dx + \text{area of first rectangle}$$

So

$$\sum_{k=1}^{100,000} \frac{1}{k} < (\ln 100,000) + 1$$

Thus

$$\ln(100,001) < \sum_{k=1}^{100,000} \frac{1}{k} < (\ln 100,000) + 1$$

$$11.5 < \sum_{k=1}^{100,000} \frac{1}{k} < 12.5 \quad \text{so} \quad \sum_{k=1}^{100,000} \frac{1}{k} \approx 12.$$

6.9 Solutions

1.

$$\text{SIMP} = \frac{1}{3}(2\,\text{MID} + \text{TRAP}) = \frac{1}{3}\left(2\,\text{MID} + \frac{\text{LEFT} + \text{RIGHT}}{2}\right) = \frac{1}{3}\left(2 \cdot 0.6857 + \frac{0.8333 + 0.5833}{2}\right) \approx 0.6932$$

The actual value is $\ln 2 \approx 0.6931$.

3. SIMP(10)$\approx$ 0.23182. When we do $n = 10$ intervals and $n = 20$ intervals, these digits match, so they must be correct digits since Simpson's rule will improve the number of digits of accuracy when we double the number of intervals. This reason also holds true for Problems 4- 10.

4. 4.2365. ($n = 10$ intervals).

5. 1.5699. ($n = 10$ intervals)

7. 1.0894. ($n = 10$ intervals)

9. 0.904524. ($n = 10$ intervals)

10. 0.6593299. ($n = 10$ intervals)

11. (a) $\int_0^1 7x^6\,dx = x^7\Big|_0^1 = 1.$

(b)

	LEFT(5)	RIGHT(5)	TRAP(5)	MID(5)	SIMP(5)
VALUE	0.438144	1.838144	1.138144	0.931623	1.0004633
ERROR	−0.561856	0.838144	0.138144	−0.068377	0.0004633

(c)

	LEFT(10)	RIGHT(10)	TRAP(10)	MID(10)	SIMP(10)
VALUE	0.6848835	1.3848835	1.0348835	0.9826019	1.000029115
ERROR	−0.3151165	0.3848835	0.0348835	−0.0173981	0.000029115

(d)

RATIOS		
	LEFT	1.78
	RIGHT	2.18
	TRAP	3.96
	MID	3.93
	SIMP	15.91.

The values are about what we would expect, in that the LEFT and RIGHT approximations improve by about the same factor, the TRAP and MID approximations improve by the square of this factor, and the SIMP approximation improves by the fourth power of this factor. This is what the discussion in the book predicts.

13. (a) For the left-hand rule, error is approximately proportional to $\frac{1}{n}$. If we let n_p be the number of subdivisions needed for accuracy to p places, then there is a constant k such that

$$\begin{aligned} 5 \times 10^{-5} &= \frac{1}{2} \times 10^{-4} \approx \frac{k}{n_4} \\ 5 \times 10^{-9} &= \frac{1}{2} \times 10^{-8} \approx \frac{k}{n_8} \\ 5 \times 10^{-13} &= \frac{1}{2} \times 10^{-12} \approx \frac{k}{n_{12}} \\ 5 \times 10^{-21} &= \frac{1}{2} \times 10^{-20} \approx \frac{k}{n_{20}} \end{aligned}$$

Thus the ratios $n_4 : n_8 : n_{12} : n_{20} \approx 1 : 10^4 : 10^8 : 10^{16}$, and assuming the computer time necessary is proportional to n_p, the computer times are approximately

4 places:	2 seconds	
8 places:	2×10^4 seconds	≈ 0.00063 years
12 places:	2×10^8 seconds	≈ 6.34 years
20 places:	2×10^{16} seconds	$\approx 634{,}000{,}000$ years

(b) For the trapezoidal rule, error is approximately proportional to $\frac{1}{n^2}$. If we let N_p be the number of subdivisions needed for accuracy to p places, then there is a constant C such that

$$\begin{aligned} 5 \times 10^{-5} &= \frac{1}{2} \times 10^{-4} \approx \frac{C}{N_4{}^2} \\ 5 \times 10^{-9} &= \frac{1}{2} \times 10^{-8} \approx \frac{C}{N_8{}^2} \\ 5 \times 10^{-13} &= \frac{1}{2} \times 10^{-12} \approx \frac{C}{N_{12}{}^2} \\ 5 \times 10^{-21} &= \frac{1}{2} \times 10^{-20} \approx \frac{C}{N_{20}{}^2} \end{aligned}$$

Thus the ratios $N_4{}^2 : N_8{}^2 : N_{12}{}^2 : N_{20}{}^2 \approx 1 : 10^4 : 10^8 : 10^{16}$, and the ratios $N_4 : N_8 : N_{12} : N_{20} \approx 1 : 10^2 : 10^4 : 10^8$. So the computer times are approximately

4 places:	2 seconds	
8 places:	2×10^2 seconds	≈ 0.0000063 years
12 places:	2×10^4 seconds	≈ 0.00063 years
20 places:	2×10^8 seconds	≈ 6.34 years

15. Since the midpoint rule is sensitive to f'', the simplifying assumption should be that f'' does not change sign in the interval of integration. Thus MID(10) and MID(20) will both be overestimates or will both be underestimates. Since the larger number, MID(10) is less accurate than the smaller number, they must both be overestimates. Then the information that ERROR(10)=4 ERROR(20) means that the the value of the integral and the two sums are arranged as follows:

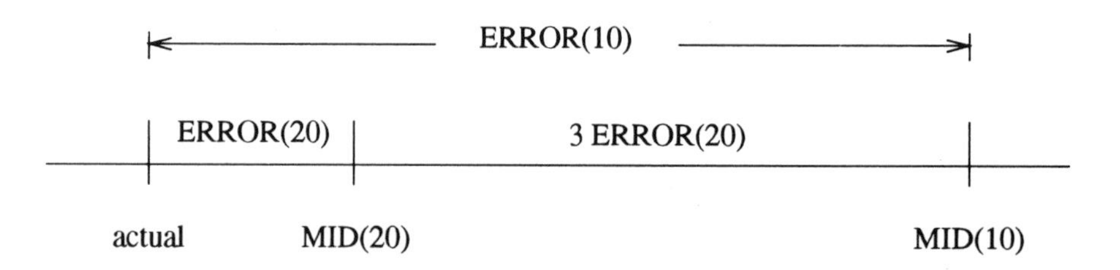

Thus 3 ERROR(20) = MID(10) - MID(20) = $35.619 - 35.415 = 0.204$, so ERROR(20) = 0.068 and ERROR(10) = 4 ERROR(20) = 0.272 .

17. False. If the function $f(x)$ is a line, then the trapezoid rule gives the exact answer to $\int_a^b f(x)\,dx$.

19. True. Since f' and g' are greater than 0, all left rectangles give underestimates. The bigger the derivative, the bigger the underestimate, so the bigger the error. (Note: if we didn't have $0 < f' < g'$, but instead just had $f' < g'$, the statement wouldn't necessarily be true. This is because some left rectangles could be overestimates and some could be underestimates–so, for example, it could be that the error in approximating g is 0! If $0 < f' < g'$, however, this can't be the case.

20. (a) If $f(x) = 1$, then

$$\int_a^b f(x)\,dx = (b-a).$$

Also,

$$\frac{h}{3}\left(\frac{f(a)}{2} + 2f(m) + \frac{f(b)}{2}\right) = \frac{b-a}{3}\left(\frac{1}{2} + 2 + \frac{1}{2}\right) = (b-a).$$

So the equation holds for $f(x) = 1$.
If $f(x) = x$, then

$$\int_a^b f(x)\,dx = \left.\frac{x^2}{2}\right|_a^b = \frac{b^2 - a^2}{2}.$$

Also,

$$\begin{aligned}
\frac{h}{3}\left(\frac{f(a)}{2} + 2f(m) + \frac{f(b)}{2}\right) &= \frac{b-a}{3}\left(\frac{a}{2} + 2\frac{a+b}{2} + \frac{b}{2}\right) \\
&= \frac{b-a}{3}\left(\frac{a}{2} + a + b + \frac{b}{2}\right) \\
&= \frac{b-a}{3}\left(\frac{3}{2}b + \frac{3}{2}a\right) \\
&= \frac{(b-a)(b+a)}{2} \\
&= \frac{b^2 - a^2}{2}.
\end{aligned}$$

So the equation holds for $f(x) = x$.

If $f(x) = x^2$, then $\int_a^b f(x)\,dx = \left.\frac{x^3}{3}\right|_a^b = \frac{b^3 - a^3}{3}$. Also,

$$\begin{aligned}
\frac{h}{3}\left(\frac{f(a)}{2} + 2f(m) + \frac{f(b)}{2}\right) &= \frac{b-a}{3}\left(\frac{a^2}{2} + 2\left(\frac{a+b}{2}\right)^2 + \frac{b^2}{2}\right) \\
&= \frac{b-a}{3}\left(\frac{a^2}{2} + \frac{a^2 + 2ab + b^2}{2} + \frac{b^2}{2}\right) \\
&= \frac{b-a}{3}\left(\frac{2a^2 + 2ab + 2b^2}{2}\right) \\
&= \frac{b-a}{3}\left(a^2 + ab + b^2\right) \\
&= \frac{b^3 - a^3}{3}.
\end{aligned}$$

So the equation holds for $f(x) = x^2$.

(b) The "Facts about Sums and Constant Multiples of Integrands" give us that when $f(x) = Ax^2 + Bx + C$, then:

$$\int_a^b f(x)\,dx = \int_a^b (Ax^2 + Bx + C)\,dx = A\int_a^b x^2\,dx + B\int_a^b x\,dx + C\int_a^b 1\,dx.$$

Now we use the results of part (a) to get:

$$\int_a^b f(x)\,dx = A\frac{h}{3}\left(\frac{a^2}{2} + 2m^2 + \frac{b^2}{2}\right) + B\frac{h}{3}\left(\frac{a}{2} + 2m + \frac{b}{2}\right) + C\frac{h}{3}\left(\frac{1}{2} + 2\cdot 1 + \frac{1}{2}\right)$$

$$= \frac{h}{3}\left(\frac{Aa^2 + Ba + C}{2} + 2(Am^2 + Bm + C) + \frac{Ab^2 + Bb + C}{2}\right)$$
$$= \frac{h}{3}\left(\frac{f(a)}{2} + 2f(m) + \frac{f(b)}{2}\right)$$

21. (a) Suppose $q_i(x)$ is the quadratic function approximating $f(x)$ on the subinterval $[x_i, x_{i+1}]$, and m_i is the midpoint of the interval, $m_i = (x_i + x_{i+1})/2$. Then, using the equation in Problem 20, with $a = x_i$ and $b = x_{i+1}$ and $h = \Delta x = x_{i+1} - x_i$:

$$\int_{x_i}^{x_{i+1}} f(x)dx \approx \int_{x_i}^{x_{i+1}} q_i(x)dx = \frac{\Delta x}{3}\left(\frac{q_i(x_i)}{2} + 2q_i(m_i) + \frac{q_i(x_{i+1})}{2}\right).$$

(b) Summing over all subintervals gives

$$\int_a^b f(x)dx \approx \sum_{i=0}^{n-1}\int_{x_i}^{x_{i+1}} q_i(x)dx = \sum_{i=0}^{n-1}\frac{\Delta x}{3}\left(\frac{q_i(x_i)}{2} + 2q_i(m_i) + \frac{q_i(x_{i+1})}{2}\right).$$

Splitting the sum into two parts:

$$= \frac{2}{3}\sum_{i=0}^{n-1} q_i(m_i)\Delta x + \frac{1}{3}\sum_{i=0}^{n-1}\frac{q_i(x_i) + q_i(x_{i+1})}{2}\Delta x$$
$$= \frac{2}{3}\text{MID}(n) + \frac{1}{3}\text{TRAP}(n)$$
$$= \text{SIMP}(n).$$

6.10 Solutions

1.

$$\int_1^\infty e^{-2x}\,dx = \lim_{b\to\infty}\int_1^b e^{-2x}\,dx = \lim_{b\to\infty} -\frac{e^{-2x}}{2}\Big|_1^b$$
$$= \lim_{b\to\infty}(-e^{-2b}/2 + e^{-2}/2) = 0 + e^{-2}/2 = e^{-2}/2,$$

where the first limit is 0 because $\lim_{x\to\infty} e^{-x} = 0$.

3. Using integration by parts with $u = x$ and $v' = e^{-x}$, we find that

$$\int xe^{-x}\,dx = -xe^{-x} - \int -e^{-x}dx = -(1+x)e^{-x}$$

so

$$\int_0^\infty \frac{x}{e^x}\,dx = \lim_{b\to\infty}\int_0^b \frac{x}{e^x}\,dx$$
$$= \lim_{b\to\infty} -1(1+x)e^{-x}\Big|_0^b$$
$$= \lim_{b\to\infty}\left[1 - (1+b)e^{-b}\right]$$
$$= 1.$$

5.

$$\begin{aligned}\int_{\pi}^{\infty} \sin y \, dy &= \lim_{b\to\infty} \int_{\pi}^{b} \sin y \, dy \\ &= \lim_{b\to\infty} (-\cos y)\Big|_{\pi}^{b} \\ &= \lim_{b\to\infty} [-\cos b - (-\cos\pi)].\end{aligned}$$

As $b \to \infty$, $-\cos b$ fluctuates between -1 and 1, so the limit fails to exist: the integral diverges. (This doesn't follow right from the fact that $\sin y$ fluctuates between -1 and 1!)

7.

$$\begin{aligned}\int_{\pi/4}^{\pi/2} \frac{\sin x}{\sqrt{\cos x}}\, dx &= \lim_{b\to\pi/2^-} \int_{\pi/4}^{b} \frac{\sin x}{\sqrt{\cos x}}\, dx \\ &= \lim_{b\to\pi/2^-} -\int_{\pi/4}^{b} (\cos x)^{-1/2}(-\sin x)\, dx \\ &= \lim_{b\to\pi/2^-} -2(\cos x)^{1/2}\Big|_{\pi/4}^{b} \\ &= \lim_{b\to\pi/2^-} [-2(\cos b)^{1/2} + 2(\cos\pi/4)^{1/2}] \\ &= \sqrt{2}.\end{aligned}$$

9. This integral is improper because $1/v$ blows up at $v = 0$. To evaluate it, we must split the region of integration up into two pieces, from 0 to 1 and from -1 to 0. But notice,

$$\int_0^1 \frac{1}{v}\, dv = \lim_{b\to 0+} \int_b^1 \frac{1}{v}\, dv = \lim_{b\to 0+} \left(\ln v\Big|_b^1\right) = -\ln b.$$

As $b \to 0^+$, this goes to infinity and the integral diverges, so our original integral also diverges.

11.

$$\begin{aligned}\int_1^{\infty} \frac{1}{\sqrt{x^2+1}}\, dx &= \lim_{b\to\infty} \int_1^{b} \frac{1}{\sqrt{x^2+1}}\, dx \\ &= \lim_{b\to\infty} \ln|x + \sqrt{x^2+1}|\Big|_1^b \\ &= \lim_{b\to\infty} \ln(b + \sqrt{b^2+1}) - \ln(1+\sqrt{2}).\end{aligned}$$

As $b \to \infty$, this limit does not exist, so the integral diverges.

13.

$$\begin{aligned}
\int_1^\infty \frac{y}{y^4+1}\,dy &= \lim_{b\to\infty} \frac{1}{2}\int_1^b \frac{2y}{(y^2)^2+1}\,dy \\
&= \lim_{b\to\infty} \frac{1}{2}\arctan(y^2)\Big|_1^b \\
&= \lim_{b\to\infty} \frac{1}{2}[\arctan(b^2) - \arctan 1] \\
&= (1/2)[\pi/2 - \pi/4] = \pi/8.
\end{aligned}$$

15. We use V- 26 with $a = 4$ and $b = -4$:

$$\begin{aligned}
\int_0^4 \frac{1}{u^2-16}\,du &= \lim_{b\to 4^-} \int_0^b \frac{1}{u^2-16}\,du \\
&= \lim_{b\to 4^-} \int_0^b \frac{1}{(u-4)(u+4)}\,du \\
&= \lim_{b\to 4^-} \frac{(\ln|u-4| - \ln|u+4|)}{8}\Big|_0^b \\
&= \lim_{b\to 4^-} \frac{1}{8}\left(\ln|b-4| + \ln 4 - \ln|b+4| - \ln 4\right).
\end{aligned}$$

As $b \to 4^-$, $\ln|b-4| \to -\infty$, so the limit does not exist and the integral diverges.

17. With the substitution $w = \ln x$, $dw = \frac{1}{x}dx$,

$$\int \frac{dx}{x\ln x} = \int \frac{1}{w}\,dw = \ln|w| + C = \ln|\ln x| + C$$

so

$$\begin{aligned}
\int_2^\infty \frac{dx}{x\ln x} &= \lim_{b\to\infty} \int_2^b \frac{dx}{x\ln x} \\
&= \lim_{b\to\infty} \ln|\ln x|\Big|_2^b \\
&= \lim_{b\to\infty} [\ln|\ln b| - \ln|\ln 2|].
\end{aligned}$$

As $b \to \infty$, the limit goes to ∞ and hence the integral diverges.

19. Using the substitution $w = -x^{\frac{1}{2}}$, $-2dw = x^{-\frac{1}{2}}\,dx$,

$$\int e^{-x^{\frac{1}{2}}} x^{-\frac{1}{2}}\,dx = -2\int e^w\,dw = -2e^{-x^{\frac{1}{2}}} + C.$$

So

$$\begin{aligned}
\int_0^\pi \frac{1}{\sqrt{x}} e^{-\sqrt{x}}\,dx &= \lim_{b\to 0+} \int_b^\pi \frac{1}{\sqrt{x}} e^{-\sqrt{x}}\,dx \\
&= \lim_{b\to 0+} -2e^{-\sqrt{x}}\Big|_b^\pi \\
&= 2 - 2e^{-\sqrt{\pi}}.
\end{aligned}$$

21. As in Problem 17, $\int \frac{dx}{x\ln x} = \ln|\ln x| + C$, so

$$\begin{aligned}\int_1^2 \frac{dx}{x\ln x} &= \lim_{b\to 1+}\int_b^2 \frac{dx}{x\ln x}\\ &= \lim_{b\to 1+} \ln|\ln x|\Big|_b^2\\ &= \lim_{b\to 1+} \ln(\ln 2) - \ln(\ln b).\end{aligned}$$

As $b \to 1^+, \ln(\ln b) \to -\infty$, so the integral diverges.

23. $\int \frac{dx}{x^2-1} = \int \frac{dx}{(x-1)(x+1)} = \frac{1}{2}(\ln|x-1| - \ln|x+1|) + C = \frac{1}{2}\left(\ln\frac{|x-1|}{|x+1|}\right) + C$, so

$$\begin{aligned}\int_4^\infty \frac{dx}{x^2-1} &= \lim_{b\to\infty}\int_4^b \frac{dx}{x^2-1}\\ &= \lim_{b\to\infty}\frac{1}{2}\left(\ln\frac{|x-1|}{|x+1|}\right)\Big|_4^b\\ &= \lim_{b\to\infty}\left[\frac{1}{2}\ln\left(\frac{b-1}{b+1}\right) - \frac{1}{2}\ln\frac{3}{5}\right]\\ &= -\frac{1}{2}\ln\frac{3}{5} = \frac{1}{2}\ln\frac{5}{3}.\end{aligned}$$

25. The curve has an asymptote at $t = \frac{\pi}{2}$, and so the area integral is improper there.

$$\text{Area} = \int_0^{\frac{\pi}{2}} \frac{dt}{\cos^2 t} = \lim_{b\to\frac{\pi}{2}}\int_0^b \frac{dt}{\cos^2 t} = \lim_{b\to\frac{\pi}{2}} \tan t\Big|_0^b,$$

which diverges. Therefore the area is infinite.

27. (a) $\int \frac{1}{z^2-z}\,dz = \int \frac{1}{(z-1)z}\,dz = \ln|z-1| - \ln|z| + C$ by V- 26 of the integral table.

(b)

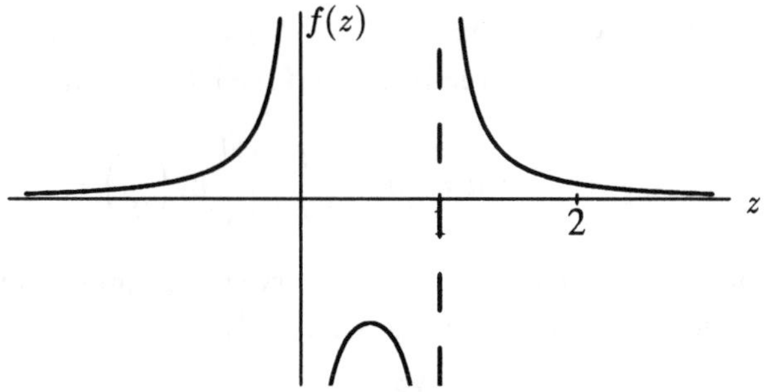

$f(z)$

z

2

The function is undefined at $z = 0$ and $z = 1$.

(c) $\int_3^\infty \frac{1}{z^2-z}\,dz = \lim_{b\to\infty} \int_3^b \frac{1}{z^2-z}\,dz = \lim_{b\to\infty}(\ln|z-1| - \ln|z|)\Big|_3^b = \lim_{b\to\infty}\left[(\ln\frac{|b-1|}{|b|}) - \ln\frac{2}{3})\right] = -\ln\frac{2}{3} = \ln\frac{3}{2}$.

(d) $\int_1^3 \frac{1}{z^2-z}\,dz = \lim_{b\to 1^+} \int_b^3 \frac{1}{z^2-z}\,dz = \lim_{b\to 1^+} \ln\frac{|z-1|}{|z|}\Big|_b^3 = \lim_{b\to 1^+}(\ln\frac{2}{3} - \ln\frac{b-1}{b})$. As $b \to 1^+, \ln\frac{b-1}{b} \to -\infty$, so the limit and the integral diverge.

(e) If $\int_a^x \frac{dz}{z^2-z}$ is not improper, then

$$\begin{aligned}\int_a^x \frac{dz}{z^2-z} &= (\ln|z-1| - \ln|z|)\Big|_a^x \\ &= \ln|x-1| - \ln|x| - (\ln|a-1| - \ln|a|) \\ &= \ln\frac{|x-1||a|}{|x||a-1|}.\end{aligned}$$

This formula won't work if the integral is improper. In fact, if 0 or 1 is between a and x, the integral diverges, as it did in (d). Thus we must have $a, x < 0, 0 < a, x < 1$, or $a, x > 1$ for the formula to work. (We assume that a and x are numbers, not $\pm\infty$.)

29. (a) The factor $\ln x$ grows slowly enough not to change the convergence or divergence of the integral, although it will change what it converges or diverges to. This, in some sense, is the moral of our story. On to the justification.

The integral is always improper, because $\ln x$ is not defined for $x = 0$. Integrating by parts (or, alternatively, the integral table) yields

$$\begin{aligned}\int_0^e x^p \ln x\,dx &= \lim_{a\to 0+} \int_a^e x^p \ln x\,dx \\ &= \lim_{a\to 0+} \left(\frac{1}{p+1}x^{p+1}\ln x - \frac{1}{(p+1)^2}x^{p+1}\right)\Big|_a^e \\ &= \lim_{a\to 0+} \left[\left(\frac{1}{p+1}e^{p+1} - \frac{1}{(p+1)^2}e^{p+1}\right)\right. \\ &\quad \left.- \left(\frac{1}{p+1}a^{p+1}\ln a - \frac{1}{(p+1)^2}a^{p+1}\right)\right].\end{aligned}$$

If $p < -1$, then as $a \to 0^+$, $a^{p+1} \to \infty$ and $\ln a \to -\infty$ so the limit diverges. If $p > -1$, then it's easy to see that $a^{p+1} \to 0$ as $a \to 0$. Also, $a^{p+1}\ln a \to 0$. This second one isn't so easy to see. It's true because if we let $t = \frac{1}{a}$ then

$$\lim_{a\to 0+} a^{p+1}\ln a = \lim_{t\to\infty} \left(\frac{1}{t}\right)^{p+1} \ln\left(\frac{1}{t}\right) = \lim_{t\to\infty} -\frac{\ln t}{t^{p+1}}.$$

This last limit is zero because $\ln t$ grows very slowly, much more slowly than t^{p+1}. So if $p > -1$, the integral converges and equals $e^{p+1}[1/(p+1) - 1/(p+1)^2] = pe^{p+1}/(p+1)^2$. What happens if $p = -1$? Then we get

$$\int_0^e \frac{\ln x}{x}\,dx = \lim_{a\to 0+} \int_a^e \frac{\ln x}{x}\,dx$$

$$= \lim_{a\to 0+} \left.\frac{(\ln x)^2}{2}\right|_a^e$$
$$= \lim_{a\to 0+} \left(\frac{1-(\ln a)^2}{2}\right).$$

Since $\ln a \to -\infty$ as $a \to 0^+$, this limit diverges.

To summarize, $\int_0^e x^p \ln x$ converges for $p > -1$ to the value $pe^{p+1}/(p+1)^2$.

(b) We can reason in a similar fashion for this part.

$$\begin{aligned}
\int_e^\infty x^p \ln x\, dx &= \lim_{b\to\infty} \int_e^b x^p \ln x\, dx \\
&= \lim_{b\to\infty} \left.\left[\frac{1}{p+1}x^{p+1}\ln x - \frac{1}{(p+1)^2}x^{p+1}\right]\right|_e^b \\
&= \lim_{b\to\infty} \left[\left(\frac{1}{p+1}b^{p+1}\ln b - \frac{1}{(p+1)^2}b^{p+1}\right)\right. \\
&\quad \left. - \left(\frac{1}{p+1}e^{p+1} - \frac{1}{(p+1)^2}e^{p+1}\right)\right].
\end{aligned}$$

As before, if $p > -1$, the limit diverges, since b^{p+1} and $\ln b$ both approach ∞ as b does. If $p < -1$, both b^{p+1} and $b^{p+1}\ln b$ approach 0 as $b \to \infty$. (Again, this follows because $\ln x$ grows more slowly than x^{p+1}.) So the value of the integral is $-pe^{p+1}/(p+1)^2$. And, like before, the case $p = -1$ has to be handled separately. For $p = -1$,

$$\int_e^\infty \frac{\ln x}{x}\, dx = \lim_{b\to\infty}\int_e^b \frac{\ln x}{x}\, dx = \lim_{b\to\infty} \left.\frac{(\ln x)^2}{2}\right|_e^b = \lim_{b\to\infty}\left(\frac{(\ln b)^2 - 1}{2}\right).$$

As $b \to \infty$, this limit diverges, so just as before, the integral diverges if $p = -1$.

To summarize, $\int_e^\infty x^p \ln x\, dx$ converges for $p > -1$ to the value $-pe^{p+1}/(p+1)^2$.

31. (a) The electrons are moved from $r = 1$ to $r = a$. The magnitude of the work done is given by

$$W = \int_a^1 \frac{kq_1q_2}{r^2}\, dr = kq_1q_2\left.\left(-\frac{1}{r}\right)\right|_a^1 = kq_1q_2\left(1 - \frac{1}{a}\right)$$

Making a the lower limit of integration makes the work positive.

(b) As $a \to 0$, the work tends to infinity.

33. (a) Using a calculator or a computer, the graph is:

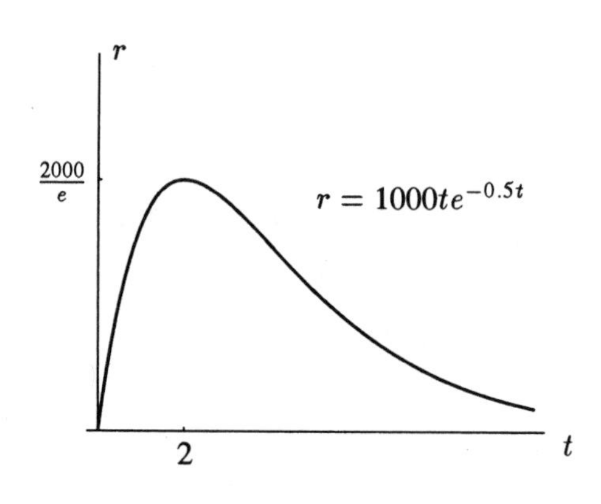

(b) People are getting sick fastest when the rate of infection is highest, i.e. when r has a maximum. Since

$$\begin{aligned} r' &= 1000e^{-0.5t} - 1000(0.5)te^{-0.5t} \\ &= 500e^{-0.5t}(2-t) \end{aligned}$$

this must occur at $t = 2$.

(c) The total number of sick people $= \int_0^\infty 1000te^{-0.5t}\,dt$.
Using integration by parts, with $u = t$, $v' = e^{-0.5t}$:

$$\begin{aligned} \text{Total} &= \lim_{b\to\infty} 1000\left(\frac{-t}{0.5}e^{-0.5t}\Big|_0^b - \int_0^b \frac{-1}{0.5}e^{-0.5t}\,dt\right) \\ &= \lim_{b\to\infty} 1000\left(-2be^{-0.5b} - \frac{2}{0.5}e^{-0.5b}\right)\Big|_0^b \\ &= \lim_{b\to\infty} 1000\left(-2be^{-0.5b} - 4e^{-0.5b} + 4\right) \\ &= 4000 \text{ people.} \end{aligned}$$

(Note: you can see $\lim_{b\to\infty} be^{-0.5b} = 0$ by looking at the graph.)

6.11 Solutions

1. It converges:

$$\int_{50}^\infty \frac{dz}{z^3} = \lim_{b\to\infty}\int_{50}^b \frac{dz}{z^3} = \lim_{b\to\infty}\left(-\frac{1}{2}z^{-2}\Big|_{50}^b\right) = \frac{1}{2}\lim_{b\to\infty}\left(\frac{1}{50^2} - \frac{1}{b^2}\right) = \frac{1}{5000}$$

3. For $\theta \geq 2$, we have $\dfrac{1}{\sqrt{\theta^3+1}} \leq \dfrac{1}{\sqrt{\theta^3}} = \dfrac{1}{\theta^{\frac{3}{2}}}$, and $\displaystyle\int_2^\infty \frac{d\theta}{\theta^{3/2}}$ converges (check by integration), so $\displaystyle\int_2^\infty \frac{d\theta}{\sqrt{\theta^3+1}}$ converges.

5. Since $\frac{1}{1+e^y} \le \frac{1}{e^y} = e^{-y}$ and $\int_0^\infty e^{-y}\,dy$ converges, the integral $\int_0^\infty \frac{dy}{1+e^y}$ is convergent.

7. This integral is convergent because, for $\phi \ge 1$,

$$\frac{2+\cos\phi}{\phi^2} \le \frac{3}{\phi^2},$$

and $\int_1^\infty \frac{3}{\phi^2}\,d\phi = 3\int_1^\infty \frac{1}{\phi^2}\,d\phi$ converges.

9. Since $\frac{1}{\phi^2} \le \frac{2-\sin\phi}{\phi^2}$ for $0 \le \phi \le \pi$, and since $\int_0^\pi \frac{1}{\phi^2}\,d\phi$ diverges, $\int_0^\pi \frac{2-\sin\phi}{\phi^2}\,d\phi$ must diverge.

11. Since $\frac{1}{u+u^2} < \frac{1}{u^2}$ for $u \ge 1$, and since $\int_1^\infty \frac{du}{u^2}$ converges, $\int_1^\infty \frac{du}{u+u^2}$ converges.

13. This improper integral diverges. We expect this because, for large θ, $\frac{1}{\sqrt{\theta^2+1}} \approx \frac{1}{\sqrt{\theta^2}} = \frac{1}{\theta}$ and $\int_1^\infty \frac{d\theta}{\theta}$ diverges. More precisely, for $\theta \ge 1$

$$\frac{1}{\sqrt{\theta^2+1}} \ge \frac{1}{\sqrt{\theta^2+\theta^2}} = \frac{1}{\sqrt{2}\sqrt{\theta^2}} = \frac{1}{\sqrt{2}} \cdot \frac{1}{\theta}$$

and $\int_1^\infty \frac{d\theta}{\theta}$ diverges. (The factor $\frac{1}{\sqrt{2}}$ doesn't affect the divergence.)

15. The integral diverges. Since $e^{-x} \le x$ for $x \ge 1$, we have $\frac{x}{e^{-x}+x} \ge \frac{1}{2}$. So

$$\int_1^\infty \frac{x}{e^{-x}+x}\,dx \ge \int_1^\infty \frac{1}{2}\,dx$$

Thus the integral does not converge, since $\int_1^\infty \frac{1}{2}\,dx$ diverges.

17. Approximating the integral by $\int_0^{10} e^{-x^2}\cos x\,dx$ yields 0.690 to two decimal places. This is a good approximation to the improper integral because the "tail" is small:

$$\int_{10}^{\infty} e^{-x^2}\cos x\,dx \le \int_{10}^{\infty} e^{-x}\,dx = e^{-10},$$

which is very small.

18. To find a, we first calculate $\int_0^{10} e^{-\frac{x^2}{2}}\,dx$. Since $\frac{x^2}{2} \ge x$ for $x \ge 10$, this will differ from $\int_0^{\infty} e^{-\frac{x^2}{2}}\,dx$ by at most

$$\int_{10}^{\infty} e^{-\frac{x^2}{2}}\,dx \le \int_{10}^{\infty} e^{-x}\,dx = e^{-10},$$

which is very small. Using Simpson's rule with 100 intervals (well more than necessary), we find $\int_0^{10} e^{-\frac{x^2}{2}}\,dx \approx 1.253314137$. Thus, since $e^{-\frac{x^2}{2}}$ is even, $\int_{-10}^{10} e^{-\frac{x^2}{2}}\,dx \approx 2.506628274$, and this is extremely close to $\int_{-\infty}^{\infty} e^{-\frac{x^2}{2}}\,dx$.
To find a, we need $\int_{-\infty}^{\infty} ae^{-\frac{x^2}{2}}\,dx = 1$.

$$a = \frac{1}{\int_{-\infty}^{\infty} e^{-\frac{x^2}{2}}\,dx} \approx 0.399 \text{ to three decimal places.}$$

19. (a) If we substitute $w = x - k$ and $dw = dx$, we find

$$\int_{-\infty}^{\infty} ae^{-\frac{(x-k)^2}{2}}\,dx = \int_{-\infty}^{\infty} ae^{-\frac{w^2}{2}}\,dw.$$

This integral is the same as the integral in Problem 18, so the value of a will be the same, namely 0.399.

(b) The answer is the same because $g(x)$ is the same as $f(x)$ in Problem 18 except that it is shifted by k to the right. Since we are integrating from $-\infty$ to ∞, however, this shift doesn't mean anything for the integral.

21. (a) For large x, $\frac{2x^2+1}{4x^4+4x^2-2} \approx \frac{2x^2}{4x^4} = \frac{1}{2x^2}$, and since $\int_1^{\infty} \frac{dx}{2x^2}$ converges, we expect the original integral to converge also. More precisely, we can say that for $x \ge 1$, $2x^2+1 \le 3x^2$ and $4x^4+4x^2-2 \ge 4x^4$, so

$$\int_1^{\infty} \frac{2x^2+1}{4x^4+4x^2-2}dx \le \int_1^{\infty} \frac{3x^2}{4x^4}dx = \frac{3}{4}$$

(b) For large x, $\left(\frac{2x^4+1}{4x^4+4x^2-2}\right)^{\frac{1}{4}} \approx \left(\frac{2x^2}{4x^4}\right)^{\frac{1}{4}} = \frac{1}{2^{\frac{1}{4}}x^{\frac{1}{2}}}$, and since $\int_1^{\infty} \frac{dx}{2^{\frac{1}{4}}x^{\frac{1}{2}}}$ diverges, we expect the original integral will diverge. To show this, notice that for $x \geq 1$, $2x^2+1 \geq 2x^2$ and $4x^4+4x^2-2 \leq 4x^4+4x^4 = 8x^4$, so

$$\int_1^{\infty}\left(\frac{2x^2+1}{4x^4+4x^2-2}\right)^{\frac{1}{4}} dx \geq \int_1^{\infty}\left(\frac{2x^2}{8x^4}\right)^{\frac{1}{4}} dx = \frac{1}{\sqrt{2}}\int_1^{\infty}\frac{dx}{\sqrt{x}}$$

So the original integral diverges.

23. First let's calculate the indefinite integral $\int \frac{dx}{x(\ln x)^p}$. Let $\ln x = w$, then $\frac{dx}{x} = dw$. So

$$\begin{aligned}\int \frac{dx}{x(\ln x)^p} &= \int \frac{dw}{w^p} \\ &= \begin{cases} \ln|w|, & \text{if } p=1 \\ \frac{1}{1-p}w^{1-p}, & \text{if } p \neq 1 \end{cases} \\ &= \begin{cases} \ln|\ln x|, & \text{if } p=1 \\ \frac{1}{1-p}(\ln x)^{1-p}, & \text{if } p \neq 1. \end{cases}\end{aligned}$$

Notice that $\lim_{x\to 1} \ln x = 0$, and $\lim_{x\to 0+} \ln x = -\infty$.

For this integral notice that $\ln 1 = 0$, so the integrand blows up at $x = 1$.

(a) $p = 1$:

$$\int_1^2 \frac{dx}{x\ln x} = \lim_{a\to 1+}\left(\ln|\ln 2| - \ln|\ln a|\right)$$

Since $\ln a \to 0$ as $a \to 1$, $\ln|\ln a| \to -\infty$ as $b \to 1$. So the integral is divergent.

(b) $p < 1$:

$$\begin{aligned}\int_1^2 \frac{dx}{x(\ln x)^p} &= \frac{1}{1-p}\lim_{a\to 1+}\left((\ln 2)^{1-p} - (\ln a)^{1-p}\right) \\ &= \frac{1}{1-p}(\ln 2)^{1-p}.\end{aligned}$$

(c) $p > 1$:

$$\int_1^2 \frac{dx}{x(\ln x)^p} = \frac{1}{1-p}\lim_{a\to 1+}\left((\ln 2)^{1-p} - (\ln a)^{1-p}\right)$$

As $\lim_{a\to 1+}(\ln a)^{1-p} = \lim_{a\to 1+}\frac{1}{(\ln a)^{p-1}} = +\infty$, the integral diverges.

Thus, $\int_1^2 \frac{dx}{x(\ln x)^p}$ is convergent for $p < 1$, divergent for $p \geq 1$.

25. (a)

$$\begin{aligned} b = 3: \quad & n = 20 && \text{gives} && 0.3941 \\ & n = 50 && \text{gives} && 0.3943 \\ & n = 100 && \text{gives} && 0.3943 \\ & n = 200 && \text{gives} && 0.3943 \end{aligned}$$

So $\int_1^3 e^{-x^2/2}\,dx \approx 0.3943$.

$$\begin{aligned} b = 4: \quad & n = 200 && \text{gives} && 0.3976 \\ & n = 500 && \text{gives} && 0.3976 \end{aligned}$$

So $\int_1^4 e^{-x^2/2}\,dx \approx 0.3976$.

$$\begin{aligned} b = 5: \quad & n = 200 && \text{gives} && 0.3977 \\ & n = 500 && \text{gives} && 0.3977 \end{aligned}$$

So $\int_1^5 e^{-x^2/2}\,dx \approx 0.3977$.

$$\begin{aligned} b = 6 \quad & n = 200 && \text{gives} && 0.3977 \\ & n = 500 && \text{gives} && 0.3977 \end{aligned}$$

So $\int_1^6 e^{-x^2/2}\,dx \approx 0.3977$.

Thus we conclude that

$$\int_1^\infty e^{-x^2/2}\,dx \approx 0.3977.$$

(b)

$$\begin{aligned} n = 20: \quad & b = 3 && \text{gives} && 0.3941 \\ & b = 4 && \text{gives} && 0.3970 \\ & b = 5 && \text{gives} && 0.3967 \\ & b = 6 && \text{gives} && 0.3961 \\ & b = 10 && \text{gives} && 0.3925 \\ & b = 100 && \text{gives} && 0.0118 \\ & b = 500 && \text{gives} && 9.3 \times 10^{-39} \\ & b = 1000 && \text{gives} && 0 \end{aligned}$$

The result is zero.

(c) The value of $e^{-x^2/2}$ approaches zero very rapidly. Thus, if b is large (in reality, if $b > 5$), $\int_1^b e^{-x^2/2}\,dx$ is almost the same as $\int_1^\infty e^{-x^2/2}\,dx$ since the tail end of the curve contributes almost nothing to the total area. So, by fixing b and letting n grow large (as you did in part a), you were estimating with increasing accuracy areas which approximated the integral $\int_1^\infty e^{-x^2/2}\,dx$. Now consider what happens when you fix n

and let b grow large (as you did in part b). When $n = 20$, you are approximating the curve by 20 rectangles of equal width. The larger the value of b, the larger the width of each rectangle and the smaller the height (because we're using the midpoint approximation and the integrand is decreasing). The rectangles get shorter so much faster than they get wide that their area approaches zero. Thus, the limit is 0, instead of the value of the integral.

6.12 Solutions

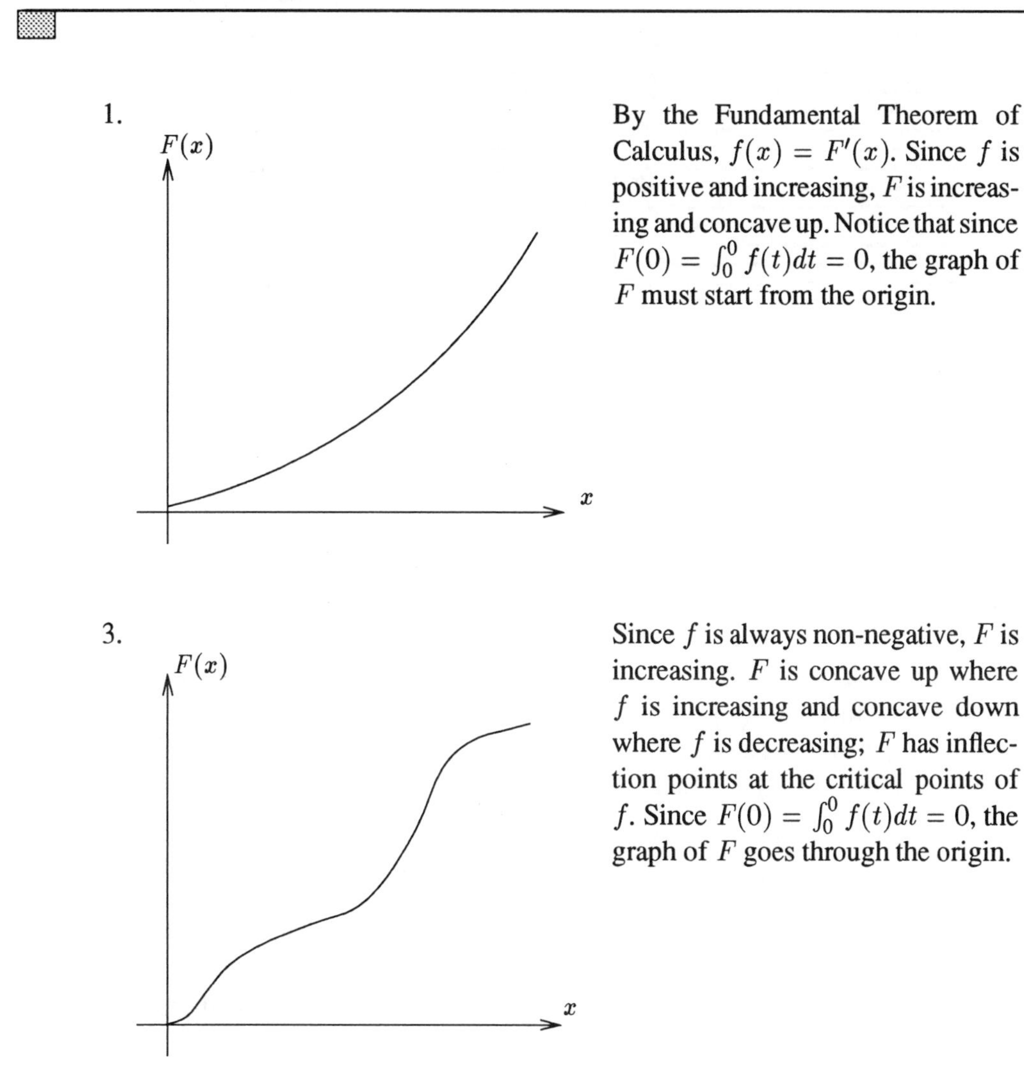

1. By the Fundamental Theorem of Calculus, $f(x) = F'(x)$. Since f is positive and increasing, F is increasing and concave up. Notice that since $F(0) = \int_0^0 f(t)dt = 0$, the graph of F must start from the origin.

3. Since f is always non-negative, F is increasing. F is concave up where f is increasing and concave down where f is decreasing; F has inflection points at the critical points of f. Since $F(0) = \int_0^0 f(t)dt = 0$, the graph of F goes through the origin.

5. (a) Again using 0.00001 as the lower limit, because the integral is improper, gives $\text{Si}(4) = 1.76$, $\text{Si}(5) = 1.55$.

   (b) $\text{Si}(x)$ decreases when the integrand is negative, which occurs when $\pi < x < 2\pi$.

7. $\sqrt{3+\cos(x^2)}$.

9. $\arctan(x^2)$.

11. $\frac{d}{dx}\int_x^1 \ln t\,dt = \frac{d}{dx}\left(-\int_1^x \ln t\,dt\right) = -\ln x$.

13.

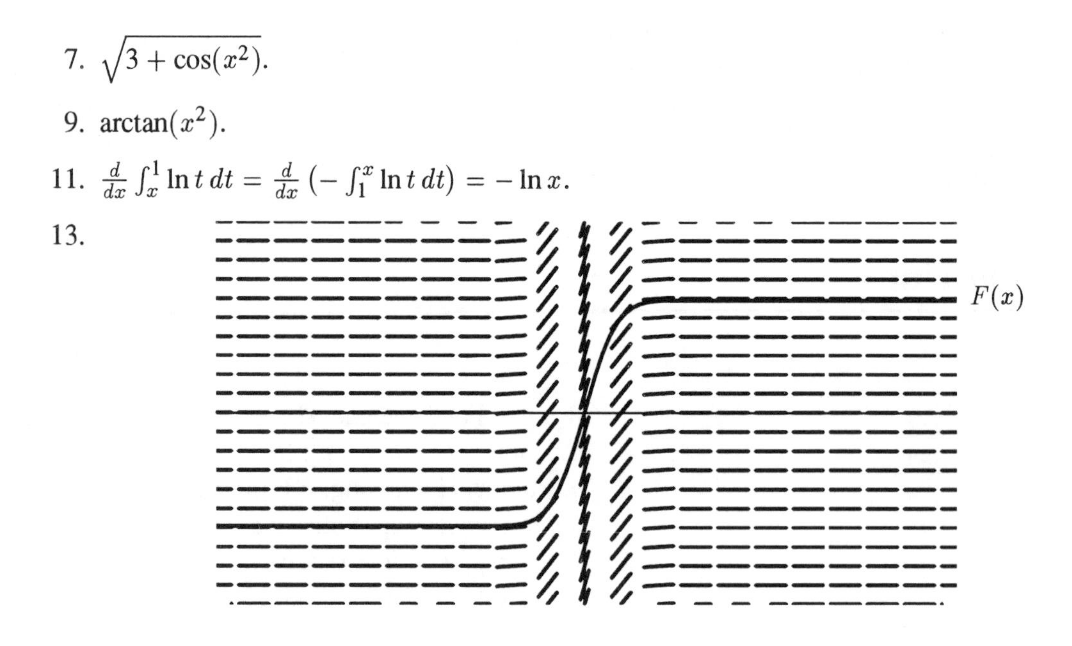

$F(x)$ is an antiderivative of e^{-x^2}; since $F(0) = 0$, we have

$$F(x) = \int_0^x e^{-t^2}\,dt.$$

Thus $\lim_{x\to\infty} F(x) = \lim_{x\to\infty}\int_0^x e^{-t^2}\,dt$. Estimating $\int_0^x e^{-t^2}\,dt$ for large x using Simpson's Rule, we find that $\lim_{x\to\infty}\int_0^x e^{-t^2}\,dt \approx 0.8862269$, or $\lim_{x\to\infty} F(x) \approx 0.8862269$.

15. (a) The most obvious feature of the graph of $y = \sin(x^2)$ is its symmetry about the y-axis. This means the function $g(x) = \sin(x^2)$ is an even function, i.e. for all x, $g(x) = g(-x)$. Since $\sin(x^2)$ is even, its antiderivative F must be odd, that is $F(-x) = -F(-x)$. This can be seen since if $F(t) = \int_0^t \sin(x^2)\,dx$,

$$F(-t) = \int_0^{-t} \sin(x^2)\,dx = -\int_{-t}^{0} \sin(x^2)\,dx = -\int_0^t \sin(x^2)\,dx,$$

since the area from $-t$ to 0 is the same as the area from 0 to t. Thus $F(t) = -F(-t)$ and F is odd.

The second obvious feature of the graph of $y = \sin(x^2)$ is that it oscillates between -1 and 1 with a "period" which goes to zero as $|x|$ increases. This implies that $F'(x)$ alternates between intervals where it is positive or negative, and increasing or decreasing, with frequency growing arbitrarily large as $|x|$ increases. Thus $F(x)$ itself similarly alternates between intervals where it is increasing or decreasing, and concave up or concave down.

Finally, since $y = \sin(x^2) = F'(x)$ passes through (0,0), and $F(0) = 0$, F is tangent to the $x-$axis at the origin.

(b)

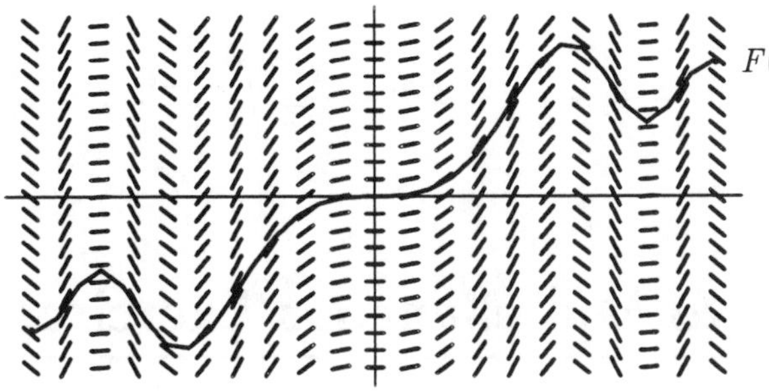

F never crosses the x-axis in the region $x > 0$, and $\lim_{x\to\infty} F(x)$ exists. One way to see these facts is to note that by the Construction Theorem,

$$F(x) = F(x) - F(0) = \int_0^x F'(t)\,dt.$$

So $F(x)$ is just the area under the curve $y = \sin(t^2)$ for $0 \le t \le x$. Now looking at the graph of curve, we see that this area will include alternating pieces above and below the $x-$axis. We can also see that the area of these pieces is approaching 0 as we go further out. So we add a piece, take a piece away, add another piece, take another piece away, and so on. It turns out that this means that the sums of the pieces converge. To see this, think of walking from point A to point B. If you walk almost to B, then go a smaller distance toward A, then a yet smaller distance back toward B, and so on, you will eventually approach some point between A and B. So we can see that $\lim_{x\to\infty} F(x)$ exists. Also, since we always subtract a smaller piece than we just added, and the first piece is added instead of subtracted, we see that we never get a negative sum; thus $F(x)$ is never negative in the region $x > 0$, so $F(x)$ never crosses the $x-$axis there.

17.

$$\begin{aligned}\frac{d}{dx}[x\,\text{erf}(x)] &= \text{erf}(x)\frac{d}{dx}(x) + x\frac{d}{dx}[\text{erf}(x)]\\ &= \text{erf}(x) + x\frac{d}{dx}\left(\frac{2}{\sqrt{\pi}}\int_0^x e^{-t^2}\,dt\right)\\ &= \text{erf}(x) + \frac{2}{\sqrt{\pi}}xe^{-x^2}.\end{aligned}$$

19. Let $w = x^2$, so $dw = 2x\,dx$. This yields

$$\int xe^{-x^4}\,dx = \frac{1}{2}\int e^{-w^2}\,dw = \frac{1}{2}\frac{\sqrt{\pi}}{2}\text{erf}(w) + C = \frac{\sqrt{\pi}}{4}\text{erf}(x^2) + C.$$

21.

$$\frac{1}{\sqrt{2\pi}}\int_{x_1}^{x_2} e^{-\frac{t^2}{2}}\,dt = -\frac{1}{\sqrt{2\pi}}\int_0^{x_1} e^{-\frac{t^2}{2}}\,dt + \frac{1}{\sqrt{2\pi}}\int_0^{x_2} e^{-\frac{t^2}{2}}\,dt = \frac{1}{\sqrt{\pi}}\left[\operatorname{erf}\left(\frac{x_2}{\sqrt{2}}\right) - \operatorname{erf}\left(\frac{x_1}{\sqrt{2}}\right)\right].$$

6.13 Answers to Miscellaneous Exercises for Chapter 6

1. True. The antiderivatives of $3x^2$ are of the form $x^3 + C$. No two such curves intersect. (If they did, then we'd have $x^3 + C = x^3 + C'$, so $C = C'$, but then the curves are the same!)

3. True. $\int_0^x f(t)\,dt = \int_0^1 f(t)\,dt + \int_1^x f(t)\,dt$, and $\int_0^1 f(t)\,dt$ is a constant given f.

5. $f(x) = \frac{1}{2}x^4 - \ln|x| + \frac{1}{x} + C.$

7. $f(x) = e^{\pi}x - \frac{2}{\sqrt{x}} + C.$

9.

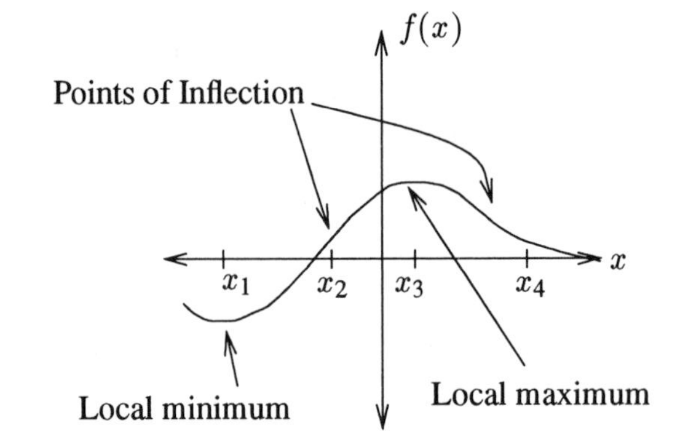

11. (a) Let $w = 2x$, so $w^2 = 4x^2$ and $dw = 2\,dx$.

$$\int \frac{dx}{\sqrt{1-4x^2}} = \frac{1}{2}\int \frac{dw}{\sqrt{1-w^2}} = \frac{1}{2}\arcsin w + C \text{ by VI- 28}$$
$$= \frac{1}{2}\arcsin 2x + C.$$

(b) $\int_0^{\frac{\pi}{8}} \frac{dx}{\sqrt{1-4x^2}} = \frac{1}{2}\arcsin 2x\Big|_0^{\frac{\pi}{8}} \approx 0.45167.$

(c) Simpson's rule with 100 intervals also yields ≈ 0.45167.

13. Since the definition of f is different on $0 \le x \le 1$ than it is on $1 \le x \le 2$, break the definite integral at $x = 1$.

$$\begin{aligned}
\int_0^2 f(t)\,dt &= \int_0^1 f(t)\,dt + \int_1^2 f(t)\,dt \\
&= \int_0^1 t^2\,dt + \int_1^2 (2-t)\,dt \\
&= \frac{t^3}{3}\bigg|_0^1 + \left(2t - \frac{t^2}{2}\right)\bigg|_1^2 \\
&= 1/3 + 1/2 \\
&= 5/6 \approx 0.833
\end{aligned}$$

15. $x \sin x + \cos x + C$.

(Integrate by parts: $u = x$, $v' = \cos x$, or use III- 16 with $p(x) = x$ and $a = 1$ in the integral table.)

17. $\frac{2}{5}(1-x)^{\frac{5}{2}} - \frac{2}{3}(1-x)^{\frac{3}{2}} + C$.

(Let $w = 1 - x$.)

19. We integrate by parts, with $u = y$, $v' = \sin y$. We have $u' = 1$, $v = -\cos y$, and

$$\int y \sin y\,dy = -y\cos y - \int (-\cos y)\,dy = -y\cos y + \sin y + C.$$

Check:

$$\frac{d}{dy}(-y\cos y + \sin y + C) = -\cos y + y\sin y + \cos y = y\sin y.$$

21.

$$\int \frac{dz}{z^2+z} = \int \frac{dz}{z(z+1)} = \int \left(\frac{1}{z} - \frac{1}{z+1}\right) dz = \ln|z| - \ln|z+1| + C.$$

(One can also look this up in the integral table.) Check:

$$\frac{d}{dz}(\ln|z| - \ln|z+1| + C) = \frac{1}{z} - \frac{1}{z+1} = \frac{1}{z^2+z}.$$

23. We integrate by parts, using $u = (\ln x)^2$ and $v' = 1$. Then $u' = 2\frac{\ln x}{x}$ and $v = x$, so

$$\int (\ln x)^2\,dx = x(\ln x)^2 - 2\int \ln x\,dx.$$

But, by the integral table, $\int \ln x\, dx = x\ln x - x + C$. Therefore,

$$\int (\ln x)^2\, dx = x(\ln x)^2 - 2x\ln x + 2x + C.$$

Check:

$$\frac{d}{dx}\left[x(\ln x)^2 - 2x\ln x + 2x + C\right] = (\ln x)^2 + x\frac{2\ln x}{x} - 2\ln x - 2x\frac{1}{x} + 2 = (\ln x)^2.$$

25.

$$\int e^{0.5-0.3t}\, dt = e^{0.5}\int e^{-0.3t}\, dt = -\frac{e^{0.5}}{0.3}e^{-0.3t} + C.$$

27. The integral table yields

$$\begin{aligned}\int \frac{5x+6}{x^2+4}\, dx &= \frac{5}{2}\ln|x^2+4| + \frac{6}{2}\arctan\frac{x}{2} + C\\ &= \frac{5}{2}\ln|x^2+4| + 3\arctan\frac{x}{2} + C.\end{aligned}$$

Check:

$$\begin{aligned}\frac{d}{dx}\left(\frac{5}{2}\ln|x^2+4| + \frac{6}{2}\arctan\frac{x}{2} + C\right) &= \frac{5}{2}\left(\frac{1}{x^2+4}(2x) + 3\frac{1}{1+(x/2)^2}\frac{1}{2}\right)\\ &= \frac{5x}{x^2+4} + \frac{6}{x^2+4} = \frac{5x+6}{x^2+4}.\end{aligned}$$

29. By VI- 30 in the table of integrals, we have

$$\int \sqrt{4-x^2}\, dx = \frac{x\sqrt{4-x^2}}{2} + 4\int \frac{1}{\sqrt{4-x^2}}\, dx.$$

The same table informs us in formula VI- 28 that

$$\int \frac{1}{\sqrt{4-x^2}}\, dx = \arcsin\frac{x}{2} + C.$$

Thus

$$\int \sqrt{4-x^2}\, dx = \frac{x\sqrt{4-x^2}}{2} + 4\arcsin\frac{x}{2} + C.$$

31. Denote $\int \cos^2\theta\, d\theta$ by A. Let $u = \cos\theta$, $v' = \cos\theta$. Then, $v = \sin\theta$ and $u' = -\sin\theta$. Integrating by parts, we get:

$$A = \cos\theta\sin\theta - \int(-\sin\theta)\sin\theta\, d\theta.$$

Employing the identity $\sin^2\theta = 1 - \cos^2\theta$, the equation above becomes:

$$\begin{aligned} A &= \cos\theta\sin\theta + \int d\theta - \int \cos^2\theta\, d\theta \\ &= \cos\theta\sin\theta + \theta - A + C. \end{aligned}$$

Solving this equation for A, and using the identity $\sin 2\theta = 2\cos\theta\sin\theta$ we get:

$$A = \int \cos^2\theta\, d\theta = \frac{1}{4}\sin 2\theta + \frac{1}{2}\theta + C.$$

[Note: An alternate solution would have been to use the identity $\cos^2\theta = \frac{1}{2}\cos 2\theta + \frac{1}{2}$.]

33. Substitute $w = 2x - 6$. Then $dw = 2dx$ and

$$\begin{aligned} \int \tan(2x-6)\, dx &= \frac{1}{2}\int \tan w\, dw \\ &= -\frac{1}{2}\ln|\cos w| + C \text{ by the integral table I- 7} \\ &= -\frac{1}{2}\ln|\cos(2x-6)| + C. \end{aligned}$$

35. $\int_1^3 \ln(x^3)\, dx = 3\int_1^3 \ln x\, dx = 3(x\ln x - x)\Big|_1^3 = 9\ln 3 - 6 \approx 3.8875.$
This matches the approximation given by Simpson's rule with 10 intervals.

37. $\int e^{2x}\sin 2x\, dx = \frac{1}{4}e^{2x}(\sin 2x - \cos 2x) + C$ by II- 8 in the integral table.
Thus $\int_{-\pi}^{\pi} e^{2x}\sin 2x = [\frac{1}{4}e^{2x}(\sin 2x - \cos 2x)]\Big|_{-\pi}^{\pi} = \frac{1}{4}(e^{-2\pi} - e^{2\pi}) \approx -133.8724.$
We get -133.37 using Simpson's rule with 10 intervals. With 100 intervals, we get -133.8724. Thus our answer matches the approximation of Simpson's rule.

39. After the substitution $w = x + 2$, the first integral becomes

$$\int w^{-2}\,dw$$

After the substitution $w = x^2 + 1$, the first integral becomes

$$\frac{1}{2}\int w^{-2}\,dw$$

41. After the substitution $w = 1 - x^2$, the first integral becomes

$$-\frac{1}{2}\int w^{-1}\,dw.$$

After the substitution $w = \ln x$, the second integral becomes

$$\int w^{-1}\,dw.$$

43. (a) $\ln(1+e^x)+C$. (Let $w = 1 + e^x$.)

(b) $\arctan(e^x) + C$. (Let $w = e^x$, getting $\int \frac{dw}{1+w^2}$, and use formula V- 24 .)

(c) $x - \ln(1+e^x) + C$.
[Note that $\frac{1}{1+e^x} = 1 - \frac{e^x}{1+e^x}$ and use part (a).]

45. Let $C(y)$ =consumption of petroleum from 1991 through the year $1991 + y$. Let $a = 1.02$ and $K = 1.4 \times 10^{20}$. We are told that in the year $1990 + t$, the annual rate of consumption will be Ka^t Joules/year. Thus

$$C(y) = \sum_{t=1}^{y} Ka^t.$$

Since $a > 1$, the function $u = Ka^t$ is increasing and $C(y)$ can be viewed as a right-hand Riemann sum overestimate for a definite integral

$$C(y) \geq \int_0^y Ka^t\,dt = \frac{K}{\ln a}(a^y - 1).$$

Thus we seek y such that

$$\frac{K}{\ln a}(a^y - 1) = 10^{22},$$

or

$$a^y = \frac{10^{22}\ln a}{K} + 1 \approx 2.414.$$

Taking logarithms, we get $y \ln a \approx \ln 2.414$, which gives $y \approx 45$. So in about 45 years, we will run out of petroleum!

47. (a) $f(t) = Q - \frac{Q}{A}t$

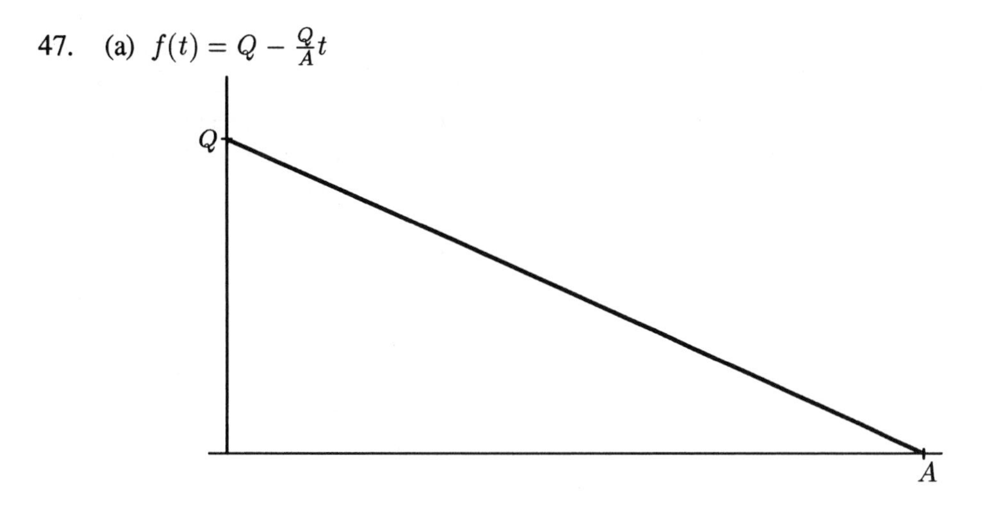

(b)

$$\begin{aligned} \text{Average level} &= \frac{\int_0^A f(t)\,dt}{A} \\ &= \frac{1}{A}\int_0^A (Q - \frac{Q}{A}t)\,dt \\ &= \frac{1}{A}(Qt - \frac{Q}{2A}t^2)\Big|_0^A = \frac{1}{A}(QA - \frac{QA}{2}) = \frac{1}{2}Q. \end{aligned}$$

Graphically, the average will be the y coordinate of the midpoint of the line in the graph above. Thus, the answer should be $\frac{Q+0}{2} = \frac{Q}{2}$.
Common sense tells us that since the rate is constant, the average amount should equal the amount at the midpoint of the interval, which is indeed $\frac{Q}{2}$.

48. The point of intersection of the two curves $y = x^2$ and $y = 6 - x$ is at (2,4). The average height of the shaded area is the average value of the difference between the functions:

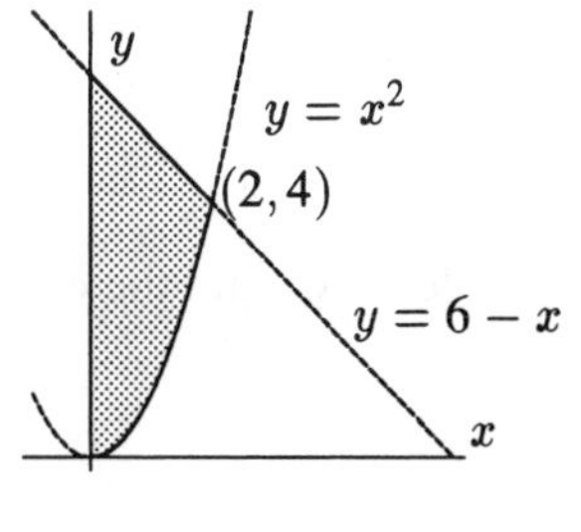

$$\begin{aligned} \frac{1}{(2-0)}\int_0^2 ((6-x) - x^2)\,dx &= (3x - \frac{x^2}{4} - \frac{x^3}{6})\Big|_0^2 \\ &= \frac{11}{3}. \end{aligned}$$

49. The average width of the shaded area is the average value of the horizontal distance between the two functions. If we call this horizontal distance $h(y)$, then the average width is

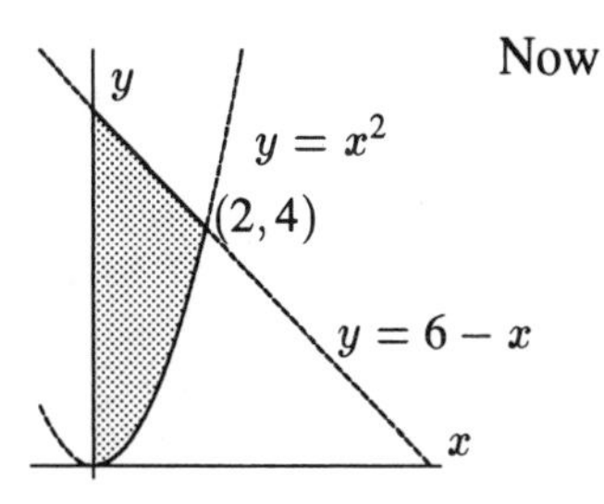

$$\frac{1}{(6-0)}\int_0^6 h(y)\,dy.$$

Now

we could compute this integral if we wanted to, but we don't need to. We can simply note that the integral (without the $\frac{1}{6}$ term) is just the area of the shaded region; similarly, the integral in Problem 48 is *also* just the area of the shaded region. So they are the same. Now we know that our average width is just $\frac{1}{3}$ as much as the average height, since we divide by 6 instead of 2. So the answer is $\frac{11}{9}$.

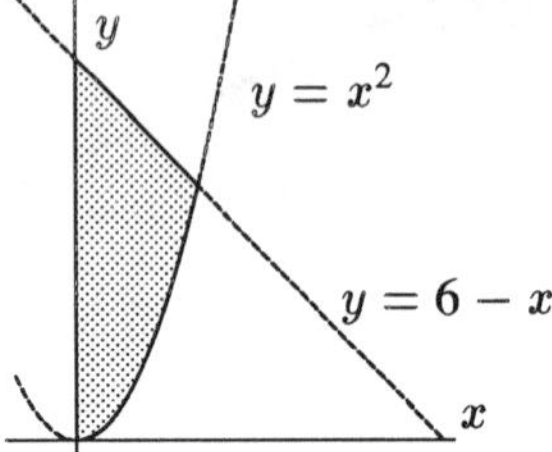

Figure 6.1: Figure for Problems 48–49.

51. $\int_4^\infty \frac{dt}{t^{\frac{3}{2}}}$ should converge, since $\int_1^\infty \frac{dt}{t^n}$ converges for $n > 1$.
We calculate its value.
$\int_4^\infty \frac{dt}{t^{\frac{3}{2}}} = \lim_{b\to\infty} \int_4^b t^{-\frac{3}{2}}\, dt = \lim_{b\to\infty} -2t^{-\frac{1}{2}}\Big|_4^b = \lim_{b\to\infty} (1 - \frac{2}{\sqrt{b}}) = 1.$

53. $\int \frac{dx}{x\ln x} = \ln|\ln x| + C$. (Substitute $w = \ln x,\ dw = \frac{1}{x}\, dx$).
Thus

$$\int_{10}^\infty \frac{dx}{x\ln x} = \lim_{b\to\infty} \int_{10}^b \frac{dx}{x\ln x} = \lim_{b\to\infty} \ln|\ln x|\Big|_{10}^b = \lim_{b\to\infty} \ln(\ln b) - \ln(\ln 10).$$

As $b \to \infty, \ln(\ln b) \to \infty$, so this diverges. Indeed, we would suspect it diverges since for large enough $x, \ln x$ is smaller than x^p, for any $p > 0$. Thus $\int_{10}^\infty \frac{dx}{x\ln x}$ seems like it would act like $\int_{10}^\infty \frac{dx}{x}$, which diverges.

55. We find the exact value:

$$\begin{aligned}
\int_{10}^\infty \frac{1}{4-z^2}\, dz &= \int_{10}^\infty \frac{1}{4-z^2}\, dz \\
&= -\int_{10}^\infty \frac{1}{z^2-4}\, dz \\
&= -\int_{10}^\infty \frac{1}{(z+2)(z-2)}\, dz \\
&= -\lim_{b\to\infty} \int_{10}^b \frac{1}{(z+2)(z-2)}\, dz \\
&= -\lim_{b\to\infty} \frac{1}{4} (\ln|z-2| - \ln|z+2|)\Big|_{10}^b
\end{aligned}$$

$$\begin{aligned} &= -\frac{1}{4}\lim_{b\to\infty}[(\ln|b-2| - \ln|b+2|) - (\ln 8 - \ln 12)] \\ &= -\frac{1}{4}\lim_{b\to\infty}[(\ln\frac{b-2}{b+2}) + \ln\frac{3}{2}] \\ &= -\frac{1}{4}(\ln 1 + \ln 3/2) = -\frac{\ln 3/2}{4} \approx -0.10. \end{aligned}$$

57. Since $\tan\theta$ is between -1 and 1 on the interval $-\pi/4 \le \theta \le \pi/4$, our integral converges. Moreover, since $\tan\theta$ is an odd function, we have

$$\begin{aligned} \int_{-\frac{\pi}{4}}^{\frac{\pi}{4}} \tan\theta\, d\theta = \int_{-\frac{\pi}{4}}^{0} \tan\theta\, d\theta + \int_{0}^{\frac{\pi}{4}} \tan\theta\, d\theta &= -\int_{-\frac{\pi}{4}}^{0} \tan(-\theta)\, d\theta + \int_{0}^{\frac{\pi}{4}} \tan\theta\, d\theta \\ &= -\int_{0}^{\frac{\pi}{4}} \tan\theta\, d\theta + \int_{0}^{\frac{\pi}{4}} \tan\theta\, d\theta = 0. \end{aligned}$$

59. Substituting $w = t+5$, we see that our integral is just $\int_0^{15} \frac{dw}{\sqrt{w}}$. This will converge, since $\int_0^b \frac{dw}{w^p}$ converges for $0 < p < 1$. We find its exact value:

$$\int_0^{15} \frac{dw}{\sqrt{w}} = \lim_{a\to 0+} \int_a^{15} \frac{dw}{\sqrt{w}} = \lim_{a\to 0+} 2w^{\frac{1}{2}}\Big|_a^{15} = 2\sqrt{15}.$$

61. This function is difficult to integrate, so instead we try to compare it with some other function. Since $\frac{\sin^2\theta}{\theta^2+1}$ is always ≥ 0, we see that $\int_0^\infty \frac{\sin^2\theta}{\theta^2+1}\, d\theta \ge 0$.
Also, since $\sin^2\theta \le 1$,

$$\int_0^\infty \frac{\sin^2\theta}{\theta^2+1}\, d\theta \le \int_0^\infty \frac{1}{\theta^2+1}\, d\theta = \lim_{b\to\infty} \arctan\theta\Big|_0^b = \frac{\pi}{2}.$$

Thus $\int_0^\infty \frac{\sin^2\theta}{\theta^2+1}\, d\theta$ converges, and its value is between 0 and $\frac{\pi}{2}$.

63. Since $0 \le \sin x < 1$ for $0 \le x \le 1$, we have

$$\begin{aligned} (\sin x)^{\frac{3}{2}} &< (\sin x) \\ \text{so}\quad \frac{1}{(\sin x)^{\frac{3}{2}}} &> \frac{1}{(\sin x)} \\ \text{or}\quad (\sin x)^{-\frac{3}{2}} &> (\sin x)^{-1} \end{aligned}$$

Thus $\int_0^1 (\sin x)^{-1} dx = \lim_{a\to 0} \ln\left|\frac{1}{\sin x} - \frac{1}{\tan x}\right|_a^1$, which is infinite.
Hence, $\int_0^1 (\sin x)^{-\frac{3}{2}}\, dx$ is infinite.

65. (a) $\int_0^\infty \sqrt{x}e^{-x}\,dx \approx 0.8862269\ldots$ [It turns out that $\int_0^\infty \sqrt{x}e^{-x}\,dx = \frac{\sqrt{\pi}}{2}$]

(b) $\int_1^\infty \ln\left(\frac{e^x+1}{e^x-1}\right)dx = 0.747402\ldots$

67. It will be helpful to have the following:

$$\int x^n e^{-x}\,dx = -x^n e^{-x} + n\int x^{n-1}e^{-x}\,dx.$$

To show this, integrate by parts with $u = x^n$ and $v' = e^{-x}$, so $u' = nx^{n-1}$ and $v = -e^{-x}$. Then

$$\begin{aligned}\int x^n e^{-x} &= x^n(-e^{-x}) - \int nx^{n-1}(-e^{-x})\,dx \\ &= -x^n e^{-x} + \int nx^{n-1}e^{-x}\,dx.\end{aligned}$$

Thus

$$\begin{aligned}\int e^{-x}\,dx &= -e^{-x} + C \\ \int xe^{-x}\,dx &= -xe^{-x} - e^{-x} + C \\ \int x^2e^{-x}\,dx &= -x^2e^{-x} - 2xe^{-x} - 2e^{-x} + C \\ \int x^3e^{-x}\,dx &= -x^3e^{-x} - 3x^2e^{-x} - 6xe^{-x} - 6e^{-x} + C.\end{aligned}$$

Note that $\lim_{x\to\infty}\frac{x^n}{e^x} = 0$ for any positive integer n. We now calculate the improper integrals:

$$\begin{aligned}\int_0^\infty e^{-x}\,dx &= \lim_{b\to\infty}\left[-e^{-x}\right]\Big|_0^b = 1. \\ \int_0^\infty xe^{-x}\,dx &= \lim_{b\to\infty}\left[-xe^{-x} - e^{-x}\right]\Big|_0^b = 1. \\ \int_0^\infty x^2e^{-x}\,dx &= \lim_{b\to\infty}\left[-x^2e^{-x} - 2xe^{-x} - 2e^{-x}\right]\Big|_0^b = 2. \\ \int_0^\infty x^3e^{-x}\,dx &= \lim_{b\to\infty}\left[-x^3e^{-x} - 3x^2e^{-x} - 6xe^{-x} - 6e^{-x}\right] = 6.\end{aligned}$$

It seems that $\int_0^\infty x^n e^{-x}\,dx = n!$.

69. The average value of f on the interval $a \le x \le b$ is

$$\frac{1}{b-a}\int_a^b f(x)\,dx.$$

If f is linear, that is, if $f(x) = k + mx$, then

$$\begin{aligned}
\frac{1}{b-a}\int_a^b f(x)\,dx &= \frac{1}{b-a}\int_a^b (k+mx)\,dx \\
&= \frac{1}{b-a}\left(kx + mx^2/2\right)\Big|_a^b \\
&= \frac{1}{b-a}\left[(kb+mb^2/2) - (ka+ma^2/2)\right] \\
&= \frac{1}{b-a}\left[k(b-a) + \frac{m}{2}(b^2-a^2)\right] \\
&= k + (b+a)m/2 \\
&= \frac{(ma+k)+(mb+k)}{2} = \frac{f(a)+f(b)}{2}.
\end{aligned}$$

This is as we would expect. The area under the curve f from a to b is the area of a trapezoid (see sketch) with bases of length $f(a)$, $f(b)$, and height $b-a$. That area is $(b-a)\frac{f(a)+f(b)}{2}$. The average value of f is that area divided by the length of the interval: that's $b-a$. So the average value is $\frac{f(a)+f(b)}{2}$.

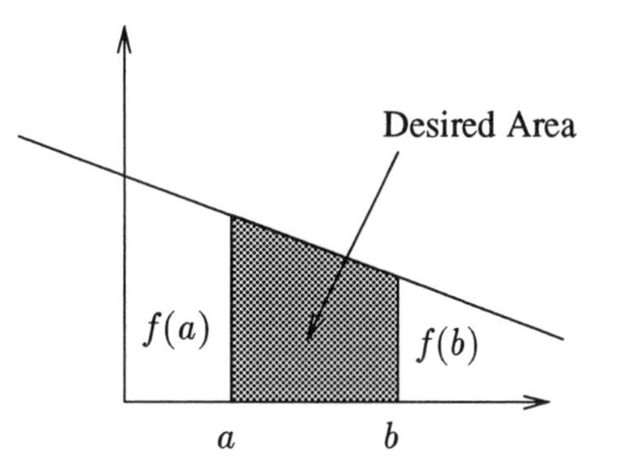

71. Let us assume that SIMP(5) and (10) are both overestimates or both underestimates. Then since SIMP(10) is more accurate and bigger than SIMP(5), they are both underestimates. Since SIMP(10) is 16 times more accurate,

$$I - \text{SIMP}(5) = 16(I - \text{SIMP}(10)).$$

Solving for I, we have

$$I = \frac{16\text{SIMP}(10) - \text{SIMP}(5)}{15} \approx 7.4175.$$

73. (a) $F'(x) = \frac{1}{\ln x}$ by the Fundamental Construction Theorem.

(b) For $x \geq 2$, $F'(x) > 0 \Rightarrow F(x)$ is increasing for $x \geq 2$. $F''(x) = -\frac{1}{x(\ln x)^2} < 0$ for $x \geq 2 \Rightarrow F(x)$ is concave down.

(c) See graph:

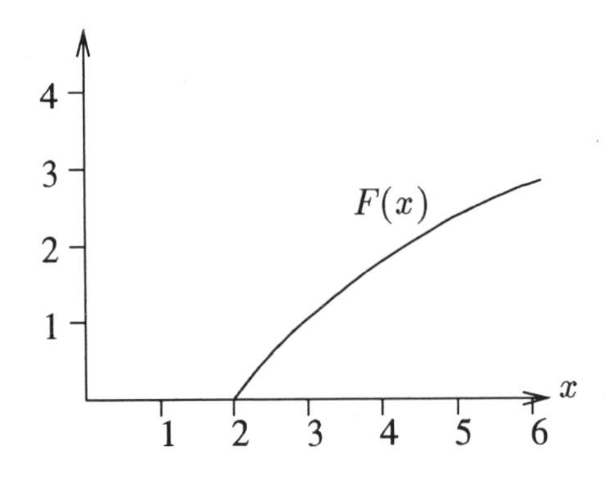

75. We note that $\sin x = \int_0^x \cos t\, dt$ and $\cos x = 1 - \int_0^x \sin t\, dt$. Thus, since $\cos t \leq 1$, we have

$$\begin{aligned} \sin x &\leq \int_0^x 1\, dt \\ &= x. \end{aligned}$$

Now using $\sin t \leq t$, we have

$$\begin{aligned} \cos x &\leq 1 - \int_0^x t\, dt \\ &= 1 - \frac{1}{2}x^2. \end{aligned}$$

Then we just keep going:

$$\begin{aligned} \sin x &\leq \int_0^x (1 - \frac{1}{2}t^2)\, dt \\ &= x - \frac{1}{6}x^3. \end{aligned}$$

Therefore

$$\cos x \leq 1 - \int_0^x (t - \frac{1}{6}t^3)\, dt = 1 - \frac{1}{2}x^2 + \frac{1}{24}x^4.$$

Chapter 7

7.1 Solutions

1. (a) One small box on the graph corresponds to moving at 25 ft/min for 15 seconds. Thus one box corresponds to a distance of 6.25 ft. Estimating the area beneath the velocity curves, we find:
Distance traveled by car 1 $\approx$ 6 boxes = 37.5 ft.
Distance traveled by car 2 $\approx$ 4 boxes = 25 ft.

(b) The two cars will have gone the same distance when the area beneath their velocity curves are equal. Since the two areas overlap, they are equal when the two shaded regions have equal areas, at $t \approx 1.5$ minutes.

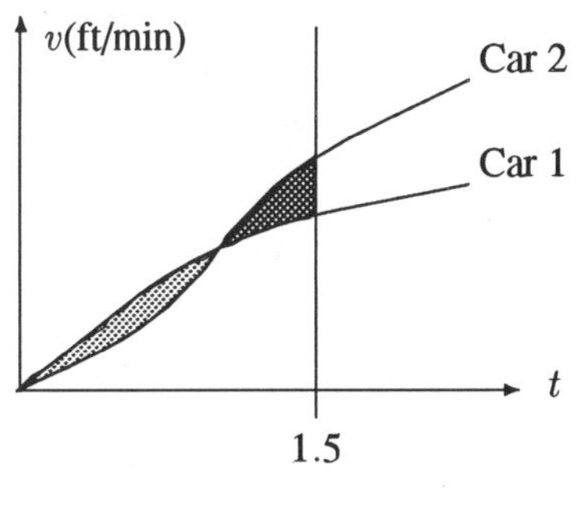

3. (a)

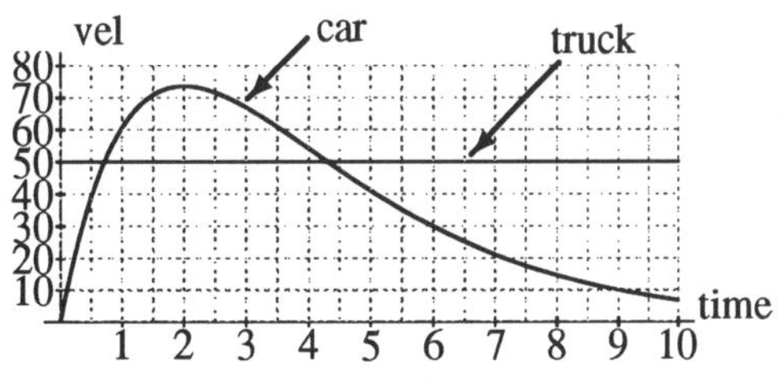

(b) The graphs intersect twice, at about 0.7 hours and 4.3 hours. At each intersection point, the velocity of the car is equal to the velocity of the truck. If the distance apart is written as $d_{\text{car}} - d_{\text{truck}}$, then its derivative is (distance apart)$' = (d_{\text{car}})' - (d_{\text{truck}})' = v_{\text{car}} - v_{\text{truck}}$. At our intersection points $v_{\text{car}} = v_{\text{truck}}$, so (distance apart)$' = 0$. Thus these points are where the distance between the two vehicles is at a local extremum. To see this, note that from the time they start until 0.7 hours later the truck is traveling at a greater velocity than the car, so the truck is ahead of the car and pulling farther away. At 0.7 hours they are traveling at the same velocity, and after 0.7 hours the car is traveling faster than the truck, so that the car begins to gain on the truck. Thus, at 0.7 hours the truck is farther from the car than it is immediately before or after 0.7 hours. Note that this says nothing about what the distance between the two is doing outside of a small interval around 0.7 hours; later, perhaps, the truck could be even farther from the car. This is only a local extremum. Similarly, because the car's velocity is greater than the truck's after 0.7 hours, it will pass the truck and pull away from the truck until 4.3 hours, at which point the two are again traveling at the same velocity. After 4.3 hours the truck travels faster than the car, so that it now gains on the car. Thus, 4.3 hours represents the point where the car is farthest ahead of the truck.

5. Name the slanted line $y = f(x)$. Then the triangle is the region under the line $y = f(x)$ and between the lines $y = 0$ and $x = b$.

   Thus Area of triangle $= \int_0^b f(x)\, dx$.

   $f(x)$ is a line of slope h/b; since this passes through the origin, $f(x) = hx/b$.

   Thus Area of triangle $= \int_0^b \frac{hx}{b}\, dx = \frac{hx^2}{2b}\Big|_0^b = \frac{hb^2}{2b} = \frac{hb}{2}$.

7. (a) The equation $v = 6 - 2t$ implies that $v > 0$ (the car is moving forwards) if $0 \leq t < 3$ and that $v < 0$ (the car is moving backwards) if $t > 3$. When $t = 3$, $v = 0$, so the car is not moving at the instant $t = 3$. The car is decelerating when $|v|$ is decreasing; since v decreases (from 6 to 0) on the interval $0 \leq t < 3$, the car decelerates on that interval. The car accelerates when $|v|$ is increasing, which occurs on the domain $t > 3$.

   (b) Let us take right as the positive direction. The car moves to the right (that is, forward) on the interval $0 \leq t < 3$, so it is furthest to the right at $t = 3$. For all $t > 3$, the car is accelerating to the left. There is no upper bound on the car's distance to the left of its starting point since it is accelerating to the left for all $t > 3$.

   (c) Let $s(t)$ be the position of the car at time t. Then

$$v(t) = \frac{d}{dt}s(t),$$

   so $s(t)$ is an antiderivative of $v(t)$. Thus,

$$s(t) = \int v(t)\, dt = \int (6 - 2t)\, dt = 6t - t^2 + C.$$

   Since the car's position is measured from its starting point, we have $s(0) = 0$, so $C = 0$. Thus, $s(t) = 6t - t^2$.

9. Let $y'(t) = \frac{dy}{dt}$. Then y is the antiderivative of y' such that $y(0) = 0$. We know that

$$y(x) = \int_0^x y'(t)\, dt.$$

   Thus, $y(x)$ is the area under the graph of $\frac{dy}{dt}$ between the y-axis and the line $t = x$ (note: we interpret "area" to be negative if a region lies below the t-axis). We therefore know that $y(t_1) = 2$, $y(t_3) = 2 - 2 = 0$, and $y(t_5) = 2$.

The function y' is positive on the intervals $(0, t_1)$ and (t_3, ∞), so y is increasing on those intervals. y' is negative on the interval (t_1, t_3), so y is decreasing on that interval. y' is increasing on the interval (t_2, t_4), so y is concave up on that interval; y' is decreasing on $(0, t_2)$, so y is concave down there. t_2, the point where the concavity changes, is an inflection point. Finally, since y' is constant on the interval (t_4, ∞), y has a linear graph with positive slope on this interval. $y(t_1) = 2$ is a local maximum, and $y(t_3) = 0$ is a local minimum.

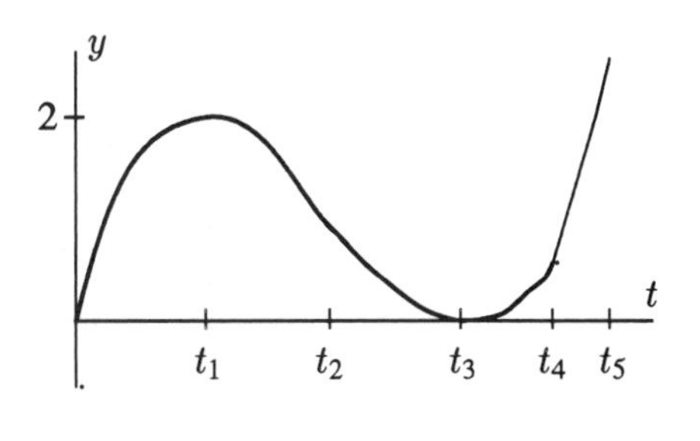

11. Let C be the rate of the flow through the hose. At $t = t_0$, the volume of water in the tank is equal to the area under the lower curve (flow rate through the hose) minus the area under the upper curve (flow rate through the hole) in the region to the left of the vertical line $t = t_0$. Since the overlap of these regions cancels, the volume is also equal to $5C$ (that's the area under the lower curve from $t = 0$ to $t = 5$) minus the region bounded by the upper curve, the horizontal line of height C, the vertical line $t = t_0$, and the vertical line $t = 5$.

If $t_0 > 15$, movement of the vertical line $t = t_0$ doesn't change the area of the latter region, so the difference becomes constant. Thus the volume of water in the tank becomes constant, and the physical system is in a steady state.

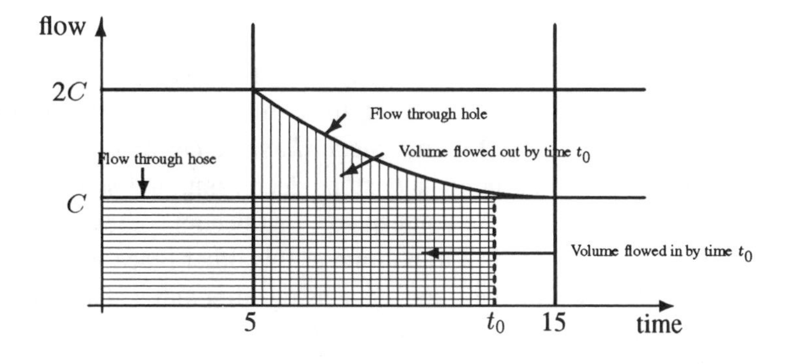

12. (a) $a(T) = \frac{1}{T}\int_0^T \sin t\, dt = \frac{1}{T}(1 - \cos T)$.

(b)

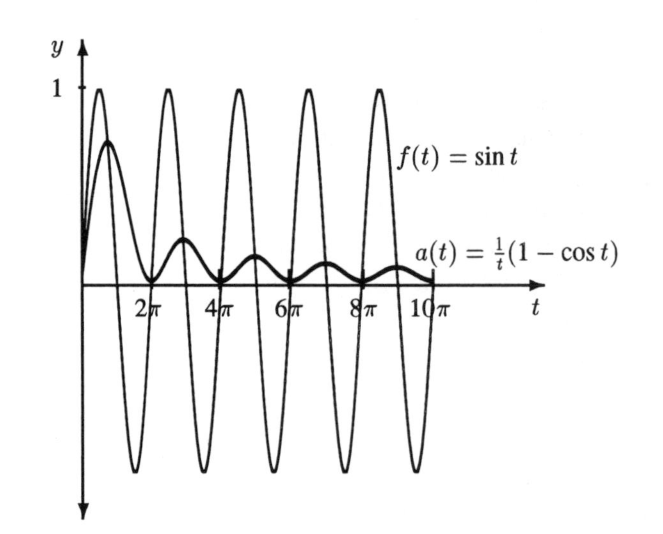

We first observe that $a(T)$ is increasing wherever $a(T)$ is smaller than $f(T)$. To see this, imagine adding a little bit ΔT to T. We choose ΔT so small that on the interval from T to $T+\Delta T$, f is greater than $a(T)$; therefore, $a(T+\Delta T)$ will be

$$\begin{aligned}
a(T+\Delta T) &= \frac{1}{T+\Delta T}\int_0^{T+\Delta T} f(t)\,dt \\
&= \frac{1}{T+\Delta T}\int_0^{T} f(t)\,dt + \frac{1}{T+\Delta T}\int_T^{T+\Delta T} f(t)\,dt \\
&> \frac{1}{T+\Delta T}\left(Ta(T) + \int_T^{T+\Delta T} a(T)\,dt\right) \\
&= \frac{1}{T+\Delta T}(Ta(T) + \Delta T a(T)) = a(T).
\end{aligned}$$

To find out where $a(t)$ is smaller than $f(t)$, we notice some things about $a(t)$. First, $a(t) = \frac{1}{t}(1-\cos t)$ is always nonnegative, because $\cos t \le 1$. So $a(t)$ is decreasing whenever $\sin t < 0 < a(t)$; that is, for t in the third or fourth quadrants. Moreover, $a(t) = 0$ when $t = 2\pi k$ for any integer k, since $\cos 2\pi k = 1$. But there $f(t) = 0$ as well. Therefore, at that point $a(t)$ stops decreasing and starts increasing. It increases until $a(t) = f(t)$ again. As you can see, that occurs somewhere near $2\pi k + \pi$, nearer for bigger k. Also, $a(t)$ is a maximum at this other place where $a(t) = f(t)$, because a is increasing before then and decreasing afterwards. One could find these points numerically by checking where $a'(t) = 0$.

13. (a) We integrate by parts. Let $u = t$, so $u' = 1$, and let $v' = \sin t$, so that $v = -\cos t$. We obtain

$$a(T) = \frac{1}{T}\int_0^T t\sin t\,dt = \frac{1}{T}\left[(-t\cos t)\Big|_0^T + \int_0^T \cos t\,dt\right] = -\cos T + \frac{\sin T}{T}.$$

(b) As in Problem 12, we know that $a(t)$ is increasing when $a(t) < f(t)$ and decreasing when $a(t) > f(t)$. We also know that $a(t)$ is maximal or minimal when $a(t) = f(t)$.

So we really only need to know where the graphs of a and f cross. Looking at the graph tells us that they cross just a little before each multiple of π. One could find the actual points by computing where $a'(t) = 0$.

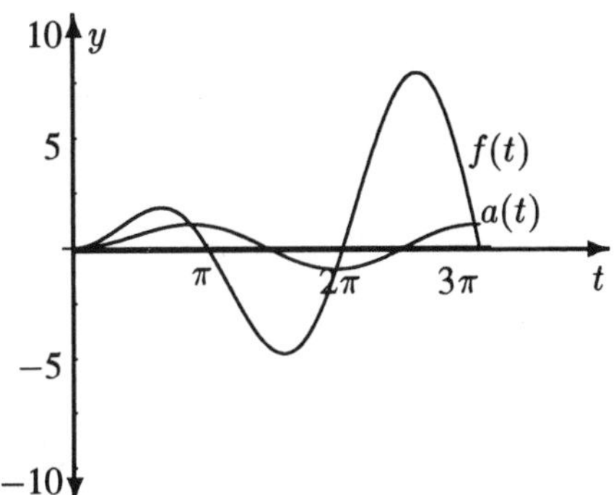

15. (a)

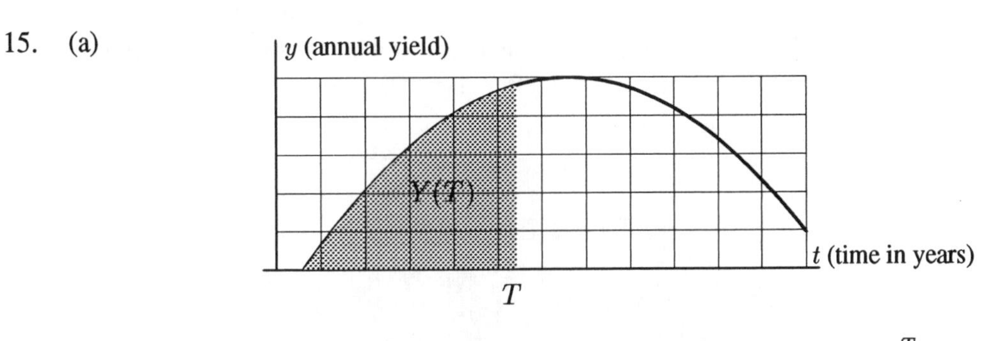

$Y(T)$ is the area of the shaded region in the picture. Thus, $Y(T) = \int_0^T y(t)\,dt$.

(b) Here is a graph of $Y(T)$. Note that the graph of y looked like the graph of a quadratic function. Thus, the graph of Y should look like a cubic, which indeed it does.

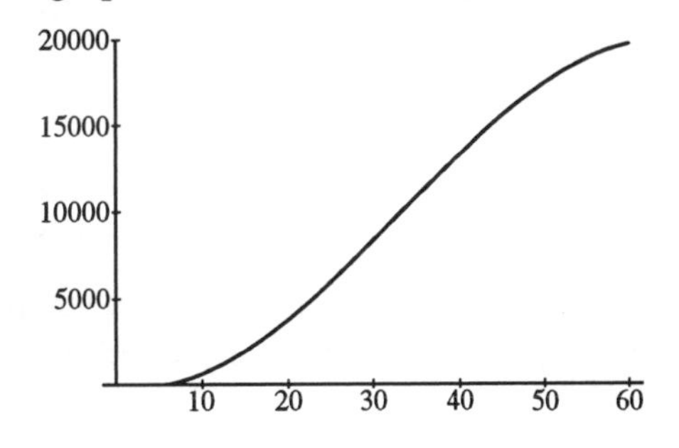

(c) $a(T) = \frac{1}{T}Y(T) = \frac{1}{T}\int_0^T y(t)\,dt$.

(d) i. If the function $a(T)$ takes on its maximum at some point T, then $a'(T) = 0$. Since $a(T) = \frac{1}{T}Y(T)$, we may differentiate using the quotient rule to show that this condition is equivalent to

$$\frac{TY'(T) - Y(T)}{T^2} = 0;$$

or, equivalently, $Y(T) = TY'(T)$.

ii. The expression above may be rewritten in terms of y, giving us

$$T\frac{d}{dT}\int_0^T y(t)\,dt = \int_0^T y(t)\,dt.$$

Simplifying, we obtain $Ty(T) = \int_0^T y(t)\,dt$, or, equivalently,

$$y(T) = \frac{1}{T}\int_0^T y(t)\,dt = a(T),$$

which is our desired condition on $y(T)$.
To find the value of T which satisfies $Ty(T) = Y(T)$, notice that $Y(T)$ is the area under the curve from 0 to T, and that $Ty(T)$ is the area of a rectangle of height $y(T)$. Thus we want the area under the curve to be equal to the area of the rectangle, or $A = B$ in the figure below. This happens when $T \approx 50$ years. In other words, the orchard should be cut down after about 50 years.

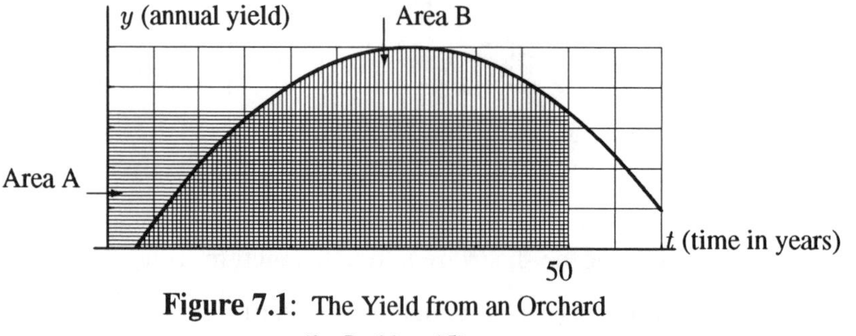

Figure 7.1: The Yield from an Orchard
(for Problem 15)

7.2 Solutions

1. (a) Suppose we choose an x, $0 \le x \le 2$. If Δx is a small fraction of a meter, then the density of the rod is approximately $\rho(x)$ anywhere from x to $x + \Delta x$ meters from the left end of the rod.

   The mass of the rod from x to $x + \Delta x$ meters is therefore $\approx \rho(x)\Delta x$.

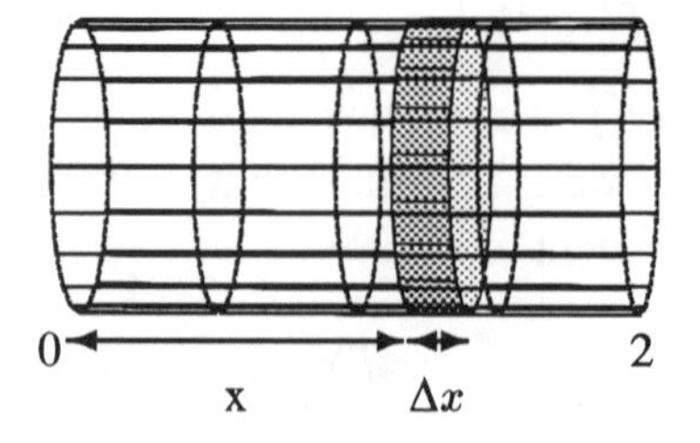

   (b) The definite integral is $M = \int_0^2 \rho(x)\,dx = \int_0^2 (2+6x)\,dx = (2x+3x^2)\Big|_0^2 = 16$ grams.

3. (a)

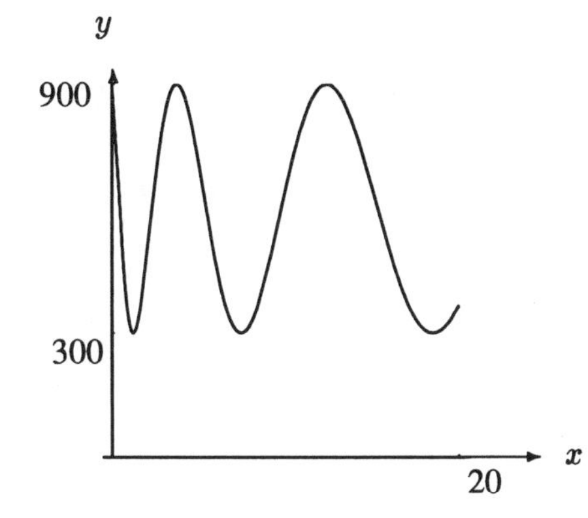

(b) Suppose we choose an $x, 0 \leq x \leq 20$. We approximate the density of the number of the cars between x and $x + \Delta x$ miles as $p(x)$ cars per mile. Therefore the number of cars between x and $x + \Delta x$ is approximately $p(x)\Delta x$.
If we slice the 20 mile strip into N slices, we get that the total number of cars is $C \approx \sum_{i=1}^{N} p(x_i)\Delta x = \sum_{i=1}^{N} \left[600 + 300\sin(4\sqrt{x_i + 0.15})\right] \Delta x$, where $\Delta x = 20/N$. (This is a right-hand approximation; the corresponding left-hand approximation is $\sum_{i=0}^{N-1} p(x_i)\Delta x$.)

(c) As $N \to \infty$, the Riemann sum above approaches the integral

$$C = \int_0^{20} (600 + 300\sin 4\sqrt{x + 0.15})\, dx.$$

If we approximate the integral using one of our approximation methods (like Simpson's rule) we find $C \approx 11513$.
We can also find the integral exactly as follows:

$$\begin{aligned}
C &= \int_0^{20} (600 + 300\sin 4\sqrt{x + 0.15})\, dx \\
&= \int_0^{20} 600\, dx + \int_0^{20} 300\sin 4\sqrt{x + 0.15}\, dx \\
&= 12000 + 300\int_0^{20} \sin 4\sqrt{x + 0.15}\, dx.
\end{aligned}$$

Let $w = \sqrt{x + 0.15}$, so $x = w^2 - 0.15$ and $dx = 2w dw$. Then

$$\begin{aligned}
\int_{x=0}^{x=20} \sin 4\sqrt{x + 0.15}\, dx &= 2\int_{w=\sqrt{0.15}}^{w=\sqrt{20.15}} w\sin 4w\, dw, \\
&\quad \text{using integral table formula 15} \\
&= 2\left[-\frac{1}{4}w\cos 4w + \frac{1}{16}\sin 4w\right]\Bigg|_{\sqrt{0.15}}^{\sqrt{20.15}} \\
&\approx -1.624
\end{aligned}$$

Using this, we have $C \approx 12000 + 300(-1.624) \approx 11513$, which matches our numerical approximation.

5. (a) Partition $0 \le r \le 8$ into eight subintervals of width $\Delta r = 1$ mile. Note that $r_i = i$. Let y_i = # of people living in the ith subinterval and y = the total population. Then $y_i \approx \rho(r_i)A_i$, where A_i = area in the ith subinterval. $A_i \approx 2\pi r_i \Delta r/2 = \pi r_i = \pi i$. So $y_i \approx \rho(i) \cdot \pi i$, and the total population will be approximately

$$y \approx \sum_{i=0}^{7} \pi i \rho(i) = \pi[0(75)+1(75)+2(67.5)+3(60)+4(52.5)+5(45)+6(37.5)+7(30)]$$

$$y \approx 3.96 \text{ million people.}$$

(b) We expect our estimate to be an underestimate for several reasons. First, intuitively, the fact that the first term in our sum is 0, when we know the population in the first mile is some positive number, leads us to believe we are underestimating. Second, the function we are approximating ($\pi r \rho(r)$), although not increasing over the whole interval $0 \le r \le 8$, is primarily increasing. We thus expect our left-hand sum to be an underestimate. Finally, A_i is actually $\pi(i + \frac{1}{2})$, not πi. (Check this.) Our underestimate in the areas (the A_i's) also causes our result to be an underestimate.

(c) For r between 1 and 8, $\rho(r) = 75 - 7.5(r-1) = 82.5 - 7.5r$. Assuming the population density is constant for $0 \le r \le 1$, the number of people living within the first mile is $\frac{\pi}{2}(75)$ thousand. In the next seven miles, the total population is approximately

$$\int_1^8 \pi r(82.5 - 7.5r)dr = \left[\frac{82.5\pi}{2}r^2 - \frac{7.5\pi}{3}r^3\right]_1^8 \approx 4.15 \text{ million.}$$

Then $y \approx \left(\frac{\pi}{2}\frac{(75)}{1000} + 4.15\right)$ million ≈ 4.27 million.

7. (a) Orient the rectangle in the coordinate plane in such a way that the side referred to in the problem — call it S — lies on the y-axis from $y = 0$ to $y = 5$, as shown in the figure. We may subdivide the rectangle into strips of width Δx and length 5. If the left side of a given strip is a distance x away from S (i.e., the y-axis), its density 2 is $1/(1+x^4)$. If Δx is small enough, the density is approximately constant on the entire strip – i.e., the density of the whole strip is about $1/(1+x^4)$. The mass of the strip is just the density times the area, or $5\Delta x/(1+x^4)$. Thus the mass of the whole rectangle is approximated by the left Riemann sum

$$\sum_{k=0}^{n-1} \frac{5\Delta x}{1 + x_k^4},$$

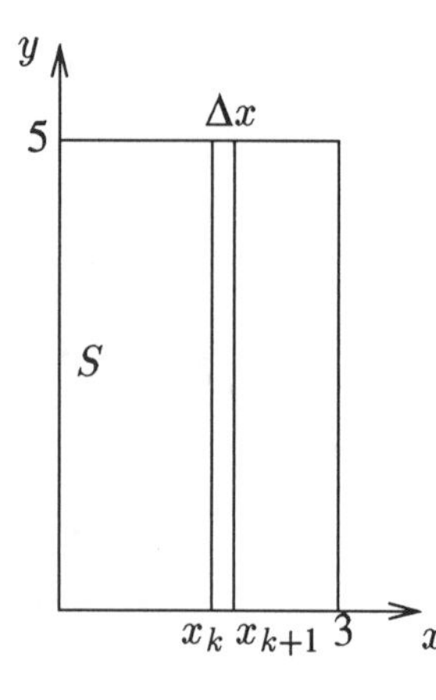

Figure 7.2: Slicing the Rectangle

where $0 = x_0 < x_1 < x_2 < \ldots < x_{n-1} < x_n = 3$ and $\Delta x = x_k - x_{k-1}$. (Since the density function $1/(1 + x^4)$ is strictly decreasing on the entire interval, the left

Riemann sum overestimates the mass.) If we had used the right hand side of each strip in approximating its density, we would have obtained the right Riemann sum

$$\sum_{k=1}^{n} \frac{5\Delta x}{1+x_k^4},$$

which is an underestimate of the mass.

(b) The exact mass of the rectangle is obtained by letting $\Delta x \to 0$ in the Riemann sums above, giving us the integral

$$\int_0^3 \frac{5\,dx}{1+x^4}.$$

Since it is not easy to find an antiderivative of $5/(1+x^4)$, we evaluate this integral numerically. Because the integrand is decreasing, we know that the value of the integral is between the left and right hand sums. For $n = 256$, the right sum is 5.46 and the left sum is 5.52; since both of these quantities are 5.5 to one decimal place, then the exact mass must be 5.5 to one decimal place as well.

9. First we rewrite the chart, listing the density with the corresponding distance from the center of the Earth (x km below the surface is equivalent to $6370 - x$ km from the center):

This gives us spherical shells whose volumes are $\frac{4}{3}\pi(r_i^3 - r_{i+1}^3)$ for any two consecutive distances from the origin. We will assume that the density of the Earth is increasing with depth. Therefore, the average density of the ith shell is between D_i and D_{i+1}, the densities at top and bottom of shell i. So $\frac{4}{3}\pi D_{i+1}(r_i^3 - r_{i+1}^3)$ and $\frac{4}{3}\pi D_i(r_i^3 - r_{i+1}^3)$ are upper and lower bounds for the mass of the shell.

i	x_i	$r_i = 6370 - x_i$	D_i
0	0	6370	3.3
1	1000	5370	4.5
2	2000	4370	5.1
3	2900	3470	5.6
4	3000	3370	10.1
5	4000	2370	11.4
6	5000	1370	12.6
7	6000	370	13.0
8	6370	0	13.0

To get a rough approximation of the mass of the Earth, we don't need to use all the data. Let's just use the densities at $x = 0$, 2900, 5000 and 6370 km. Calculating an upper bound on the mass,

$$M_U = \frac{4}{3}\pi[13.0(1370^3-0^3)+12.6(3470^3-1370^3)+5.6(6370^3-3470^3)]\cdot 10^{15} \approx 7.29\times 10^{27} \text{ g}.$$

The factor of 10^{15} may appear unusual. Remember the radius is given in kilometers and the density is given in g/cm^3, so we must convert kilometers to centimeters: 1 km $= 10^5$ cm , so 1 km$^3 = 10^{15}$ cm^3.

The lower bound is

$$M_L = \frac{4}{3}\pi[12.6(1370^3-0^3)+5.6(3470^3-1370^3)+3.3(6370^3-3470^3)]\cdot 10^{15} \approx 4.05\times 10^{27} \text{ g}.$$

Here, our upper bound is just under 2 times our lower bound.

Using all our data, we can find a more accurate estimate. The upper and lower bounds are

$$M_U = \frac{4}{3}\pi \sum_{i=0}^{7} D_{i+1}(r_i^3 - r_{i+1}^3) \cdot 10^{15} \text{ g}$$

and

$$M_L = \frac{4}{3}\pi \sum_{i=0}^{7} D_i(r_i^3 - r_{i+1}^3) \cdot 10^{15} \text{ g}.$$

We have

$$\begin{aligned}
M_U &= \frac{4}{3}\pi[4.5(6370^3 - 5370^3) + 5.1(5370^3 - 4370^3) + 5.6(4370^3 - 3470^3)\\
&+ 10.1(3470^3 - 3370^3) + 11.4(3370^3 - 2370^3) + 12.6(2370^3 - 1370^3)\\
&+ 13.0(1370^3 - 370^3) + 13.0(370^3 - 0^3)] \cdot 10^{15} \text{ g}\\
&\approx 6.50 \times 10^{27} \text{ g}.
\end{aligned}$$

and

$$\begin{aligned}
M_L &= \frac{4}{3}\pi[3.3(6370^3 - 5370^3) + 4.5(5370^3 - 4370^3) + 5.1(4370^3 - 3470^3)\\
&+ 5.6(3470^3 - 3370^3) + 10.1(3370^3 - 2370^3) + 11.4(2370^3 - 1370^3)\\
&+ 12.6(1370^3 - 370^3) + 13.0(370^3 - 0^3)] \cdot 10^{15} \text{ g}\\
&\approx 5.46 \times 10^{27} \text{ g}.
\end{aligned}$$

10. (a) Partition $0 \le h \le 100$ into N subintervals of width $\Delta h = \frac{100}{N}$. The density is taken to be approximately $\rho(h_i)$ on the ith spherical shell, and the volume is approximately the surface area of a sphere of radius $r_e + h_i$ meters times Δh, where $r_e = 6.378 \times 10^6$ meters is the radius of the Earth. If the volume of the ith shell is V_i, then $V_i \approx 4\pi(r_e + h_i)^2\Delta h$, and a left-hand Riemann sum for the total mass is $M \approx \sum_{i=0}^{N-1} 4\pi(r_e + h_i)^2 \times 1.28e^{-0.000124h_i}\Delta h.$

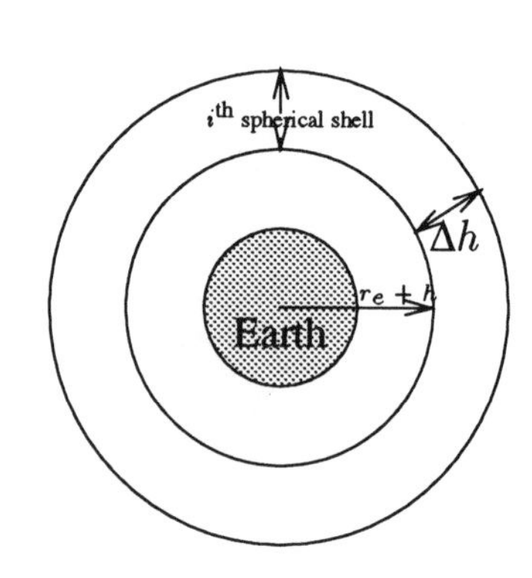

(b) This Riemann sum becomes the integral

$$\begin{aligned}
M &= 4\pi \int_0^{100} (r_e + h)^2 \times 1.28e^{-0.000124h}\, dh\\
&= 5.12\pi \int_0^{100} (r_e + h)^2 e^{-0.000124h}\, dh\\
&= 5.12\pi \int_0^{100} (r_e + h)^2 e^{ah}\, dh,
\end{aligned}$$

where $a = -0.000124$.

Using integration by parts, $\int \underbrace{(r_e + h)^2}_{u} \underbrace{e^{ah}}_{v'} \, dh = \underbrace{(r_e + h)^2}_{u} \underbrace{\frac{e^{ah}}{a}}_{v} - \int \underbrace{2(r_e + h)}_{u'} \underbrace{\frac{e^{ah}}{a}}_{v} \, dh$,

and $\int \underbrace{2(r_e + h)}_{u} \underbrace{\frac{e^{ah}}{a}}_{v'} \, dh = \underbrace{2(r_e + h)}_{u} \underbrace{\frac{e^{ah}}{a^2}}_{v} - \int \underbrace{2}_{u'} \underbrace{\frac{e^{ah}}{a^2}}_{v} \, dh = \frac{2(r_e + h)e^{ah}}{a^2} - \frac{2}{a^3}e^{ah} +$

C.

Thus $\int (r_e + h)^2 e^{ah} \, dh = \frac{(r_e + h)^2 e^{ah}}{a} - \frac{2(r_e + h)e^{ah}}{a^2} + \frac{2e^{ah}}{a^3} + C$

and

$$M = 5.12\pi \left[e^{ah} \left(\frac{(r_e + h)^2}{a} - \frac{2(r_e + h)}{a^2} + \frac{2}{a^3} \right) \right] \Bigg|_0^{100}.$$

Let this bracketed antiderivative be $F(h)$. Substituting back in for a and r_e and evaluating, we get $M = 5.12\pi[F(100) - F(0)] \approx 6.48 \times 10^{16}$ kg.

(c) Assuming the exponential model is correct at all heights, the total mass of the atmosphere is given by $\lim_{H\to\infty} 5.12\pi \int_0^H (r_e + h)^2 e^{ah} \, dh = \lim_{H\to\infty} 5.12\pi[F(H) - F(0)]$. Notice that $\lim_{H\to\infty} F(H) = 0$ because of the factor $e^{-0.000124h}$ in the antiderivative. Therefore the value of the improper integral giving the total mass of the atmosphere is simply $-5.12\pi = \left[\frac{r_e^2}{a} - \frac{2r_e}{a^2} + \frac{2}{a^3} \right] = 5.26 \times 10^{18}$ kg.

11. (a) The ratio of the mass computed in Problem 10, part (b), to the total mass is $\frac{6.48\times10^{16}}{5.26\times10^{18}} \approx 0.0123 = 1.23\%$.

(b) For any height H, we have that the mass of the atmosphere from the Earth's surface to the height H is $5.12\pi(F(H) - F(0)) = 5.12\pi F(H) + 5.26 \times 10^{18}$ kilograms, where $F(H) = [-\frac{(6.378\times10^6+H)^2}{0.000124} - \frac{2(6.378\times10^6+H)}{(0.000124)^2} - \frac{2}{(0.000124)^3}]e^{-0.000124H}$.

Due to the squared term, the bracketed factor is dominated by the first term: $F(H) \approx -\frac{(6.378\times10^6+H)^2}{0.000124}e^{-0.000124H}$. We want to find H such that

$$5.12\pi F(H) + 5.26 \times 10^{18} = (0.9)(5.26 \times 10^{18}).$$

So, $F(H) = -\frac{(6.378\times10^6+H)^2}{0.000124}e^{-0.000124H} = -3.27 \times 10^{16}$. Using our approximation for $F(H)$, we want

$$-\frac{(6.378 \times 10^6 + H)^2}{0.000124}e^{-0.000124H} = -3.27 \times 10^{16}.$$

Consequently, $(6.378 \times 10^6 + H)^2 e^{-0.000124H} = 4.055 \times 10^{12}$. By trial and error, $H \approx 18650$ meters.

7.3 Solutions

1. Vertical slices are circular. Horizontal slices would be similar to ellipses in cross-section, or at least ovals (a word derived from *ovum*, the Latin word for egg).

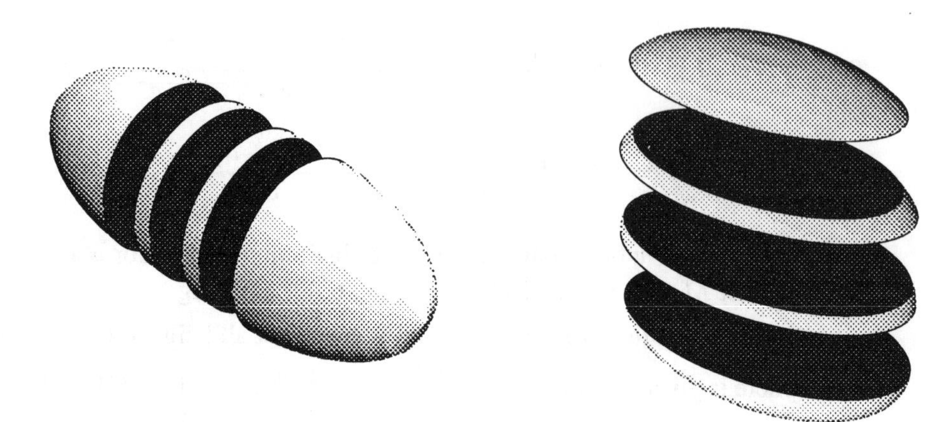

3.

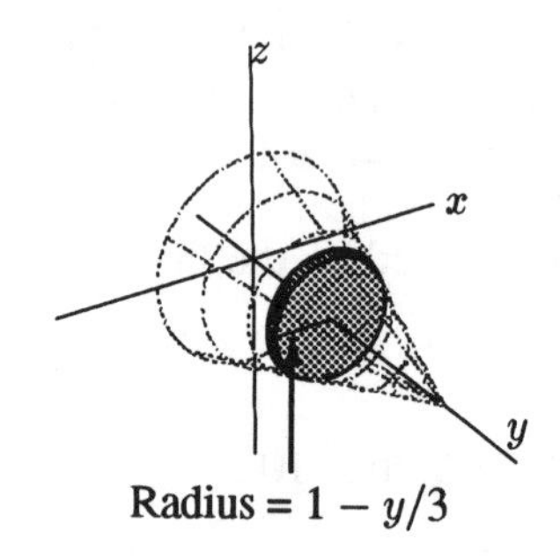

Radius = $1 - y/3$

Slice parallel to the base of the cone, or, equivalently, rotate the line $x = (3 - y)/3$ about the y–axis. (One can also slice the other way.) The the volume V is given by

$$\begin{aligned} V &= \int_{y=0}^{y=3} \pi x^2 \, dy = \int_0^3 \pi \left(\frac{3-y}{3} \right)^2 \, dy \\ &= \int_0^3 \left(1 - \frac{2y}{3} + \frac{y^2}{9} \right) \, dy = \pi \left(y - \frac{y^2}{3} + \frac{y^3}{27} \right) \Big|_0^3 = \pi. \end{aligned}$$

5.

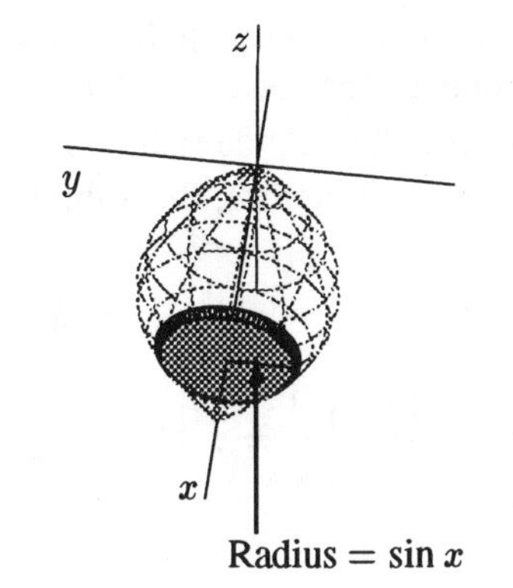

We take slices perpendicular to the x–axis. The Riemann sum for approximating the volume is $\sum \pi \sin^2 x \Delta x$. The volume is the integral corresponding to that sum, namely

$$
\begin{aligned}
V &= \int_0^{\pi} \pi \sin^2 x \, dx \\
&= \pi \left[-\frac{1}{2} \sin x \cos x + \frac{1}{2} x \right] \Bigg|_0^{\pi} = \frac{\pi^2}{2} \approx 4.935.
\end{aligned}
$$

7.

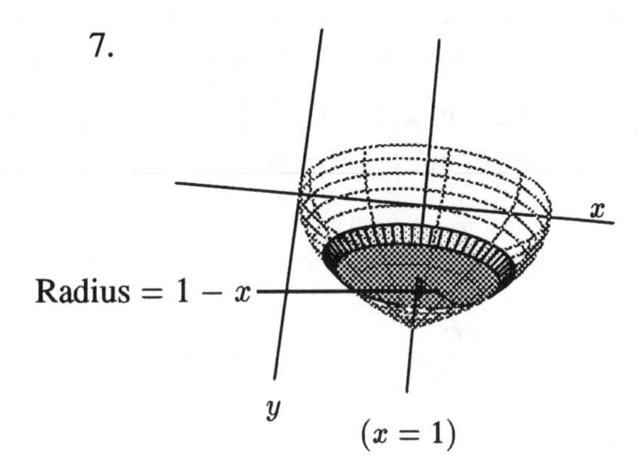

We slice the region perpendicular to the y–axis. The Riemann sum we get is $\sum \pi(1-x)^2 \Delta y = \sum \pi(1-y^2)^2 \Delta y$. So the volume V is the integral

$$
\begin{aligned}
V &= \int_0^1 \pi(1-y^2)^2 \, dy \\
&= \pi \int_0^1 (1 - 2y^2 + y^4) \, dy \\
&= \pi \left(y - \frac{2y^3}{3} + \frac{y^5}{5} \right) \Bigg|_0^1 \\
&= (8/15)\pi \approx 1.68.
\end{aligned}
$$

9.

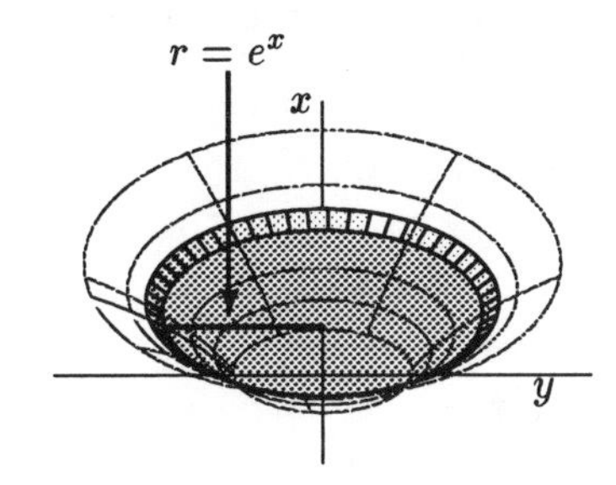

This is the volume of revolution gotten from the rotating the curve $y = e^x$. Take slices perpendicular to the x-axis. The slices will be circles, and we get

$$
\begin{aligned}
V &= \int_{x=0}^{x=1} \pi y^2 \, dx = \pi \int_0^1 e^{2x} \, dx \\
&= \frac{\pi e^{2x}}{2} \Bigg|_0^1 = \frac{\pi(e^2 - 1)}{2} \approx 10.036.
\end{aligned}
$$

11.

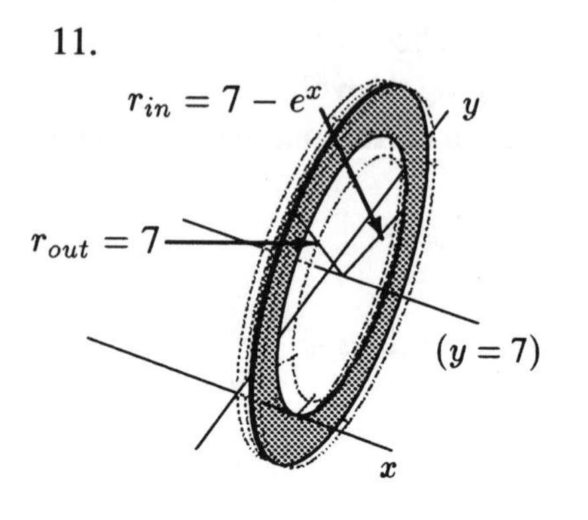

This problem can be done by slicing the volume into washers with planes perpendicular to the axis of rotation, $y = 7$, just like in Example 5. This time the outside radius of a washer is 7, and the inside radius is $7 - e^x$. Therefore, the volume V is

$$
\begin{aligned}
V &= \int_{x=0}^{x=1} [\pi 7^2 - \pi(7 - e^x)^2]\, dx = \pi \int_0^1 (14e^x - e^{2x})\, dx \\
&= \pi \left[14e^x - \frac{1}{2}e^{2x}\right]\Big|_0^1 = \pi\left[14e - \frac{1}{2}e^2 - \left(14 - \frac{1}{2}\right)\right] \\
&\approx 65.54.
\end{aligned}
$$

13.

We slice perpendicular to the x-axis. As stated in the problem, the cross-sections obtained thereby will be semicircles, with radius $\frac{e^x}{2}$. The volume of one semicircular slice is $\frac{1}{2}\pi\left(\frac{e^x}{2}\right)^2 dx$. (Look at the picture.) Adding up the volumes of the slices yields

$$
\begin{aligned}
\text{Volume} &= \int_{x=0}^{x=1} \pi \frac{y^2}{2}\, dx = \frac{\pi}{8}\int_0^1 e^{2x}\, dx \\
&= \frac{\pi e^{2x}}{16}\Big|_0^1 = \frac{\pi(e^2 - 1)}{16} \approx 1.25.
\end{aligned}
$$

15.

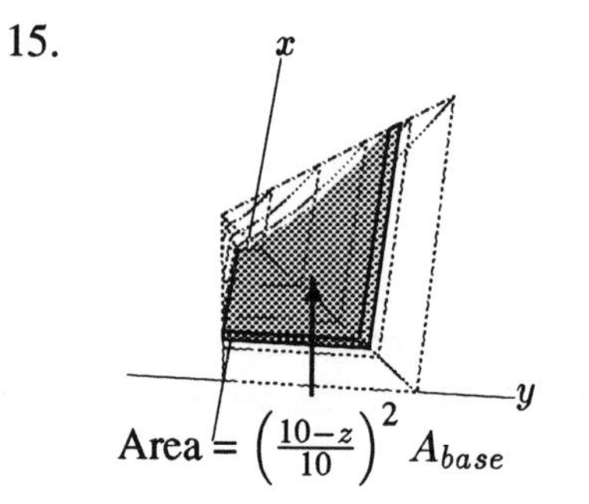

We use the hint. Slice the "pyramid" with planes parallel to the xy-plane. The area of the horizontal cross-section at height z is $\left(\frac{10 - z}{10}\right)^2 \cdot$ (the area of the base), that is, the area of the original figure; this is a formalization of the hint. The area of the base, A, is

$$A = \int_0^1 e^x\, dx = e - 1.$$

So the "pyramid" has volume

$$
\begin{aligned}
\int_{z=0}^{z=10} \left(\frac{10 - z}{10}\right)^2 (e - 1)\, dz &= \frac{e - 1}{100}\int_0^{10} (10 - z)^2\, dz \\
&= \frac{e - 1}{100}\int_0^{10} (z^2 - 20z + 100)\, dz \\
&= \frac{e - 1}{100}\left(\frac{z^3}{3} - 10z^2 + 100z\right)\Big|_0^{10}
\end{aligned}
$$

$$= \frac{10}{3}(e-1) \approx 5.73.$$

Note: Since the base area is $b = e - 1$, and the height is 10, this volume is just $bh/3$, a formula that works for normal pyramids and "odd pyramids".

17. We want to approximate $\int_0^{120} A(h)\,dh$, where h is height, and $A(h)$ represents the cross-sectional area of the tree at height h. Since $A = \pi r^2$ (circular cross-sections), and $c = 2\pi r$, where c is the circumference, we have $A = \pi r^2 = \pi[c/(2\pi)]^2 = c^2/(4\pi)$. We make a table of $A(h)$ based on this:

height (feet)	0	20	40	60	80	100	120
Area (square feet)	53.79	38.52	28.73	15.60	2.865	0.716	0.080

We now form left & right sums using the chart:

$$\begin{aligned} \text{LEFT}(6) &= 53.79 \cdot 20 + 38.52 \cdot 20 + 28.73 \cdot 20 + 15.60 \cdot 20 + 2.865 \cdot 20 + 0.716 \cdot 20 \\ &= 2804.42. \\ \text{RIGHT}(6) &= 38.52 \cdot 20 + 28.73 \cdot 20 + 15.60 \cdot 20 + 2.865 \cdot 20 + 0.716 \cdot 20 + 0.080 \cdot 20 \\ &= 1730.22. \end{aligned}$$

So

$$\text{TRAP}(6) = \frac{\text{RIGHT}(6) + \text{LEFT}(6)}{2} = \frac{2804.42 + 1730.22}{2} = 2267.32 \text{ cubic feet.}$$

19. The problem appears complicated, because we are now working in three dimensions. However, if we take one dimension at a time, we will see the solution is not too difficult.
For example, let's just work at a constant depth, say 0. We apply the trapezoid rule to find the approximate area along the length of the boat. For example, by the trapezoid rule the approximate area at depth 0 from the front of the boat to 10 feet toward the back is $\frac{(2+8)\cdot 10}{2} = 50$. Overall, at depth 0 we have that the area for each length span as follows:

length span:	0–10	10–20	20–30	30–40	40–50	50–60
depth 0	50	105	145	165	165	130

We can fill in the whole chart the same way:

		Area					
length span:		0–10	10–20	20–30	30–40	40–50	50–60
	0	50	105	145	165	165	130
	2	25	60	90	105	105	90
depth	4	15	35	50	65	65	50
	6	5	15	25	35	35	25
	8	0	5	10	10	10	10

Now, to find the volume, we just apply the trapezoid rule to the depths and areas. For example, according to the trapezoid rule the approximate volume as the depth goes from 0 to 2 and the length goes from 0 to 10 is $\frac{(50+25)\cdot 2}{2} = 75$. Again, we fill in a chart:

Volume

length span:	0–10	10–20	20–30	30–40	40–50	50–60
0–2	75	165	235	270	270	220
depth 2–4	40	95	140	170	170	140
span 4–6	20	50	75	100	100	75
6–8	5	20	35	45	45	35

Adding all this up, we find the volume is approximately 2595 cubic feet.

You might wonder what would have happened if we had done our trapezoids along the depth axis first instead of along the length axis. If you try this, you'd find that you come up with the same answers in the volume chart! For the trapezoid rule, it doesn't matter which axis you choose first.

21. Note that this function is actually $x^{3/2}$ in disguise. So

$$\begin{aligned} L &= \int_0^2 \sqrt{1 + \left[\frac{3}{2}x^{\frac{1}{2}}\right]^2}\, dx = \int_{x=0}^{x=2} \sqrt{1 + \frac{9}{4}x}\, dx \\ &= \frac{4}{9}\int_{w=1}^{w=\frac{11}{2}} w^{\frac{1}{2}}\, dw \\ &= \frac{8}{27}w^{\frac{3}{2}}\Big|_1^{\frac{11}{2}} = \frac{8}{27}\left(\left(\frac{11}{2}\right)^{\frac{3}{2}} - 1\right) \approx 3.526, \end{aligned}$$

where we set $w = 1 + \frac{9}{4}x$, so $dx = \frac{4}{9}dw$.

23. Since $y = (e^x + e^{-x})/2$, $y' = (e^x - e^{-x})/2$. The length of the catenary is

$$\begin{aligned} \int_{-1}^{1} \sqrt{1 + (y')^2}\, dx &= \int_{-1}^{1} \sqrt{1 + \left[\frac{e^x - e^{-x}}{2}\right]^2}\, dx = \int_{-1}^{1} \sqrt{1 + \frac{e^{2x}}{4} - \frac{1}{2} + \frac{e^{-2x}}{4}}\, dx \\ &= \int_{-1}^{1} \sqrt{\left[\frac{e^x + e^{-x}}{2}\right]^2}\, dx = \int_{-1}^{1} \frac{e^x + e^{-x}}{2}\, dx \\ &= \left[\frac{e^x - e^{-x}}{2}\right]\Bigg|_{-1}^{1} = e - e^{-1} \approx 2.35. \end{aligned}$$

25. Since the ellipse is symmetric about its axes, we can just find its arclength in the first quadrant and multiply that by 4. To determine the arclength of this section, we first solve for y in terms of x: since $x^2/4 + y^2 = 1$ is the equation for the ellipse, we have $y^2 = 1 - x^2/4$, so $y = \sqrt{1 - x^2/4}$. We also need to find dy/dx; we can do this by differentiating $y^2 = 1 - x^2/4$ implicitly, obtaining $2y dy/dx = -x/2$, whence $dy/dx = -x/(4y)$. We now set up the integral:

$$\begin{aligned} \text{length} &= \int_0^2 \sqrt{1 + \left(-\frac{x}{4y}\right)^2}\, dx = \int_0^2 \sqrt{1 + \frac{x^2}{16y^2}}\, dx \\ &= \int_0^2 \sqrt{1 + \frac{x^2}{16 - 4x^2}}\, dx = \int_0^2 \sqrt{\frac{16 - 3x^2}{16 - 4x^2}}\, dx. \end{aligned}$$

This is an improper integral, since $16 - 4x^2 = 0$ for $x = 2$. Hence, integrating it numerically is somewhat tricky. None of the methods we have studied will work in this case. However, what we *can* do is numerically integrate from 0 to 1.999, and then use a vertical line to approximate the last section. The upper point of the line is $(1.999, 0.016)$; the lower point is $(2, 0)$. The length of the line connecting these two points is $\sqrt{(2 - 1.999)^2 + (0 - 0.016)^2} \approx 0.016$. Approximating the integral from 0 to 1.999 gives 2.391; hence the total arclength of the first quadrant is approximately 2.391 + 0.016 = 2.407. So the arclength of the whole ellipse is about 9.63.

27. (a) If $f(x) = \int_0^x \sqrt{g'(t)^2 - 1}\, dt$, then, by the Fundamental Theorem of Calculus, $f'(x) = \sqrt{g'(x)^2 - 1}$. So the arclength of f from 0 to x is

$$\begin{aligned} \int_0^x \sqrt{1 + (f'(t))^2}\, dt &= \int_0^x \sqrt{1 + (\sqrt{g'(t)^2 - 1})^2}\, dt \\ &= \int_0^x \sqrt{1 + g'(t)^2 - 1}\, dt \\ &= \int_0^x g'(t)\, dt = g(x) - g(0) = g(x). \end{aligned}$$

(b) If g is the arclength of any function f, then by the Fundamental Theorem of Calculus, $g'(x) = \sqrt{1 + f'(x)^2} \geq 1$. So if $g'(x) < 1$, g cannot be the arclength of a function.

(c) We find a function f whose arclength from 0 to x is $g(x) = 2x$. Using part (a), we see that

$$f(x) = \int_0^x \sqrt{(g'(t))^2 - 1}\, dt = \int_0^x \sqrt{2^2 - 1}\, dt = \sqrt{3}x.$$

This is the equation of a line. Does it make sense to you that the arclength of a line segment depends linearly on its right endpoint?

7.4 Solutions

1. Let x be the distance measured from the bottom the tank. It follows that $0 \leq x \leq 10$. To pump a layer of water of thickness Δx at x feet from the bottom, the work needed is $62.4\pi 6^2(20-x)\Delta x$. Therefore, the total work is

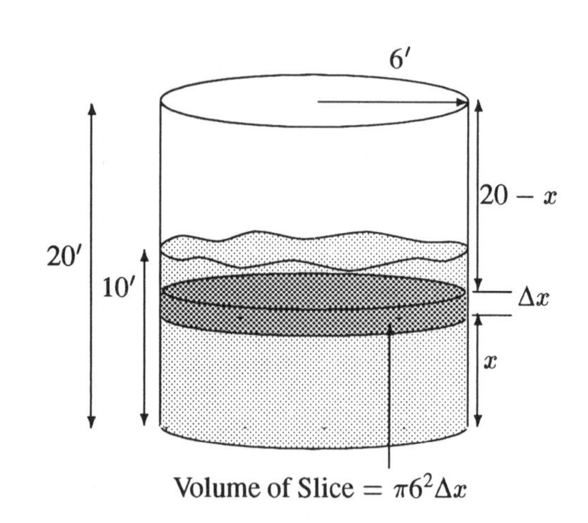

Volume of Slice = $\pi 6^2 \Delta x$

$$\begin{aligned} W &= \int_0^{10} 36 \cdot 62.4\pi(20-x)dx \\ &= 36 \cdot 62.4\pi(20x - \frac{1}{2}x^2)\Big|_0^{10} \\ &= 36 \cdot 62.4\pi(200-50) \\ &\approx 1{,}058{,}591.1 \quad \text{ft-lb.} \end{aligned}$$

3. Consider lifting a rectangular slab of water h feet from the top up to the top. The area of such a slab is $(10)(20) = 200$ square feet; if the thickness is dh, then the volume of such a slab is $200\,dh$ cubic feet. This much water weighs 62.4 pounds per ft^3, so the weight of such a slab is $(200\,dh)(62.4) = 12480\,dh$ pounds. To lift that much water h feet requires $12480h\,dh$ foot-pounds of work. To lift the whole tank, we lift one plate at a time; integrating over the slabs yields

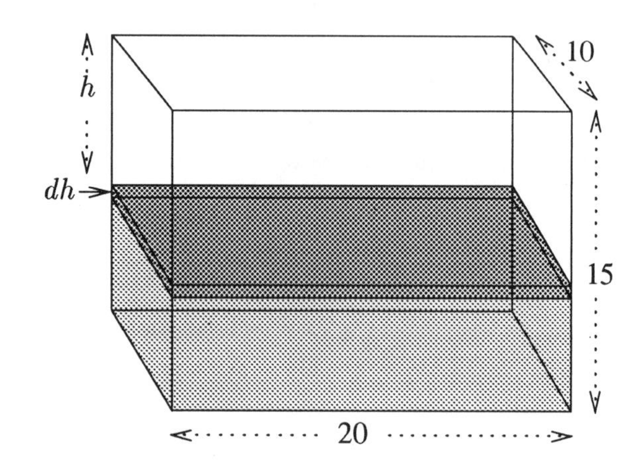

$$\int_0^{15} 12480h\,dh = \frac{12480h^2}{2}\Big|_0^{15} = \frac{12480 \cdot 15^2}{2} = 1{,}404{,}000 \text{ foot-pounds.}$$

5. Let x be the height (in feet) from ground to the cube of ice. It follows that $0 \leq x \leq 100$. At height x, the ice cube weighs $2000 - 4x$ since it's being lifted at a rate 1 ft./min. and it's melting at a rate of 4 lb/min. To lift it Δx more the work required is $(2000-4x)\Delta x$. So the total work done is

$$\begin{aligned} W &= \int_0^{100} (2000-4x)dx \\ &= \left(2000x - 2x^2\right)\Big|_0^{100} \\ &= 2000(100) - 2 \cdot (100)^2 \\ &= 180{,}000 \quad \text{ft-lb.} \end{aligned}$$

7.

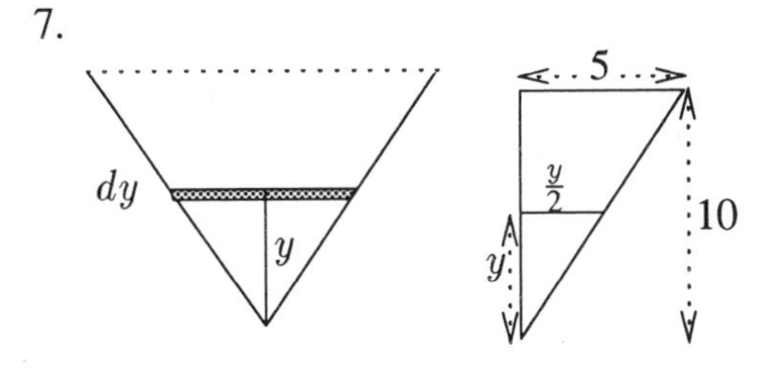

Figure 7.3: A Cold Look at Wine

Cut the wine into horizontal pieces. Let y denote height above the top of the glass stem. In this case, each layer is a cylinder of height dy and base area πx^2, where $x = y/2$ by the similar triangles in Figure 7.3. The volume of the cylinder is $\pi x^2\, dy$, so the mass of the cylinder of wine is $1.2\pi x^2\, dy$ grams. This means that the force due to gravity acting on this mass (ie, its weight) is $980(1.2\pi x^2 dy)$ dynes.

To get that layer of liquid to the top of the straw, we must raise it a height of $15 - y$. Thus the total work is

$$\begin{aligned}\int_0^8 980(1.2\pi x^2)(15-y)\,dy &= 980(1.2\pi)\int_0^8 \frac{y^2}{4}(15-y)\,dy \\ &= 294\pi\int_0^8 (15y^2 - y^3)\,dy = 294\pi\left(5y^3 - \frac{y^4}{4}\right)\Bigg|_0^8 \\ &\approx 1{,}418{,}693 \text{ ergs (about 0.142 joules).}\end{aligned}$$

9. Setting the initial kinetic energy and escape work equal to each other gives

$$\frac{1}{2}mv^2 = \frac{GMm}{R}, \text{ or } \quad v^2 = \frac{2GM}{R}.$$

Since the planet is assumed to be a sphere of radius R and density ρ, we have $M = \rho(\frac{4}{3}\pi)R^3$. Hence

$$v^2 = \frac{2G\rho(\frac{4}{3}\pi)R^3}{R}$$

and therefore

$$v = k\sqrt{\rho}R$$

where $k = \sqrt{\frac{8\pi G}{3}}$. That is, the escape velocity is proportional to R and $\sqrt{\rho}$.

11. (a) Divide the wall into N horizontal strips, each of which is of height Δh. The area of each strip is $1000\Delta h$, and the pressure at depth h_i is $62.4h_i$, so we approximate the pressure on the strip as $1000(62.4h_i)\Delta h$.

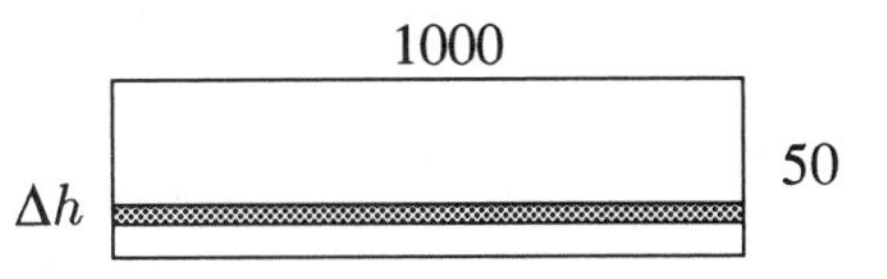

Therefore,

$$\text{Force on the Dam} \approx \sum_{i=0}^{N-1} 1000(62.4h_i)\Delta h.$$

(b) As $N \to \infty$, the Riemann sum becomes the integral, so the force on the dam is

$$\int_0^{50} (1000)(62.4h)\,dh = 62400\frac{h^2}{2}\bigg|_0^{50} = 78{,}000{,}000 \text{ pounds}.$$

13. We need to divide the disk up into circular rings of charge and integrate their contributions to the potential (at P) from 0 to a. These rings, however, are not uniformly distant from the point P. A ring of radius z is $\sqrt{R^2+z^2}$ away from point P (See picture).

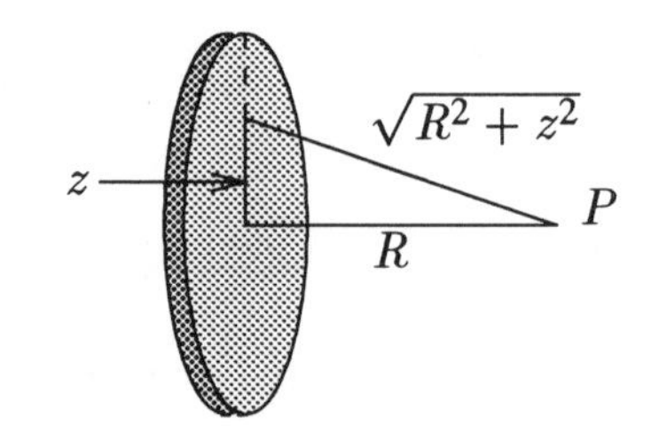

The ring has area $2\pi z\,\Delta z$, and charge $2\pi z\sigma\,\Delta z$. The potential of the ring is then $\dfrac{2\pi z\sigma\,\Delta z}{\sqrt{R^2+z^2}}$ and the total potential at point P is

$$\int_0^a \frac{2\pi z\sigma\,dz}{\sqrt{R^2+z^2}} = \pi\sigma\int_0^a \frac{2z\,dz}{\sqrt{R^2+z^2}}.$$

We make the substitution $u = z^2$. Then $du = 2z\,dz$. We obtain

$$\begin{aligned}\pi\sigma\int_0^a \frac{2z\,dz}{\sqrt{R^2+z^2}} &= \pi\sigma\int_0^{a^2} \frac{du}{\sqrt{R^2+u}} = \pi\sigma(2\sqrt{R^2+u})\bigg|_0^{a^2} \\ &= \pi\sigma(2\sqrt{R^2+z^2})\bigg|_0^a = 2\pi\sigma(\sqrt{R^2+a^2}-R).\end{aligned}$$

15. We slice the record into rings in such a way that every point has approximately the same speed: use concentric circles around the hole. We assume the record is a flat disk of uniform density: since its mass is 50 grams, and has area $\pi(10\text{cm})^2 = 100\pi\text{ cm}^2$, the record has density $\frac{50}{100\pi} = \frac{1}{2\pi}\frac{\text{gm}}{\text{cm}^2}$. So a ring of width dr, having area about $2\pi r\,dr\text{ cm}^2$, has mass approximately $(2\pi r\,dr)(1/2\pi) = r\,dr$ gm. At radius r, the velocity of the ring is

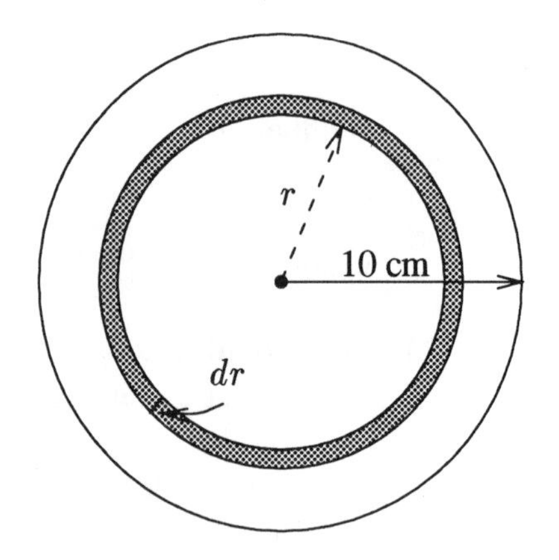

$$33\tfrac{1}{3}\frac{\text{rev}}{\text{min}}\cdot\frac{1\text{ min}}{60\text{ sec}}\cdot\frac{2\pi r\text{ cm}}{1\text{ rev}} = \frac{10\pi r}{9}\frac{\text{cm}}{\text{sec}}.$$

The kinetic energy of the ring is

$$\frac{1}{2}mv^2 = \frac{1}{2}(r\,dr\text{ gm})\left(\frac{10\pi r}{9}\frac{\text{cm}}{\text{sec}}\right)^2 = \frac{50\pi^2 r^3\,dr}{81}\frac{\text{gm}\cdot\text{cm}^2}{\text{sec}^2}.$$

So the kinetic energy of the record, summing the energies of all these rings, is

$$\int_0^{10} \frac{50\pi^2 r^3\,dr}{81} = \frac{25\pi^2 r^4}{162}\Bigg|_0^{10} \approx 15231 \frac{\text{gm}\cdot\text{cm}}{\text{sec}^2} = 15231 \text{ ergs}.$$

16.

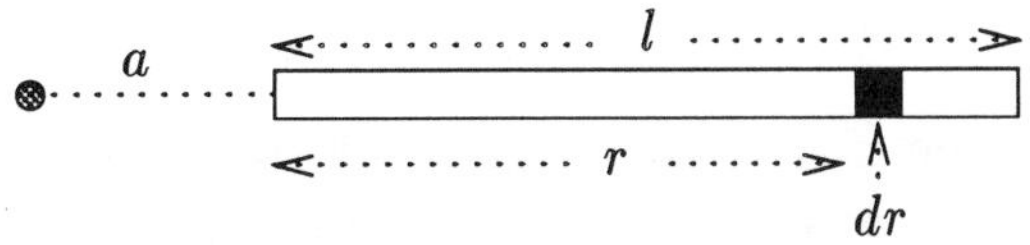

The density of the rod, in mass per unit length, is M/l. So a slice of size dr has mass $\frac{M\,dr}{l}$. It pulls the small mass m with force $Gm\frac{M\,dr}{l}/r^2 = \frac{GmM\,dr}{lr^2}$. So the total gravitational attraction between the rod and point is

$$\begin{aligned}
\int_a^{a+l} \frac{GmM\,dr}{lr^2} &= \frac{GmM}{l}\left(-\frac{1}{r}\right)\Bigg|_a^{a+l} \\
&= \frac{GmM}{l}\left(\frac{1}{a} - \frac{1}{a+l}\right) \\
&= \frac{GmM}{l}\frac{l}{a(a+l)} = \frac{GmM}{a(a+l)}.
\end{aligned}$$

17.

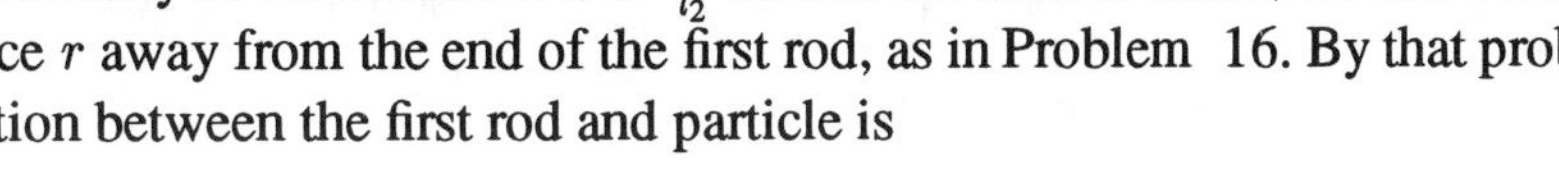

This time, let's split the second rod into small slices of length dr. Each slice is of mass $\frac{M_2}{l_2}\,dr$, since the density of the second rod is $\frac{M_2}{l_2}$. Since the slice is small, we can treat it as a particle at distance r away from the end of the first rod, as in Problem 16. By that problem, the force of attraction between the first rod and particle is

$$\frac{GM_1\frac{M_2}{l_2}\,dr}{(r)(r+l_1)}.$$

So the total force of attraction between the rods is

$$\int_a^{a+l_2} \frac{GM_1\frac{M_2}{l_2}\,dr}{(r)(r+l_1)} = \frac{GM_1M_2}{l_2}\int_a^{a+l_2} \frac{dr}{(r)(r+l_1)}$$

$$\begin{aligned}
&= \frac{GM_1M_2}{l_2}\int_a^{a+l_2}\frac{1}{l_1}\left(\frac{1}{r}-\frac{1}{r+l_1}\right)dr.\\
&= \frac{GM_1M_2}{l_1l_2}\left(\ln|r|-\ln|r+l_1|\right)\Big|_a^{a+l_2}\\
&= \frac{GM_1M_2}{l_1l_2}\left[\ln|a+l_2|-\ln|a+l_1+l_2|-\ln|a|+\ln|a+l_1|\right]\\
&= \frac{GM_1M_2}{l_1l_2}\ln\left[\frac{(a+l_1)(a+l_2)}{a(a+l_1+l_2)}\right]
\end{aligned}$$

This result is symmetric: if you switch l_1 and l_2 or M_1 and M_2, you get the same answer. That means it's not important which rod is "first", and which is "second".

7.5 Solutions

1.

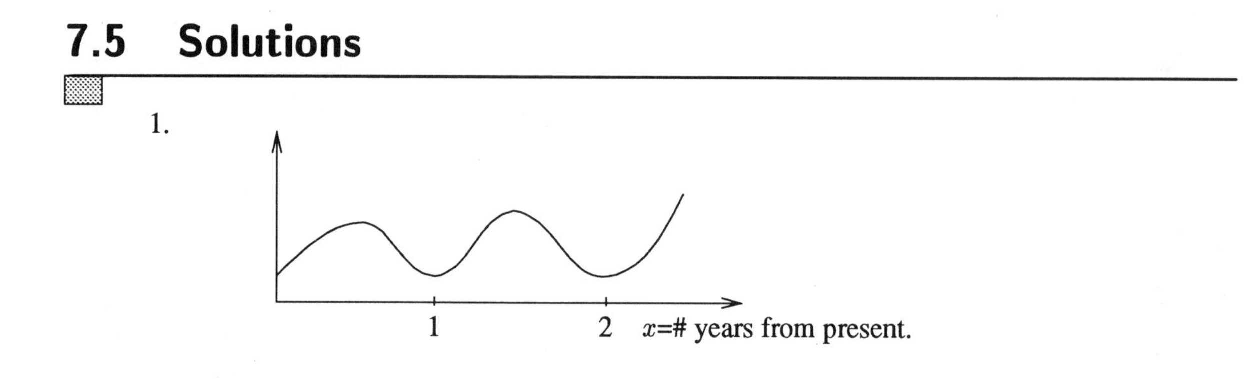

The graph reaches a peak each summer, and a trough each winter. Sunscreen sales are probably increasing from cycle to cycle, as evidenced by the upward trend of the graph. This gradual increase may be due in part to inflation, in part to population growth, and in part to a (predicted) increase in skin cancer rates thanks to the destruction of the ozone layer.

3.

$$\begin{aligned}
\text{Future Value} &= \int_0^{15} 3000e^{.06(15-t)}dt = 3000e^{.9}\int_0^{15} e^{-.06t}dt\\
&= 3000e^{.9}\left(\frac{1}{-.06}e^{-.06t}\right)\Big|_0^{15} = 3000e^{.9}\left(\frac{1}{-.06}e^{-.9}+\frac{1}{.06}e^{0}\right)\\
&\approx \$72{,}980.16
\end{aligned}$$

$$\begin{aligned}
\text{Present Value} &= \int_0^{15} 3000e^{-.06t}dt = 3000\left(-\frac{1}{.06}\right)e^{-.06t}\Big|_0^{15}\\
&\approx \$29{,}671.52.
\end{aligned}$$

There's a quicker way to calculate the present value of the income stream, since the future value of the income stream is (as we've shown) \$72,980.16, the present value of the income stream must be the present value of $72,980.16. Thus,

$$\begin{aligned}\text{Present Value} &= 72{,}980.16(e^{-.06\cdot 15}) \\ &\approx 29{,}671.52,\end{aligned}$$

which is what we got before.

5. (a) Solve for $P(t) = P$.

$$\begin{aligned}100000 &= \int_0^{10} Pe^{0.10(10-t)}dt = Pe\int_0^{10} e^{-0.10t}dt \\ &= \frac{Pe}{-0.10}e^{-0.10t}\bigg|_0^{10} = Pe(-3.678+10) \\ &= P\cdot 17.183\end{aligned}$$

So, $P \approx \$5820$ per year.

(b) To answer this, we'll calculate the present value of $100,000:

$$\begin{aligned}100000 &= Pe^{0.10(10)} \\ P &\approx \$36787.94.\end{aligned}$$

7. (a) Let L be the number of years for the balance to reach $10,000. Since our income stream is $1000 per year, the future value of this income stream should equal (in L years) $10,000. Thus

$$\begin{aligned}10000 &= \int_0^{L} 1000e^{0.05(L-t)}dt = 1000e^{0.05L}\int_0^{L} e^{-0.05t}dt \\ &= 1000e^{0.05L}\left(-\frac{1}{0.05}\right)e^{-0.05t}\bigg|_0^{L} = 20000e^{0.05L}\left(1-e^{-0.05L}\right) \\ &= 20000e^{0.05L} - 20000 \\ \text{so}\quad e^{0.05L} &= \frac{3}{2} \\ L &= 20\ln\left(\frac{3}{2}\right) \\ &\approx 8.11 \text{ years.}\end{aligned}$$

(b) We want

$$10000 = 2000e^{0.05L} + \int_0^{L} 1000e^{0.05(L-t)}dt$$

The first term on the right hand side is the future value of our initial balance. The second term is the future value of our income stream. We want this sum to equal $10,000 in L years. We solve for L:

$$10,000 = 2000e^{0.05L} + 1000e^{0.05L}\int_0^{L} e^{-0.05t}dt$$

$$
\begin{aligned}
&= 2000e^{0.05L} + 1000e^{0.05L}\left(\frac{1}{-0.05}\right)e^{-0.05t}\Big|_0^L \\
&= 2000e^{0.05L} + 20000e^{0.05L}\left(1 - e^{-0.05L}\right) \\
&= 2000e^{0.05L} + 20000e^{0.05L} - 20000
\end{aligned}
$$

So,

$$
\begin{aligned}
22000e^{0.05L} &= 30000 \\
e^{0.05L} &= \frac{30000}{22000} \\
L &= 20\ln\frac{15}{11} \\
L &\approx 6.203 \text{ years.}
\end{aligned}
$$

9. One good way to approach the problem is in terms of present values. In 1980, the present value of Germany's loan was 20 billion DM. Now let's figure out the rate that the Soviet Union would have to give money to Germany to pay off 10% interest on the loan by using the formula for the present value of a continuous stream. Since the Soviet Union sends gas at a constant rate, the rate of deposit, $P(t)$, is a constant c. Since they don't start sending the gas until after 5 years have passed, the present value of the loan is given by:

$$\text{Present Value} = \int_5^\infty P(t)e^{-rt}\,dt.$$

We want to find c so that

$$
\begin{aligned}
20{,}000{,}000{,}000 &= \int_5^\infty ce^{-rt}\,dt = c\int_5^\infty e^{-rt}\,dt \\
&= c\lim_{b\to\infty}\left(-10e^{-0.10t}\right)\Big|_5^b = ce^{-0.10(5)} \approx 6.065c.
\end{aligned}
$$

Dividing, we see that c should be about 3.3 billion DM per year. At 0.10 DM per m^3 of natural gas, the Soviet Union must deliver gas at the constant, continuous rate of about 33 billion m^3 per year.

11.

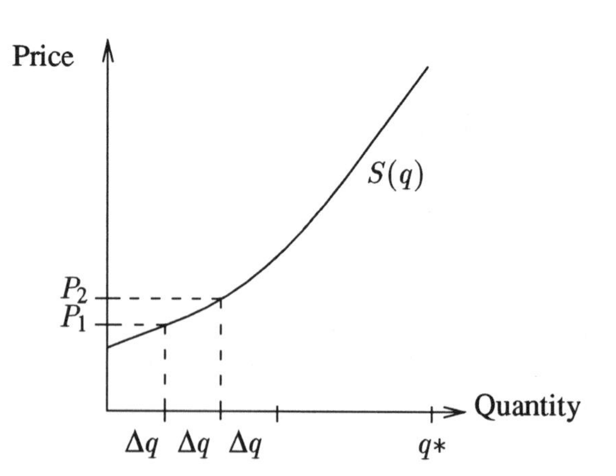

The supply curve, $S(q)$, represents the minimum price p per unit that the suppliers will be willing to supply some quantity q of the good for. If the suppliers have q^* of the good and q^* is divided into subintervals of size Δq, then if the consumers could offer the suppliers for each Δq a price increase just sufficient to induce the suppliers to sell an additional Δq of the good, the consumers' total expenditure on q^* goods would be

$$p_1\Delta q + p_2\Delta q + \cdots = \sum p_i\Delta q.$$

As $\Delta q \to 0$ the Riemann sum becomes the integral $\int_0^{q^*} S(q)\,dq$. Thus $\int_0^{q^*} S(q)\,dq$ is the amount the consumers would pay if suppliers could be forced to sell at the lowest price they would be willing to accept.

13. (a) p^*q^* = the total amount paid for q^* of the good at equilibrium.

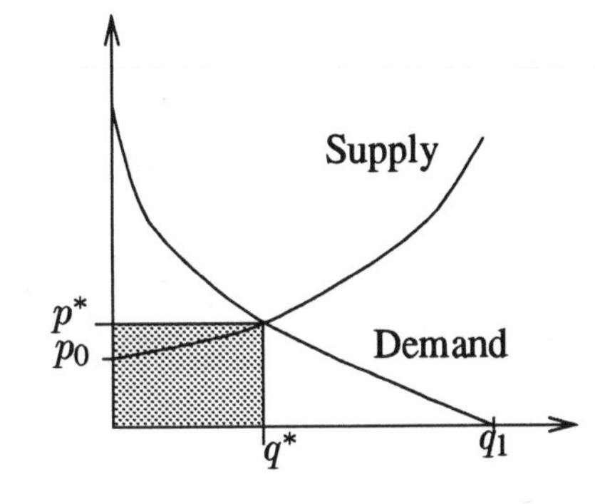

(b) $\int_0^{q^*} D(q)\,dq$ = the maximum consumers would be willing to pay if they had to pay the highest price acceptable to them for each additional unit of the good.

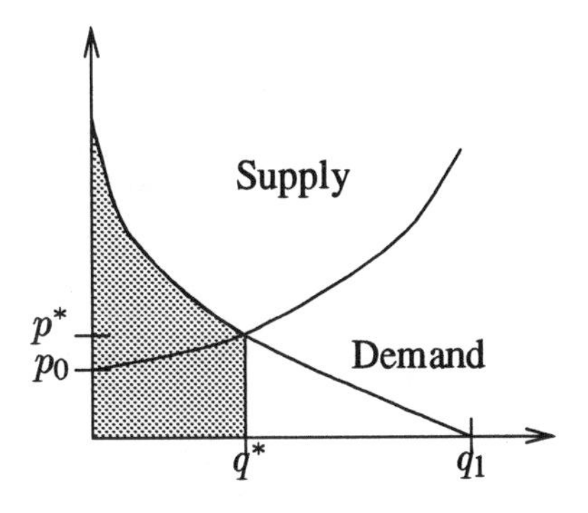

(c) $\int_0^{q^*} S(q)\,dq$ = the minimum suppliers would be willing to accept if they were paid the minimum price acceptable to them for each additional unit of the good.

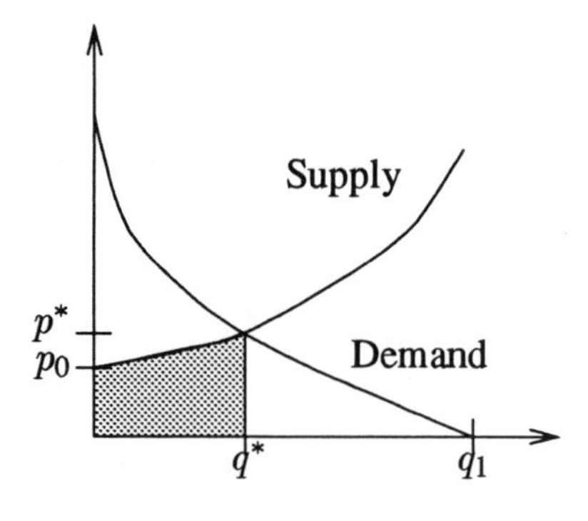

(d) $\int_0^{q^*} D(q)\,dq - p^*q^* =$ consumer surplus.

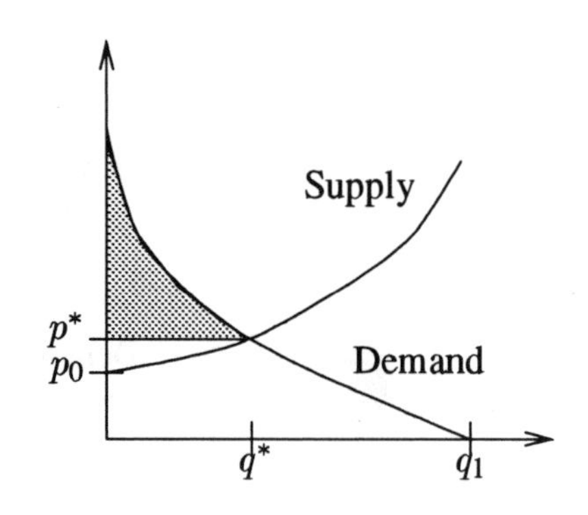

(e) $p^*q^* - \int_0^{q^*} S(q)\,dq =$ producer surplus.

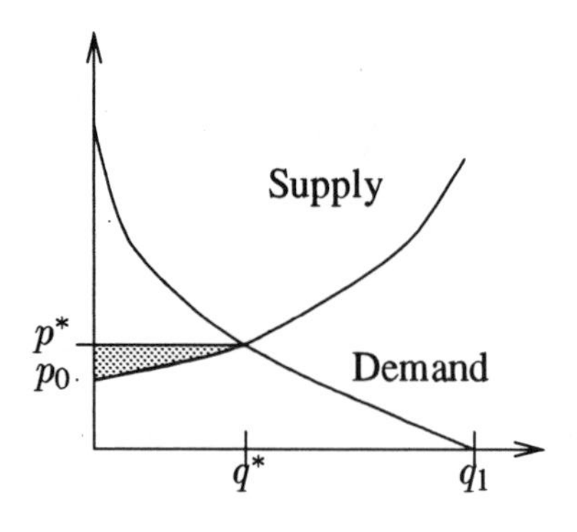

(f) $\int_0^{q^*} (D(q) - S(q))\,dq =$ producer surplus and consumer surplus.

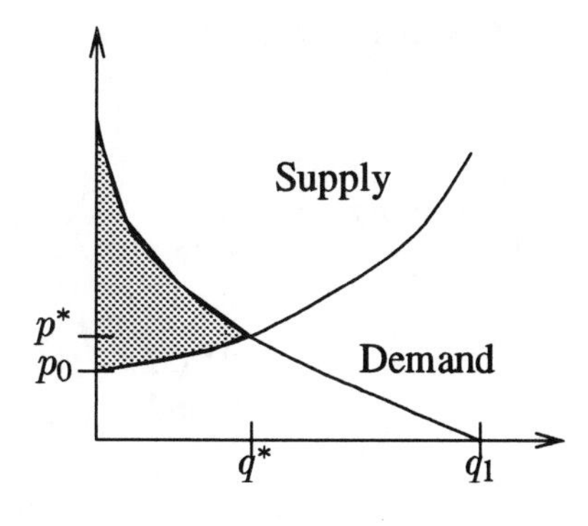

7.6 Solutions

1.

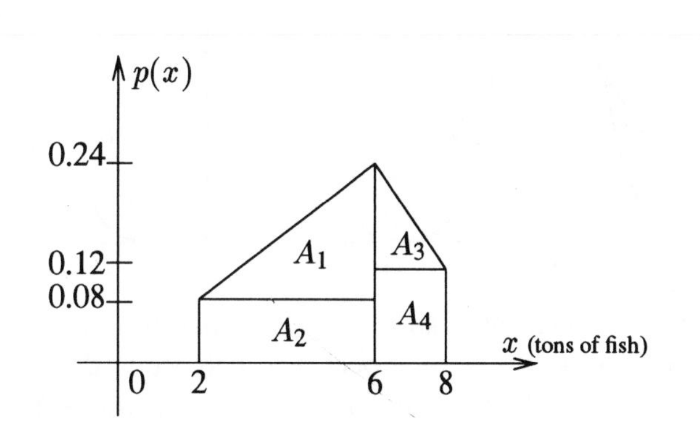

Splitting the figure into four pieces, we see that

$$\begin{aligned}\text{Area under the curve} &= A_1 + A_2 + A_3 + A_4 \\ &= \frac{1}{2}(0.16)4 + 4(0.08) + \frac{1}{2}(0.12)2 + 2(0.12) \\ &= 1.\end{aligned}$$

We expect the area to be 1, since $\int_{-\infty}^{\infty} p(x)\,dx = 1$ for any probability density function, and $p(x)$ is 0 except when $2 \le x \le 8$.

3. (a) Since $\mu = 100$ and $\sigma = 15$:

$$p(x) = \frac{1}{15\sqrt{2\pi}} e^{-\frac{1}{2}\left(\frac{x-100}{15}\right)^2}$$

(b) The fraction of the population with IQ scores between 115 and 120 is (integrating numerically)

$$\begin{aligned}\int_{115}^{120} p(x)\,dx &= \int_{115}^{120} \frac{1}{15\sqrt{2\pi}} e^{-\frac{(x-100)^2}{450}}\,dx \\ &= \frac{1}{15\sqrt{2\pi}} \int_{115}^{120} e^{-\frac{(x-100)^2}{450}}\,dx \\ &\approx 0.067 = 6.7\% \text{ of the population.}\end{aligned}$$

5. (a) i.

ii.

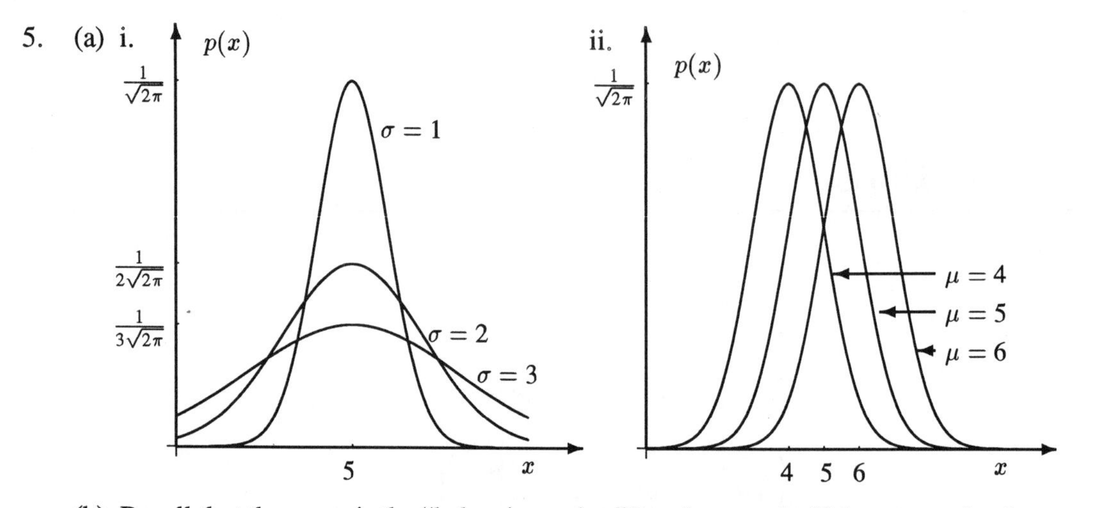

(b) Recall that the mean is the "balancing point." In other words, if the area under the curve was made of cardboard, we'd expect it to balance at the mean. All of the graphs are symmetric across the line $x = \mu$, so μ is the "balancing point" and hence the mean. As the graphs also show, increasing σ flattens out the graph, in effect lessening the concentration of the data near the mean. Thus, the smaller the σ value, the more data is clustered around the mean.

7. (a)

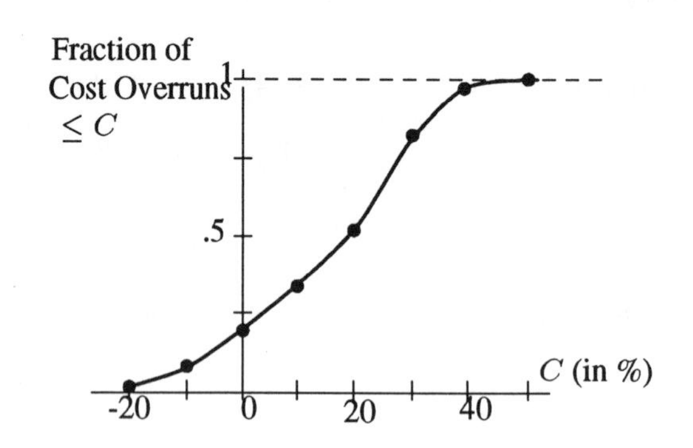

This is a cumulative distribution function.

(b) The density function is the derivative of the cumulative distribution function.

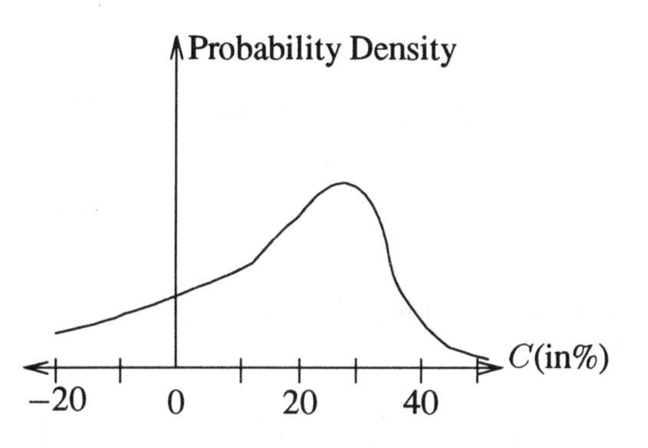

(c) Let's call the cumulative distribution function $F(C)$. The probability that there will be a cost overrun of more than 50% is $1 - F(50) = 0.01$, a 1% chance. The probability that it will between 20% and 50% is $F(50) - F(20) = 0.99 - 0.50 = 0.49$, or 49%. The most likely amount of cost overrun occurs when the slope of the tangent line to the cumulative distribution function is a maximum. This occurs at the inflection point of the cumulative distribution graph, at about $C = 28\%$.

9. (a) The fraction of students passing is given by the area under the curve from 2 to 4 divided by the total area under the curve. This appears to be about $\frac{2}{3}$.

   (b) The fraction with honor grades corresponds to the area under the curve from 3 to 4 divided by the total area. This is about $\frac{1}{3}$.

   (c) The peak around 2 probably exists because many students work to get just a passing grade.

   (d)

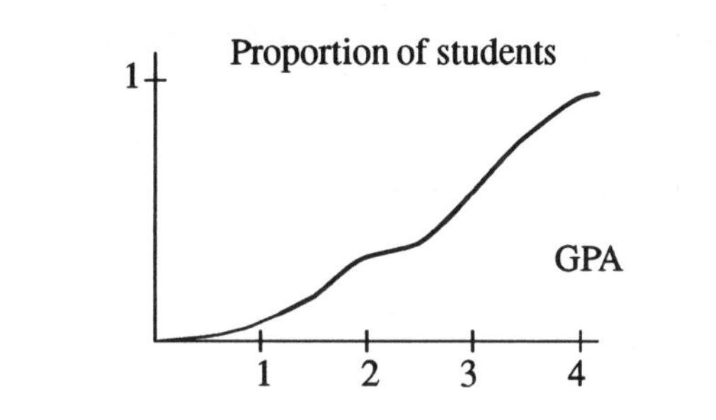

11.

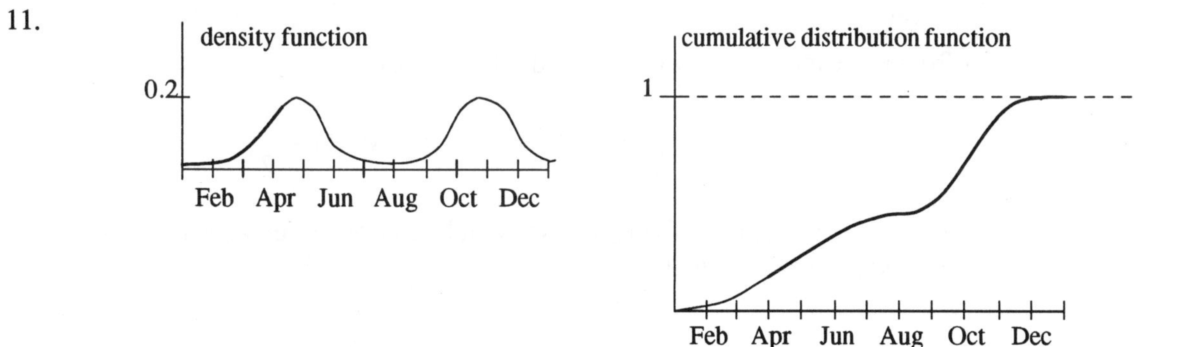

13. (a) We want to find a such that $\int_0^\infty p(v)\,dv = \lim_{r\to\infty} a\int_0^r v^2 e^{-mv^2/2kT}\,dv = 1$. Therefore,

$$\frac{1}{a} = \lim_{r\to\infty}\int_0^r v^2 e^{-mv^2/2kT}\,dv. \tag{7.1}$$

To evaluate the integral, use integration by parts with the substitutions $u = v$ and $w' = ve^{-mv^2/2kT}$:

$$\begin{aligned}\int_0^r \underbrace{v}_{u}\underbrace{ve^{-mv^2/2kT}}_{w'}\,dv &= \underbrace{v}_{u}\underbrace{\frac{e^{-mv^2/2kT}}{-m/kT}}_{w}\Bigg|_0^r - \int_0^r \underbrace{1}_{u'}\underbrace{\frac{e^{-mv^2/2kT}}{-m/kT}}_{w}\,dv \\ &= -\frac{kTr}{m}e^{-mr^2/2kT} + \frac{kT}{m}\int_0^r e^{-mv^2/2kT}\,dv.\end{aligned}$$

From the normal distribution we know that $\int_0^\infty \frac{1}{\sqrt{2\pi}}e^{-x^2/2}\,dx = \frac{1}{2}$, so

$$\int_0^\infty e^{-x^2/2}\,dx = \frac{\sqrt{2\pi}}{2}.$$

Therefore in the above integral, make the substitution $x = \sqrt{\frac{m}{kT}}v$, so that $dx = \sqrt{\frac{m}{kT}}\,dv$, or $dv = \sqrt{\frac{kT}{m}}\,dx$. Then

$$\frac{kT}{m}\int_0^r e^{-mv^2/2kT}\,dv = \left(\frac{kT}{m}\right)^{3/2}\int_0^{\sqrt{\frac{m}{kT}}r} e^{-x^2/2}\,dx.$$

Substituting this into Equation 7.1 we get

$$\frac{1}{a} = \lim_{r\to\infty}\left(-\frac{kTr}{m}e^{-mr^2/2kT} + \left(\frac{kT}{m}\right)^{3/2}\int_0^{\sqrt{\frac{m}{kT}}r} e^{-x^2/2}\,dx\right) = 0 + \left(\frac{kT}{m}\right)^{3/2}\cdot\frac{\sqrt{2\pi}}{2}.$$

Therefore, $a = \frac{2}{\sqrt{2\pi}}(\frac{m}{kT})^{3/2}$. Substituting the values for k, T, and m gives $a \approx 3.4\times10^{-8}$ SI units.

(b) To find the median, we wish to find the x such that

$$\int_0^x p(v)\,dv = \int_0^x av^2 e^{-\frac{mv^2}{2kT}}\,dv = \frac{1}{2},$$

where $a = \frac{2}{\sqrt{2\pi}}(\frac{m}{kT})^{3/2}$. Using a calculator, by trial and error we get $x \approx 441$ m/sec. To find the mean, we find

$$\int_0^\infty vp(v)\,dv = \int_0^\infty av^3 e^{-\frac{mv^2}{2kT}}\,dv.$$

This integral can be done by substitution. Let $u = v^2$, so $du = 2vdv$. Then

$$\begin{aligned}
\int_0^\infty av^3 e^{-\frac{mv^2}{2kT}}\,dv &= \frac{a}{2}\int_{v=0}^{v=\infty} v^2 e^{-\frac{mv^2}{2kT}} 2v\,dv \\
&= \frac{a}{2}\int_{u=0}^{u=\infty} ue^{-\frac{mu}{2kT}}\,du \\
&= \lim_{r\to\infty}\frac{a}{2}\int_0^r ue^{-\frac{mu}{2kT}}\,du.
\end{aligned}$$

Now, using the integral table, we have

$$\begin{aligned}
\int_0^\infty av^3 e^{-\frac{mv^2}{2kT}}\,dv &= \lim_{r\to\infty}\frac{a}{2}\left[-\frac{2kT}{m}ue^{-\frac{mu}{2kT}} - \left(-\frac{2kT}{m}\right)^2 e^{-\frac{mu}{2kT}}\right]\Bigg|_0^r \\
&= \frac{a}{2}\left(-\frac{2kT}{m}\right)^2 \\
&\approx 457.7 \text{ m/sec.}
\end{aligned}$$

The maximum for $p(v)$ will be at a point where $p'(v) = 0$.

$$\begin{aligned}
p'(v) &= a(2v)e^{-\frac{mv^2}{2kT}} + av^2\left(-\frac{2mv}{2kT}\right)e^{-\frac{mv^2}{2kT}} \\
&= ae^{-\frac{mv^2}{2kT}}\left(2v - v^3\frac{m}{kT}\right).
\end{aligned}$$

Thus $p'(v) = 0$ at $v = 0$ and at $v = \sqrt{\frac{2kT}{m}} \approx 405$. It's obvious that $p(0) = 0$. So $v = 405$ gives us a maximum: $p(405) \approx 0.002$.

(c) Notice that if T changes, a has to change along with it, since

$$a = \frac{2}{\sqrt{2\pi}}\left(\frac{m}{kT}\right)^{3/2}.$$

The median, we recall, is the value of x such that $\int_0^x p(v)\,dv = \int_0^x av^2 e^{-\frac{mv^2}{2kT}}\,dv = \frac{1}{2}$. By experimentation, again with a calculator, we find that as T increases, so does the median.

The mean, as we found in part (b), is $\frac{a}{2}\frac{4k^2T^2}{m^2} = \frac{4}{\sqrt{2\pi}}\frac{k^{1/2}T^{1/2}}{m^{1/2}}$. It is clear, then, that as T increases so does the mean. (And thus we also intuitively feel that as T increases so does the median.)

We found in part (b) that $p(v)$ reached its maximum at $v = \sqrt{\frac{2kT}{m}}$. Thus

$$\begin{aligned}
\text{the maximum value of } p(v) &= \frac{2}{\sqrt{2\pi}}\left(\frac{m}{kT}\right)^{3/2}\frac{2kT}{m}e^{-1} \\
&= \frac{4}{e\sqrt{2\pi}}\frac{m^{1/2}}{kT^{1/2}}.
\end{aligned}$$

Thus as T increases, the maximum value decreases.

15. (a) 

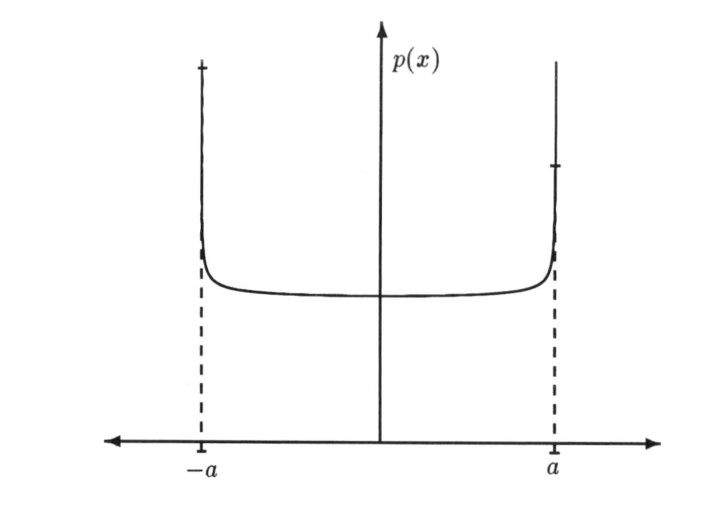

(b) The graphs should look similar.

(c) We expect $\int_{-a}^{a} \frac{dx}{\pi\sqrt{a^2 - x^2}} = 1$, since

$$f(x) = \begin{cases} \frac{1}{\pi\sqrt{a^2-x^2}} & -a < x < a; \\ 0 & |x| \geq a, \end{cases}$$

and thus $\int_{-a}^{a} \frac{dx}{\pi\sqrt{a^2 - x^2}} = 1$ by the definition of a probability density function. Indeed,

$$\begin{aligned} \int_{-a}^{a} \frac{dx}{\pi\sqrt{a^2 - x^2}} &= \frac{1}{\pi} \arcsin \frac{x}{a} \Big|_{-a}^{a} \\ &= \frac{1}{\pi}(\arcsin 1 - \arcsin(-1)) \\ &= 1. \end{aligned}$$

(d) It does make sense, physically speaking. The fact that $f(x) \to \infty$ as $x \to a$ does not mean that the ball spends an infinite amount of time at a, but just that the ratio of the time spent near $-a$ and a to the time spent elsewhere goes to ∞. This makes sense—if we watch a pendulum, we note that more time is spent near the ends of its path (where its velocity is small) than in the middle of the path (where its velocity is largest).

7.7 Answers to Miscellaneous Exercises for Chapter 7

1. (a) In the beginning, both birth and death rates are small; this is consistent with a very small population. Both rates begin climbing, the birth rate faster than the death rate, which is consistent with a growing population. The death rate is then high, but it begins to decrease as the population decreases.

(b) The bacteria population is growing most quickly when $B - D$, the rate of change of population, is maximal; that happens when B is farthest above D, which is at a point where the slopes of both graphs are equal. The point on this graph satisfying that criterion is $t \approx 6$, so the greatest rate of increase occurs about 6 hours after things have begun.

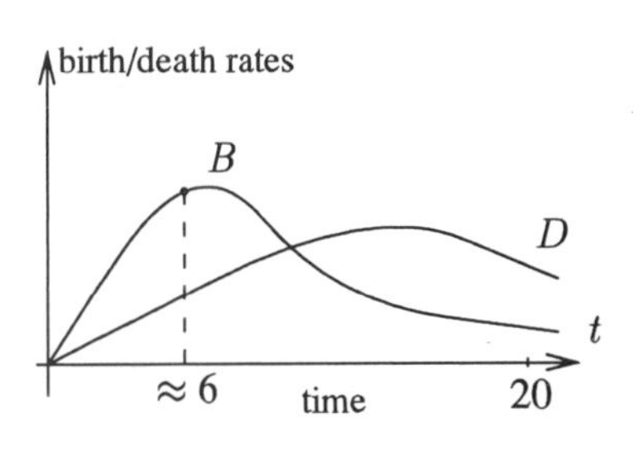

(c)

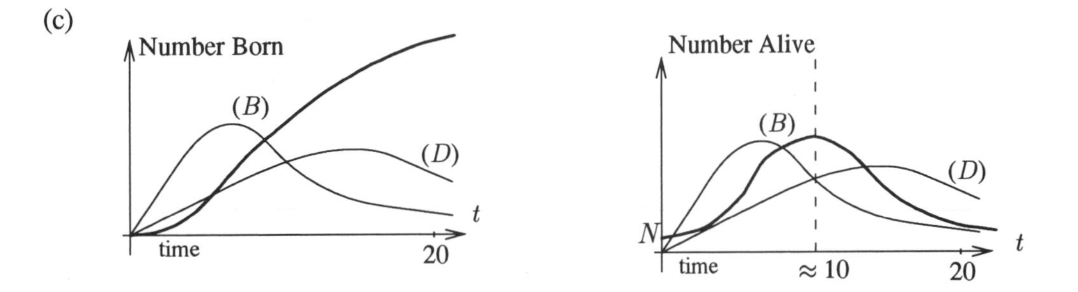

From the first figure, we see that the population is at a maximum when $B = D$, that is, after 10 hours. This stands to reason, because $B - D$ is the rate of change of population, so population is maximized when $B - D = 0$, that is, when $B = D$.

3. (a)

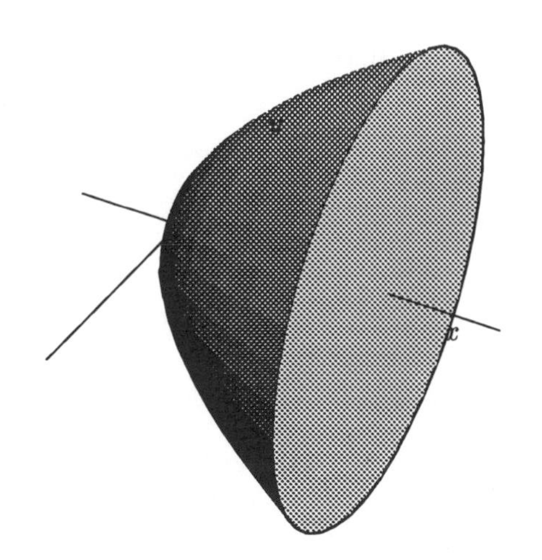

Figure 7.4: Rotated Region

(b) Partition [0,1] into N subintervals of width $\Delta x = \frac{1}{N}$. The volume of the ith disc is $\pi(\sqrt{x_i})^2\Delta x = \pi x_i \Delta x$. So, $V \approx \sum_{i=1}^{N} \pi x_i \Delta x$.

(c)

$$\text{Volume} = \int_0^1 \pi x\, dx = \frac{\pi}{2}x^2\Big|_0^1 = \frac{\pi}{2} \approx 1.57.$$

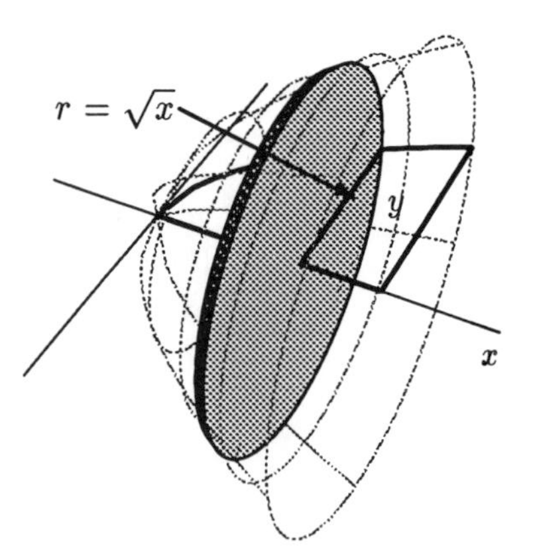

Figure 7.5: Cutaway View

5. (a)

$$c\int_0^6 e^{-ct}dt = -e^{-ct}\Big|_0^6 = 1 - e^{-6c} = 0.1 \text{ so } c = -\frac{1}{6}\ln 0.9 \approx 0.0176.$$

(b)

$$c\int_6^{12} e^{-ct}dt = -e^{-ct}\Big|_6^{12} = e^{-6c} - e^{-12c} = 0.9 - 0.81 = 0.09,$$

so the probability is 9%.

7. (a) Slice the mountain horizontally into N cylinders of height Δh. The sum of the volumes of the cylinders will be

$$\sum_{i=1}^{N} \pi r^2 \Delta h = \sum_{i=1}^{N} \pi \left(3 \cdot \frac{10^5}{\sqrt{h_i}}\right)^2 \Delta h.$$

(b)

$$\begin{aligned} \text{Volume} &= \int_{400}^{14400} \pi \left(3 \cdot \frac{10^5}{\sqrt{h_i}}\right)^2 dh = 9 \cdot 10^{10}\pi \int_{400}^{14400} \frac{1}{h}\, dh \\ &= 9 \cdot 10^{10}\pi\, [\ln 14400 - \ln 400] = 9 \cdot 10^{10}\pi \ln(14400/400) \\ &= 9 \cdot 10^{10}\pi \ln 36 \approx 1.01 \times 10^{12} \text{ cubic feet.} \end{aligned}$$

9.

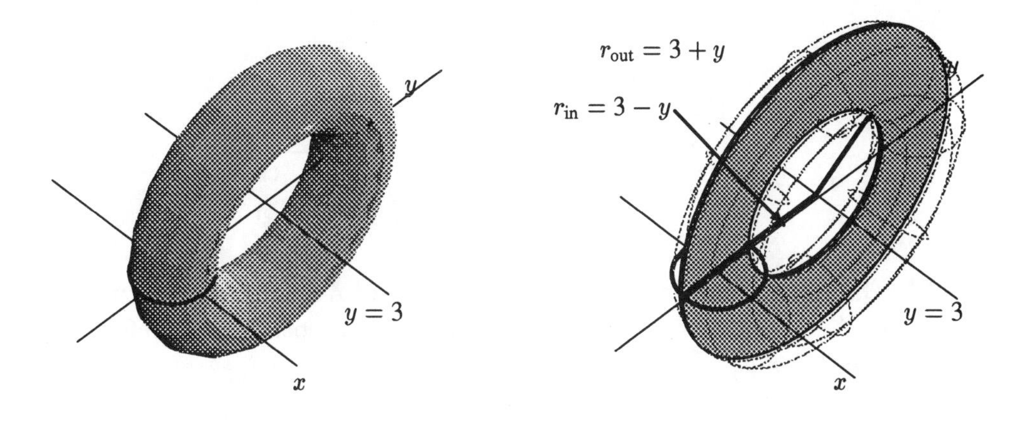

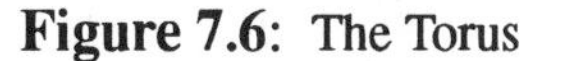
Figure 7.6: The Torus

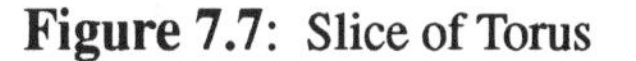
Figure 7.7: Slice of Torus

As shown in Figure 7.7, we slice the torus perpendicular to the line $y = 3$. We obtain washers with width dx, inner radius $r_{\text{in}} = 3 - y$, and outer radius $r_{\text{out}} = 3 + y$. Therefore, the area of the washer is $\pi r_{\text{out}}^2 - \pi r_{\text{in}}^2 = \pi[(3+y)^2 - (3-y)^2] = 12\pi y$. Since $y = \sqrt{1-x^2}$, the volume is gotten by summing up the volumes of the washers: we get

$$\int_{-1}^{1} 12\pi\sqrt{1-x^2}\,dx = 12\pi\int_{-1}^{1}\sqrt{1-x^2}\,dx.$$

But $\int_{-1}^{1}\sqrt{1-x^2}\,dx$ is the area of a semicircle of radius 1, which is $\frac{\pi}{2}$. So we get $12\pi \cdot \frac{\pi}{2} = 6\pi^2 \approx 59.22$. (Or, you could use $\int \sqrt{1-x^2}\,dx = \left[x\sqrt{1-x^2} + \arcsin(x)\right]$ by Formula 31 and Formula 28.)

11. The arclength of $\sqrt{1-x^2}$ from $x = 0$ to $x = 1$ is one quarter the perimeter of the unit circle. Hence the length is $\dfrac{2\pi}{4} = \dfrac{\pi}{2}$.

13.

$$\begin{aligned} L &= \int_1^2 \sqrt{1 + \left(x^4 + \frac{1}{(16x^4)} - \frac{1}{2}\right)}\,dx = \int_1^2 \sqrt{\left(x^2 + \frac{1}{(4x^2)}\right)^2}\,dx \\ &= \int_1^2 \left(x^2 + \frac{1}{4}x^{-2}\right)dx = \left[\frac{x^3}{3} - \frac{1}{4x}\right]_1^2 = \frac{59}{24}. \end{aligned}$$

15. Since $f(x) = \sin x$, $f'(x) = \cos(x)$, so

$$\text{Arc Length} = \int_0^{\pi} \sqrt{1 + \cos^2 x}\, dx.$$

17. The average horizontal width of A is the area of A divided by the vertical span of A. The area of A is $\int_0^{\pi} \sin x\, dx = -\cos x|_0^{\pi} = 2$. The vertical span is 1. Therefore, the average width of A is $2/1 = 2$.

19. (a)

$$\begin{aligned}
\text{Future Value} &= \int_0^{20} 100e^{.10(20-t)} dt = 100 \int_0^{20} e^2 e^{-.10t} dt \\
&= \frac{100e^2}{-.10} e^{-.10t}\Big|_0^{20} = \frac{100e^2}{.10}(1 - e^{-.10(20)}) \\
&\approx \$6389.06.
\end{aligned}$$

The present value of the income stream is

$$\begin{aligned}
\int_0^{20} 100e^{-.10t} dt &= 100\left(\frac{1}{-.10}\right) e^{-.10t}\Big|_0^{20} \\
&= 1000\left(1 - e^{-2}\right) \\
&= \$864.66.
\end{aligned}$$

Note that this is also the present value of the sum \$6389.06.

(b) Let T be the number of years for the balance to reach \$5000. Then

$$\begin{aligned}
5000 &= \int_0^T 100e^{.10(T-t)} dt \\
50 &= e^{.10T} \int_0^T e^{-.10t} dt \\
&= \frac{e^{.10T}}{-.10} e^{-.10t}\Big|_0^T = 10e^{.10T}(1 - e^{-.10T}) \\
&= 10e^{.10}T - 10.
\end{aligned}$$

So, $60 = 10e^{.10T}$, and $T = 10 \ln 6 \approx 17.92$ years.

21. We'll divide up time between 1971 and 1992 into intervals of length dt, and figure out how much of the Strontium-90 produced during that time interval is still around.

First, Strontium-90 decays exponentially, so if a quantity S_0 was produced t years ago, and S is the quantity around today, $S = S_0 e^{-kt}$. Since the half-life is 28 years, $\frac{1}{2} = e^{-k(28)}$, giving $k = \frac{-\ln(\frac{1}{2})}{28} \approx 0.025$.

Suppose we measure t in years from 1971, so that 1992 is $t = 21$.

1971 $(t = 0)$ dt 1992 $(t = 21)$

t $(21 - t)$

Since Strontium-90 is produced at a rate of 1 kg/year, during the interval dt we know that a quantity $1dt$ kg was produced. Since this was $(21 - t)$ years ago, the quantity remaining now is $1dte^{-0.025(21-t)}$. Summing over all such intervals gives

$$\text{Strontium in 1992} \approx \int_0^{21} e^{-0.025(21-t)}\,dt = \left.\frac{e^{-0.025(21-t)}}{0.025}\right|_0^{21} = 16.34 \text{ kg}.$$

[Note: This is exactly like a future value problem from economics, with a negative interest rate.]

23. Let x be the distance from the bucket to the surface of the water. It follows that $0 \le x \le 40$. At x feet, the bucket weighs $30 - \frac{1}{4}x$, where the $\frac{1}{4}x$ term is due to the leak. When the bucket is x feet from the surface of the water, the work done by raising it Δx feet is $(30 - \frac{1}{4}x)\Delta x$. So the total work required ot raise the bucket to the top is

$$\begin{aligned} W &= \int_0^{40} (30 - \frac{1}{4}x)dx \\ &= \left(30x - \frac{1}{8}x^2\right)\Big|_0^{40} \\ &= 30(40) - \frac{1}{8}40^2 \\ &= 1000 \quad \text{ft.lb.} \end{aligned}$$

25. Let x be the depth of the water measured from the bottom of the tank. It follows that $0 \le x \le 15$. Let r be the radius of the section of the cone with height x. By similar triangles, $\frac{r}{x} = \frac{12}{18}$, so $r = \frac{2}{3}x$. Then the work required to pump a layer of water with thickness of Δx at depth x over the top of the tank is $62.4\pi\left(\frac{2}{3}x\right)^2\Delta x(18 - x)$. So the total work done by pumping the water over the top of the tank is

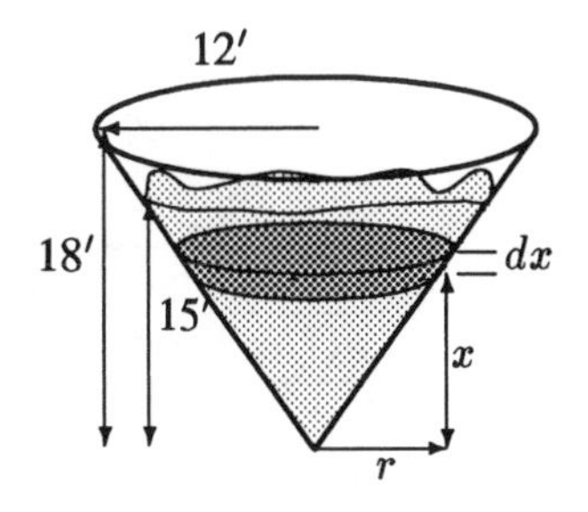

$$\begin{aligned} W &= \int_0^{15} 62.4\pi\left(\frac{2}{3}x\right)^2(18 - x)dx = \frac{4}{9}62.4\pi\int_0^{15} x^2(18 - x)dx \\ &= \frac{4}{9}62.4\pi\left(6x^3 - \frac{1}{4}x^4\right)\Big|_0^{15} = \frac{4}{9}62.4\pi(7593.75) \\ &\approx 661{,}619.41 \quad \text{ft.lb.} \end{aligned}$$

27. (a) If you slice the apple perpendicular to the core, you expect that the cross section will be approximately a circle.

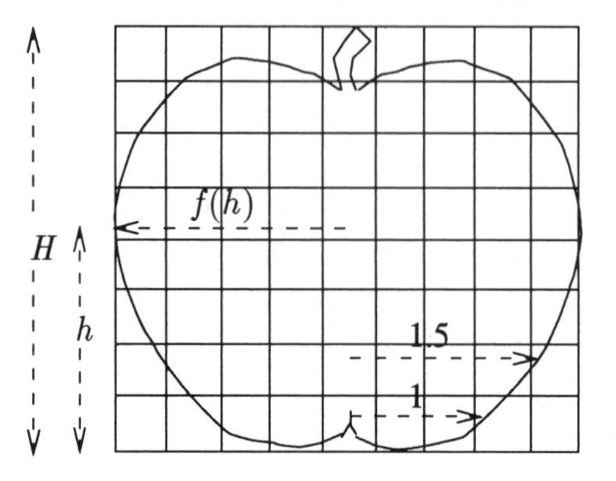

If $f(h)$ is the radius of the apple at height h above the bottom, and H is the height of the apple, then

$$\text{Volume} = \int_0^H \pi f(h)^2 \, dh.$$

Ignoring the stem, $H \approx 3.5$. Although we do not have a formula for $f(h)$, we can estimate it at various points. (Remember, we measure here from the bottom of the apple, which is not quite the bottom of the graph.)

h = Height (in inches)	0	0.5	1	1.5	2	2.5	3	3.5
$f(h)$ = Radius (in inches)	1	1.5	2	2.1	2.3	2.2	1.8	1.2

Now let $g(h) = \pi f(h)^2$, the area of the cross-section at height h. From our approximations above, we get the following table.

h = Height (in inches)	0	0.5	1	1.5	2	2.5	3	3.5
$g(h)$ = Area (sq. in.)	3.14	7.07	12.57	13.85	16.62	13.85	10.18	4.52

We can now take left- and right-hand sum approximations. Note that $\Delta h = 0.5$ inches. Thus

$$\begin{aligned}
\text{LEFT}(9) &= (3.14 + 7.07 + 12.57 + 13.85 + 16.62 + 13.85 + 10.18)(0.5) = 38.64. \\
\text{RIGHT}(9) &= (7.07 + 12.57 + 13.85 + 16.62 + 13.85 + 10.18 + 4.52)(0.5) = 39.33.
\end{aligned}$$

Thus the volume of the apple is ≈ 39 cu.in.

(b) The apple weighs $0.03 \times 39 \approx 1.17$ pounds, so it costs about 94¢. (Expensive apple!)

29. Look at the disc-shaped slab of water at height y and of thickness dy. The rate at which water is flowing out when it is at depth y is $k\sqrt{y}$ (Torricelli's Law, with k constant). Then, if $x = g(y)$, we have

$$dt = \begin{pmatrix}\text{Time for water to}\\ \text{drop by this amount}\end{pmatrix} = \frac{\text{Volume}}{\text{Rate}} = \frac{\pi(g(y))^2 dy}{k\sqrt{y}}.$$

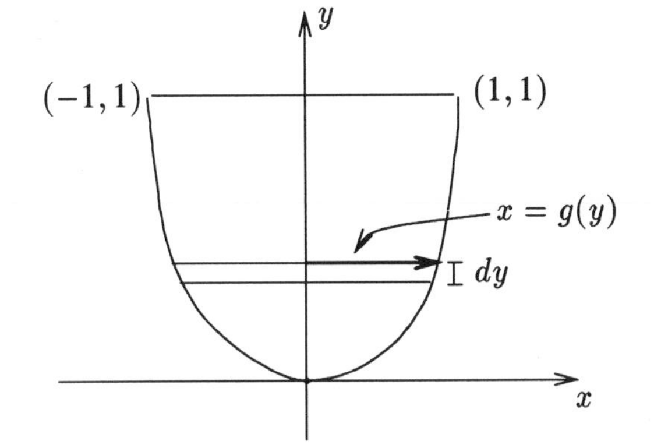

If the rate at which the depth of the water is dropping is constant, then $\frac{dy}{dt}$ is constant, so we want

$$\frac{\pi(g(y))^2}{k\sqrt{y}} = \text{constant},$$

or $\frac{(g(y))^2}{\sqrt{y}} = \text{constant}$, so $g(y) = c\sqrt[4]{y}$, for some constant c. Since $x = 1$ when $y = 1$, we have $c = 1$ and so $x = \sqrt[4]{y}$, or $y = x^4$.

31. Any small piece of mass ΔM on either of the two hoops has kinetic energy $\frac{1}{2}v^2\Delta M$. Since the angular velocity of the two hoops is the same, the actual velocity of the piece ΔM will depend on how far away it is from the axis of revolution. The further away a piece is from the axis, the faster it must be moving and the larger its velocity v. This is because if ΔM is at a distance r from the axis, in one revolution it must trace out a circular path of length $2\pi r$ about the axis. Since every piece in either hoop takes 1 minute to make 1 revolution, pieces farther from the axis must move faster, as they travel a greater distance.

The hoop rotating about the cylindrical axis has all of its mass at a distance R from the axis, whereas the other hoop has a good bit of its mass close (or on) the axis of rotation. So, since the bulk of the hoop rotating about the cylindrical axis is travelling faster than the bulk of the other hoop, it must have the higher kinetic energy.

33. Let us make coordinate axes with the origin at the center of the box. The x and y axes will lie along the central axes of the cylinders, and the (height) axis will extend vertically to the top of the box. If one slices the cylinders horizontally, one gets a cross. The cross is what you get if you cut out four corner squares from a square of side length 2. If h is the height of the cross above (or below) the xy plane, the equation of a cylinder is $h^2 + y^2 = 1$ (or $h^2 + x^2 = 1$). Thus the "armpits" of the cross occur where $y^2 - 1 = -h^2 = x^2 - 1$ for some fixed height h—that is, out $\sqrt{1-h^2}$ units from the center, or $1 - \sqrt{1-h^2}$ units away from the edge. Each corner square has area $(1-\sqrt{1-h^2})^2 = 2 - h^2 - 2\sqrt{1-h^2}$. The whole big square has area 4. Therefore, the area of the cross is

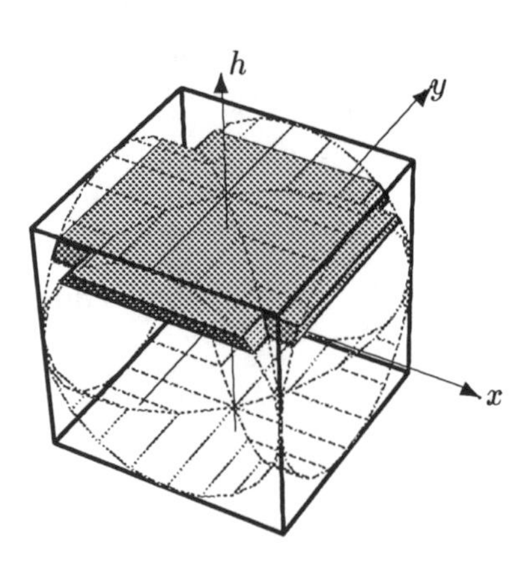

$$4 - 4(2 - h^2 - 2\sqrt{1-h^2}) = -4 + 4h^2 + 8\sqrt{1-h^2}.$$

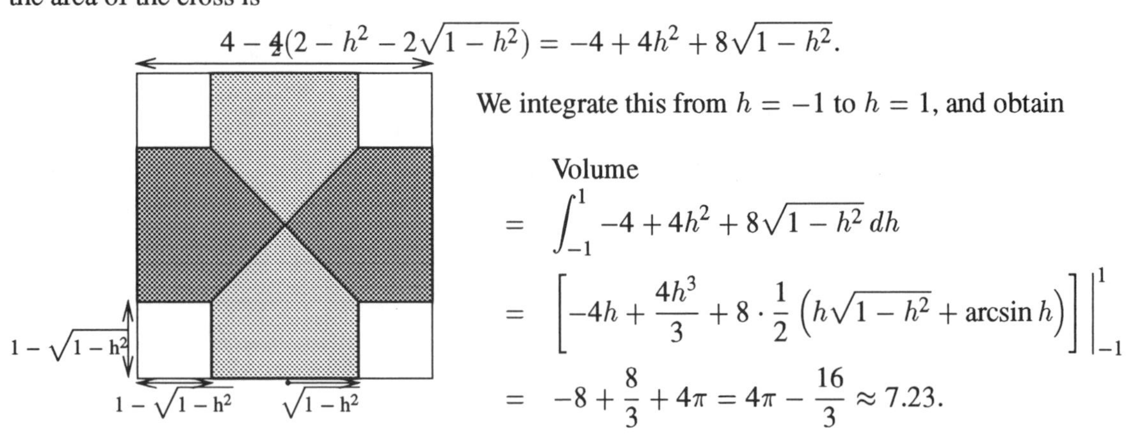

We integrate this from $h = -1$ to $h = 1$, and obtain

$$\begin{aligned}
&\text{Volume} \\
&= \int_{-1}^{1} -4 + 4h^2 + 8\sqrt{1-h^2}\, dh \\
&= \left[-4h + \frac{4h^3}{3} + 8 \cdot \frac{1}{2}\left(h\sqrt{1-h^2} + \arcsin h\right)\right]\Bigg|_{-1}^{1} \\
&= -8 + \frac{8}{3} + 4\pi = 4\pi - \frac{16}{3} \approx 7.23.
\end{aligned}$$

This is a reasonable answer, as the volume of the cube is 8, and the volume of one cylinder alone is $2\pi \approx 6.28$.

35. (a) First, we find the critical points of $p(x)$

$$\begin{aligned}
\frac{d}{dx}p(x) &= \frac{1}{\sigma\sqrt{2\pi}}\left[\frac{-2(x-\mu)}{2\sigma^2}\right] e^{-\frac{(x-\mu)^2}{2\sigma^2}} \\
&= -\frac{(x-\mu)}{\sigma^3\sqrt{2\pi}} e^{-\frac{(x-\mu)^2}{2\sigma^2}}.
\end{aligned}$$

This implies $x = \mu$ is the only critical point of $p(x)$.

To confirm that $p(x)$ is maximized at $x = \mu$, we rely on the first derivative test. As $-\frac{1}{\sigma^3\sqrt{2\pi}} e^{-\frac{(x-\mu)^2}{2\sigma^2}}$ is always negative, the sign of $p'(x)$ is the opposite of the sign of $(x-\mu)$; thus $p'(x) > 0$ when $x < \mu$, and $p'(x) < 0$ when $x > \mu$.

(b) To find the inflection points, we need to find where $p''(x)$ changes sign; that will happen only when $p''(x) = 0$. As

$$\frac{d^2}{dx^2}p(x) = -\frac{1}{\sigma^3\sqrt{2\pi}}e^{-\frac{(x-\mu)^2}{2\sigma^2}}\left[-\frac{(x-\mu)^2}{\sigma^2}+1\right],$$

$p''(x)$ changes sign when $\left[-\frac{(x-\mu)^2}{\sigma^2}+1\right]$ does, since the sign of the other factor is always negative. This occurs when

$$\begin{aligned} -\frac{(x-\mu)^2}{\sigma^2}+1 &= 0, \\ -(x-\mu)^2 &= -\sigma^2, \\ x-\mu &= \pm\sigma. \end{aligned}$$

Thus, $x = \mu + \sigma$ or $x = \mu - \sigma$. Since $p''(x) > 0$ for $x < \mu - \sigma$ and $x > \mu + \sigma$ and $p''(x) < 0$ for $\mu - \sigma \leq x \leq \mu + \sigma$, these are in fact points of inflection.

(c) μ represents the mean of the distribution, while σ is the standard deviation. In other words, σ gives a measure of the "spread" of the distribution, i.e. how tightly the observations are clustered about the mean. A small σ tells us that most of the data are close to the mean; a large σ tells us that the data is spread out.

Chapter 8

8.1 Solutions

1. If $P = P_0 e^t$, then

$$\frac{dP}{dt} = \frac{d}{dt}(P_0 e^t) = P_0 e^t = P.$$

3. (a) If $y = \sin 2t$, then $\frac{dy}{dt} = 2\cos 2t$, and $\frac{d^2y}{dt^2} = -4\sin 2t$.
Thus $\frac{d^2y}{dt^2} + 4y = -4\sin 2t + 4\sin 2t = 0$.

(b) If $y = \cos \omega t$, then

$$\frac{dy}{dt} = -\omega \sin \omega t, \quad \frac{d^2y}{dt^2} = -\omega^2 \cos \omega t.$$

Thus, if $\frac{d^2y}{dt^2} + 9y = 0$, then

$$\begin{aligned} -\omega^2 \cos \omega t + 9\cos \omega t &= 0 \\ (9 - \omega^2)\cos \omega t &= 0. \end{aligned}$$

Thus $9 - \omega^2 = 0$, or $\omega^2 = 9$, so $\omega = \pm 3$.

5. (a) If $y = \frac{e^x + e^{-x}}{2}$, then $\frac{dy}{dx} = \frac{e^x - e^{-x}}{2}$, and $\frac{d^2y}{dx^2} = \frac{e^x + e^{-x}}{2}$.
If $k = 1$, then

$$\begin{aligned} k\sqrt{1 + \left(\frac{dy}{dx}\right)^2} = \sqrt{1 + \left(\frac{e^x - e^{-x}}{2}\right)^2} &= \sqrt{1 + \frac{e^{2x}}{4} - \frac{1}{2} + \frac{e^{-2x}}{4}} \\ &= \sqrt{\frac{e^{2x}}{4} + \frac{1}{2} + \frac{e^{-2x}}{4}} \\ &= \sqrt{\left(\frac{e^x + e^{-x}}{2}\right)^2} = \frac{e^x + e^{-x}}{2} = \frac{d^2y}{dx^2}. \end{aligned}$$

(b) $y = \frac{e^{Ax} + e^{-Ax}}{2A}$, so

$$\frac{dy}{dx} = \frac{e^{Ax} - e^{-Ax}}{2} \quad \text{and} \quad \frac{d^2y}{dx^2} = A\left(\frac{e^{Ax} + e^{-Ax}}{2}\right).$$

Therefore we have

$$\begin{aligned} 1 + \left(\frac{dy}{dx}\right)^2 &= 1 + \left(\frac{e^{Ax} - e^{-Ax}}{2}\right)^2 = 1 + \frac{1}{4}\left(e^{2Ax} + e^{-2Ax} - 2\right) \\ &= \frac{1}{4}\left(e^{2Ax} + e^{-2Ax} + 2\right) = \frac{1}{4}\left(e^{Ax} + e^{-Ax}\right)^2. \end{aligned}$$

$$\begin{aligned} k\sqrt{1+\left(\frac{dy}{dx}\right)^2} &= k\sqrt{\frac{1}{4}\left(e^{Ax}+e^{-Ax}\right)^2} = \frac{k}{2}\cdot\left|e^{Ax}+e^{-Ax}\right| \\ &= k\frac{\left(e^{Ax}+e^{-Ax}\right)}{2}, \quad \text{since } e^{Ax}+e^{-Ax} > 0. \end{aligned}$$

Since $\frac{d^2y}{dx^2} = k\sqrt{1+\left(\frac{dy}{dx}\right)^2}$, we have $A = k$.

7.

$$\begin{array}{lll} \text{(A) } y = xe^{kx}, & \frac{dy}{dx} = (kx+1)e^{kx}, & \frac{d^2y}{dx^2} = (k^2x+2k)e^{kx} \\ \text{(B) } y = x^p, & \frac{dy}{dx} = px^{p-1}, & \frac{d^2y}{dx^2} = p(p-1)x^{p-2} \\ \text{(C) } y = e^{kx}, & \frac{dy}{dx} = ke^{kx}, & \frac{d^2y}{dx^2} = k^2e^{kx} \\ \text{(D) } y = mx, & \frac{dy}{dx} = m, & \frac{d^2y}{dx^2} = 0 \end{array}$$

and so:

(a) (D)

(b) (C)

(c) (A)

(d) (B)

8.2 Solutions

1. (a)

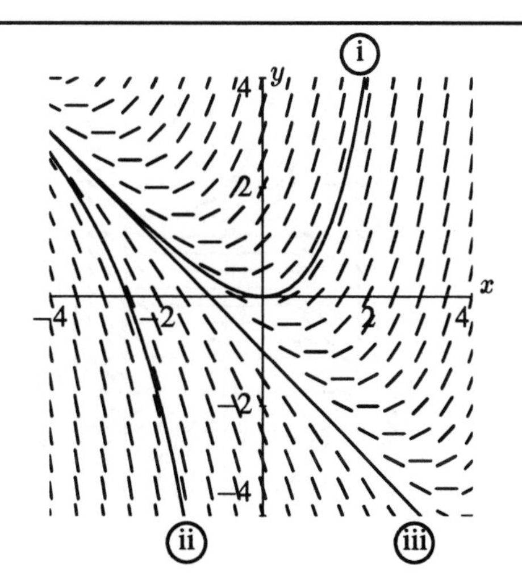

(b) The solution through $(-1, 0)$ appears to be linear and to have the equation $y = -x - 1$.

(c) If $y = -x - 1$, then $y' = -1$ and $x + y = x + (-x - 1) = -1$.

3. (a)

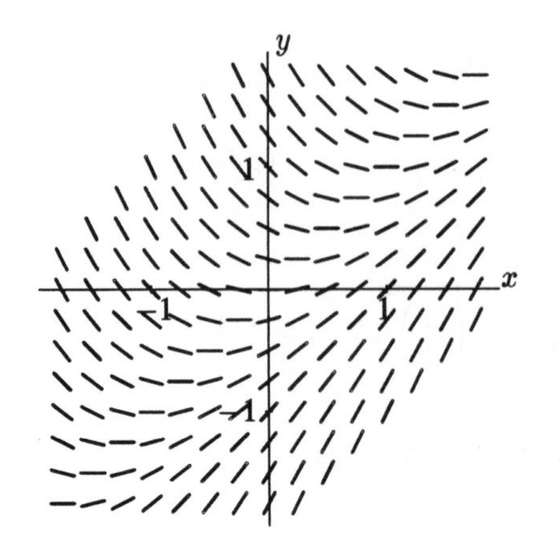

(b) From the graph, the solution through $(1,0)$ appears linear and to have the equation $y = x - 1$.

In fact, if $y = x - 1$, then $x - y = x - (x - 1) = 1 = y'$, so $y = x - 1$ is the solution through $(1,0)$.

5. The first graph has the equation $y' = x^2 - y^2$. We can see this by looking along the line $y = x$. On the first slope field, it seems that $y' = 0$ along this line, as it should if $y' = x^2 - y^2$. This is not the case for the second graph.

At $(0,1)$, $y' = -1$, and at $(1,0)$, $y' = 1$, so you are looking for points on the axes where the line is sloping at 45°. The points are distinguishable below.

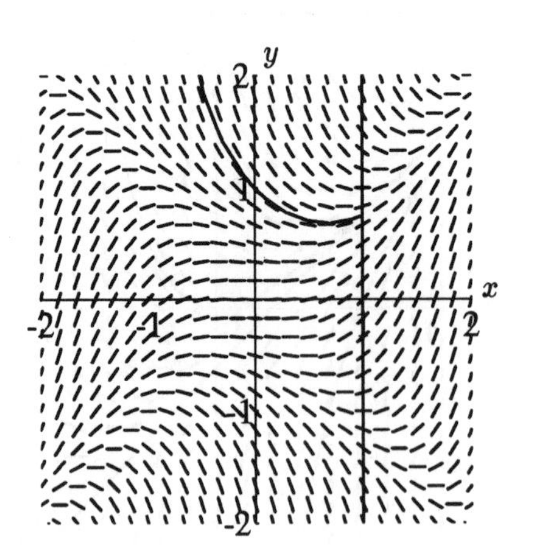

7. (a)

(b)

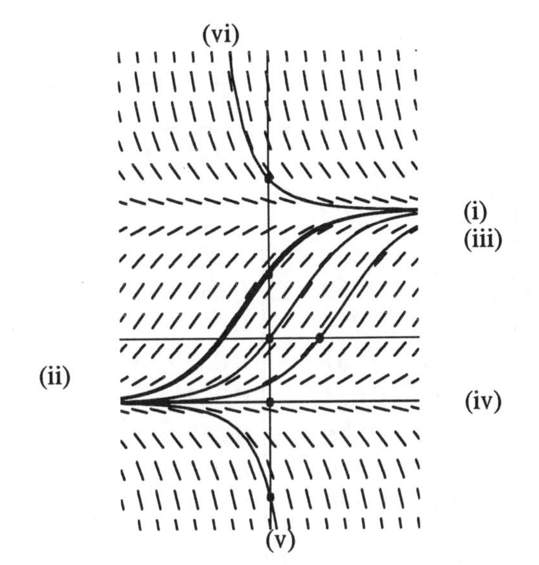

(c) The graph shows that a solution will be increasing if its y values fall in the range $-1 < y < 2$. This makes sense since $y' = 0.5(1+y)(2-y)$, so $y' > 0$ if $-1 < y < 2$. Notice that if the y value ever gets to 2, then $y' = 0$ and the function becomes constant, following the line $y = 2$. (The same is true if ever $y = -1$.)

From the graph, the solution is decreasing if $y > 2$ or $y < -1$. Again, this also follows from the equation, since in either case $y' < 0$.

The curve has a horizontal tangent if $y' = 0$, which only happens if $y = 2$ or $y = -1$. This also can be seen on the graph of the slope field.

9. (a) B (b) F (c) D (d) A (e) C (f) E

11. As $x \to \infty$, y diverges.

13. As $x \to \infty$, y seems to oscillate within a certain range. The range will depend on the starting point, but the *size* of the range appears independent of the starting point.

15. From the slope field, the function looks like a parabola of the form $y = x^2 + C$, where C depends on the starting point. In any case, $y \to \infty$ as $x \to \infty$.

8.3 Solutions

1. (a)

t	y	slope= $\frac{1}{t}$	$\Delta y = (\text{slope})\Delta t = \frac{1}{t}(0.1)$
1	0	1	0.1
1.1	0.1	0.909	0.091
1.2	0.191	0.833	0.083
1.3	0.274	0.769	0.077
1.4	0.351	0.714	0.071
1.5	0.422	0.667	0.067
1.6	0.489	0.625	0.063
1.7	0.552	0.588	0.059
1.8	0.610	0.556	0.056
1.9	0.666	0.526	0.053
2	0.719		

(b) Since $\frac{dy}{dt} = \frac{1}{t}$, $y = \ln|t| + C$.
Starting at (1,0) means $y = 0$ when $t = 1$, so $C = 0$ and $y = \ln|t|$.
After ten steps, $t = 2$, so $y = \ln 2 \approx 0.693$.

3. (a)

x	y	$\Delta y = (\text{slope})\Delta x$
0	1.000	−0.200
0.2	0.800	−0.120
0.4	0.680	−0.060
0.6	0.620	−0.005
0.8	0.615	0.052
1	0.667	

(b)

x	y	$\Delta y = (\text{slope})\Delta x$
0	1.000	−0.100
0.1	0.900	−0.080
0.2	0.820	−0.063
0.3	0.757	−0.048
0.4	0.708	−0.034
0.5	0.674	−0.020
0.6	0.654	−0.007
0.7	0.647	0.007
0.8	0.654	0.021
0.9	0.675	0.035
1	0.710	

5. Since the error is proportional to the reciprocal of the number of subintervals, the error using 10 intervals should be roughly half the error obtained using 5 intervals. Since both the estimates are underestimates, if we let A be the actual value we have:

$$\begin{aligned}\frac{1}{2}(A - 0.667) &= A - 0.710\\ A - 0.667 &= 2A - 1.420\\ A &= 0.753\end{aligned}$$

Therefore, 0.753 should be a better approximation.

7. Assume that $x > 0$ and that we use n steps in Euler's method. Label the x-coordinates we use in the process $x_0, x_1, \ldots, x_n$, where $x_0 = 0$ and $x_n = x$. Then using Euler's method to find $y(x)$, we get

	x	y	$\Delta y = (\text{slope})\Delta x$
P_0	$0 = x_0$	0	$f(x_0)\Delta x$
P_1	x_1	$f(x_0)\Delta x$	$f(x_1)\Delta x$
P_2	x_2	$f(x_0)\Delta x + f(x_1)\Delta x$	$f(x_2)\Delta x$
$\vdots$	$\vdots$	$\vdots$	$\vdots$
P_n	$x = x_n$	$\sum_{i=0}^{n-1} f(x_i)\Delta x$	

Thus the result from Euler's method is $\sum_{i=0}^{n-1} f(x_i)\Delta x$. We recognize this as the left-hand Riemann sum that approximates $\int_0^x f(t)\, dt$.

8.4 Solutions

1. $\frac{dP}{dt} = 0.02P \Rightarrow \frac{dP}{P} = 0.02\, dt$.

   $\int \frac{dP}{P} = \int 0.02\, dt \Rightarrow \ln|P| = 0.02t + C$.

   $|P| = e^{0.02t+C} \Rightarrow P = Ae^{0.02t}$, where $A = \pm e^C$.
   We are given $P(0) = 20$. Therefore, $P(0) = Ae^{(0.02)\cdot 0} = A = 20$. So the solution is $P = 20e^{0.02t}$.

2. $\frac{dQ}{dt} = \frac{Q}{5} \Rightarrow \frac{dQ}{Q} = \frac{dt}{5}$.

   $\int \frac{dQ}{Q} = \int \frac{dt}{5} \Rightarrow \ln|Q| = \frac{1}{5}t + C$.

   So $|Q| = e^{\frac{1}{5}t+C} = e^{\frac{1}{5}t}e^C \Rightarrow Q = Ae^{\frac{1}{5}t}$, where $A = \pm e^C$. From the initial conditions we know that $Q(0) = 50$, so $Q(0) = Ae^{(\frac{1}{5})\cdot 0} = A = 50$. Thus $Q = 50e^{\frac{1}{5}t}$.

3. $\frac{dm}{dt} = 3m$. As in problems 1 and 2, we get

   $$m = Ae^{3t}.$$

   Since $m = 5$ when $t = 1$, we have $5 = Ae^3$, so $A = \frac{5}{e^3}$. Thus $m = \frac{5}{e^3}e^{3t} = 5e^{3t-3}$.

5. $\frac{dy}{dx} + \frac{y}{3} = 0 \Rightarrow \frac{dy}{dx} = -\frac{y}{3} \Rightarrow \int \frac{dy}{y} = -\int \frac{1}{3}\, dx$.
   Integrating and moving terms, we have $y = Ae^{-\frac{1}{3}x}$. Since $y(0) = A = 10$, we have $y = 10e^{-\frac{1}{3}x}$.

7. $\frac{dP}{dt} = P + 4 \Rightarrow \frac{dP}{P+4} = dt.$

   $\int \frac{dP}{P+4} = \int dt \Rightarrow \ln|P+4| = t + C.$

   $P + 4 = Ae^t \Rightarrow P = Ae^t - 4$. $P = 100$ when $t = 0$, so $P(0) = Ae^0 - 4 = 100$, and $A = 104$. Therefore $P = 104e^t - 4$.

9. Factoring out the 0.1 gives $\frac{dm}{dt} = 0.1m + 200 = 0.1(m + 2000)$.
   $\frac{dm}{m+2000} = 0.1\,dt \Rightarrow \int \frac{dm}{m+2000} = \int 0.1\,dt \Rightarrow \ln|m+2000| = 0.1t + C$. So $m = Ae^{0.1t} - 2000$. Using the initial condition, $m(0) = Ae^{(0.1)\cdot 0} - 2000 = 1000$, so $A = 3000$. Thus $m = 3000e^{0.1t} - 2000$.

11. $\frac{dz}{dt} = te^z \Rightarrow e^{-z}dz = tdt \Rightarrow \int e^{-z}\,dz = \int t\,dt \Rightarrow -e^{-z} = \frac{t^2}{2} + C$.
   Since the solution passes through the origin, $z = 0$ when $t = 0$, we must have $-e^{-0} = \frac{0}{2} + C$, so $C = -1$. Thus $-e^{-z} = \frac{t^2}{2} - 1$, or $z = -\ln(1 - \frac{t^2}{2})$.

13. $\frac{dy}{dx} = \frac{5y}{x} \Rightarrow \int \frac{dy}{y} = \int 5\frac{dx}{x}. \Rightarrow \ln|y| = 5\ln|x| + C \Rightarrow |y| = e^{5\ln|x|}e^C$, and thus $y = Ax^5$ where $A = \pm e^C$. $y = 3$ when $x = 1$, so $A = 3$. Thus $y = 3x^5$.

15. $\frac{dw}{d\theta} = \theta w^2 \sin\theta^2 \Rightarrow \int \frac{dw}{w^2} = \int \theta \sin\theta^2\,d\theta \Rightarrow -\frac{1}{w} = -\frac{1}{2}\cos\theta^2 + C$. According to the initial conditions, $w(0) = 1$, so $-1 = -\frac{1}{2} + C$ and $C = -\frac{1}{2}$. Thus $-\frac{1}{w} = -\frac{1}{2}\cos\theta^2 - \frac{1}{2} \Rightarrow \frac{1}{w} = \frac{\cos\theta^2 + 1}{2} \Rightarrow w = \frac{2}{\cos\theta^2 + 1}$.

17. $e^{-\cos\theta}\frac{dz}{d\theta} = \sqrt{1 - z^2}\sin\theta \Rightarrow \int \frac{dz}{\sqrt{1-z^2}} = \int e^{\cos\theta}\sin\theta\,d\theta \Rightarrow \arcsin z = -e^{\cos\theta} + C$. According to the initial conditions: $z(0) = \frac{1}{2}$, so $\arcsin\frac{1}{2} = -e^{\cos 0} + C$, therefore $\frac{\pi}{6} = -e + C$, and $C = \frac{\pi}{6} + e$. Thus $z = \sin(-e^{\cos\theta} + \frac{\pi}{6} + e)$.

19. $(1 + t^2)y\frac{dy}{dt} = 1 - y \Rightarrow \int \frac{y\,dy}{1-y} = \int \frac{dt}{1+t^2} \Rightarrow \int \left(-1 + \frac{1}{1-y}\right)dy = \int \frac{dt}{1+t^2}$. Therefore $-y - \ln|1 - y| = \arctan t + C$. $y(1) = 0$, so $0 = \arctan 1 + C$, and $C = -\frac{\pi}{4}$, so $-y - \ln|1 - y| = \arctan t - \frac{\pi}{4}$. We cannot solve for y in terms of t.

21. $\frac{dR}{dt} = kR \Rightarrow \frac{dR}{R} = k\,dt \Rightarrow \int \frac{dR}{R} = \int k\,dt$. Integrating gives $\ln|R| = kt + C$, so $|R| = e^{kt+C} = e^{kt}e^C$. $R = Ae^{kt}$, where $A = \pm e^C$.

23. $\frac{dP}{dt} = P - a \Rightarrow \frac{dP}{P-a} = dt \Rightarrow \int \frac{dP}{P-a} = \int dt$. Integrating yields $\ln|P - a| = t + C$, so $|P - a| = e^{t+C} = e^te^C$. $P = a + Ae^t$, where $A = \pm e^C$.

25. $\frac{dP}{dt} = k(P - a) \Rightarrow \frac{dP}{P-a} = k\,dt \Rightarrow \int \frac{dP}{P-a} = \int k\,dt$. Integrating yields $\ln|P - a| = kt + C$ so $P = a + Ae^{kt}$ where $A = \pm e^C$.

27. $\frac{dy}{dt} = y(2-y) \Rightarrow \frac{dy}{y(y-2)} = -\,dt \Rightarrow \int \frac{dy}{(y-2)(y)} = -\int dt \Rightarrow -\frac{1}{2}\int (\frac{1}{y} - \frac{1}{y-2})dy = -\int dt.$
Integrating yields $\frac{1}{2}(\ln|y-2| - \ln|y|) = -t + C$, so $\ln\frac{|y-2|}{|y|} = -2t + 2C$.
Exponentiating both sides yields $|1 - \frac{2}{y}| = e^{-2t+2C} \Rightarrow \frac{2}{y} = 1 - Ae^{-2t}$, where $A = \pm e^{2C}$.
Hence $y = \frac{2}{1-Ae^{-2t}}$. But $y(0) = \frac{2}{1-A} = 1$, so $A = -1$, and $y = \frac{2}{1+e^{-2t}}$.

29. $\frac{dx}{dt} = \frac{x\ln x}{t} \Rightarrow \int \frac{dx}{x\ln x} = \int \frac{dt}{t} \Rightarrow \ln|\ln x| = \ln|t| + C \Rightarrow |\ln x| = e^C e^{\ln|t|} = e^C|t|$. Therefore $\ln x = At$, where $A = \pm e^C$, so $x = e^{At}$.

31. Separation of variables yields $\displaystyle\int \frac{dy}{y\ln y} = \int \frac{dt}{t^2}$, so $\ln|\ln y| = -\frac{1}{t} + C$.
Exponentiating both sides gives:

$$|\ln y| = e^{(-\frac{1}{t}+C)} = e^{-\frac{1}{t}}e^C.$$

So, $\ln y = Ae^{-\frac{1}{t}}$, where $A = \pm e^C$. Exponentiating once more gives $y = e^{Ae^{(-\frac{1}{t})}}$.

33. $\frac{dQ}{dt} = -t^2Q^2 - Q^2 + 4t^2 + 4 = -Q^2(t^2+1) + 4(t^2+1) = (t^2+1)(4-Q^2)$. Separating variables yields $\frac{dQ}{4-Q^2} = (t^2+1)\,dt$, so $-\int \frac{dQ}{(Q-2)(Q+2)} = -\frac{1}{4}\int (\frac{1}{Q-2} - \frac{1}{Q+2})dQ = \int (t^2+1)\,dt$
Integrating, we obtain $-\frac{1}{4}(\ln|Q-2| - \ln|Q+2|) = \frac{t^3}{3} + t + C$, so $\ln\frac{|Q-2|}{|Q+2|} = -\frac{4t^3}{3} - 4t - 4C$.
Exponentiating yields $|\frac{Q-2}{Q+2}| = e^{-\frac{4t^3}{3}-4t}e^{-4C}$. $\frac{Q-2}{Q+2} = Ae^{-\frac{4t^3}{3}-4t}$ where $A = \pm e^{4C}$. Solving for Q, $Q = \frac{4}{1-Ae^{-\frac{4t^3}{3}-4t}} - 2$. Notice that A could be any constant, including 0. In fact, we also lost the solution $Q = -2$ when we divided both sides by $4 - Q^2$. (The solution $Q = 2$ corresponds to $A = 0$, but $Q = -2$, another valid solution, is lost by our division.)

35. (a) See (b).
(b)

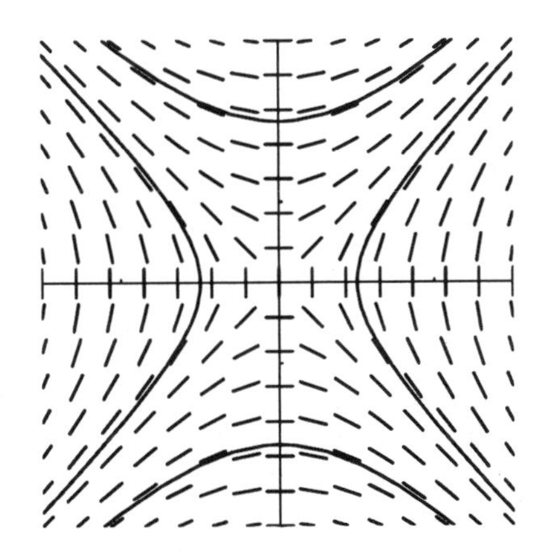

(c) $\frac{dy}{dx} = \frac{x}{y} \Rightarrow \int y\,dy = \int x\,dx \Rightarrow \frac{y^2}{2} = \frac{x^2}{2} + C$, or $y^2 - x^2 = 2C$. This is the equation of the hyperbolas in (b).

37. By looking at the slope fields, we see that any solution curve of $y' = \frac{x}{y}$ intersects any solution curve to $y' = -\frac{y}{x}$. Now if the two curves intersect at (x, y), then the two slopes at (x, y) are negative reciprocals of each other, because $-\frac{1}{x/y} = -\frac{y}{x}$. Hence, the two curves intersect at right angles.

8.5 Solutions

1. (a) If the world's population grows exponentially, satisfying $\frac{dP}{dt} = kP$, and if the arable land used is proportional to the population, then we'd expect A to satisfy $\frac{dA}{dt} = kA$. One is, of course, also assuming that the amount of arable land is large compared to the amount that is now being used.

   (b) We must solve $A = A_0e^{kt} = (1 \times 10^9)e^{kt}$, where t is the number of years after 1950. Since $2 \times 10^9 = (1 \times 10^9)e^{k(30)}$, we have $e^{30k} = 2$, so $k = \frac{\ln 2}{30} \approx 0.023$. Thus, $A \approx (1 \times 10^9)e^{0.023t}$. We want to find t such that $3.2 \times 10^9 = A = (1 \times 10^9)e^{0.023t}$ Taking logarithms yields,

$$t = \frac{\ln(3.2)}{0.023} \approx 50.6 \text{ years.}$$

   Thus the arable land will have run out by the year 2001.

3. (a) The rate of growth of the money in the account is proportional to the amount of money in the account. Thus

$$\frac{dM}{dt} = rM.$$

   (b) Solving, we have $\frac{dM}{M} = r\,dt$.

$$\int \frac{dM}{M} = \int r\,dt \Rightarrow \ln|M| = rt + C \Rightarrow M = e^{rt+C} = Ae^{rt}$$

   When $t = 0$ (in 1970), $M = 1000$, so $A = 1000$ and $M = 1000e^{rt}$.

   (c)

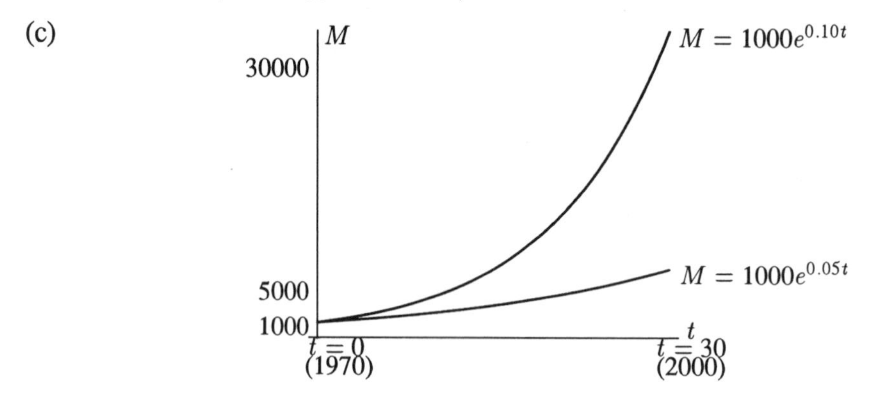

5. (a) Since the rate of change is proportional to the amount present, $\frac{dy}{dt} = ky$ for some constant k.

(b) Solving the differential equation, we have $y = Ae^{kt}$, where A is the initial amount. Since 100 grams become 54.9 grams in one hour, $54.9 = 100e^k$, so $k = \ln\frac{54.9}{100} \approx -0.6$. Thus, after 10 hours, there remains $100e^{(-0.6)10} \approx 0.248$ grams.

7. (a) If I is intensity and l is the distance traveled through the water, then for some $k > 0$,

$$\frac{dI}{dl} = -kI$$

(the proportionality constant is negative because intensity decreases with distance) Thus $I = Ae^{-kl}$. Since $I = A$ when $l = 0$, A represents the initial intensity of the light.

(b) If 50% of the light is absorbed in 10 feet, then $0.50A = Ae^{-10k}$, so $e^{-10k} = \frac{1}{2}$, giving

$$k = \frac{-\ln\frac{1}{2}}{10} = \frac{\ln 2}{10}.$$

In 20 feet, the percentage of light left is

$$e^{-\frac{\ln 2}{10}\cdot 20} = e^{-2\ln 2} = (e^{\ln 2})^{-2} = 2^{-2} = \frac{1}{4},$$

so $\frac{3}{4}$ or 75% of the light has been absorbed. Similarly, after 25 feet,

$$e^{-\frac{\ln 2}{10}\cdot 25} = e^{-2.5\ln 2} = (e^{\ln 2})^{-\frac{5}{2}} = 2^{-\frac{5}{2}} \approx 0.177.$$

Approximately 17.7% of the light is left, so 82.3% of the light has been absorbed.

9. (a) If $C' = -kC$, and then $C = C_0e^{-kt}$. Since the half-life is 5700 years, $\frac{1}{2}C_0 = C_0e^{-5700k}$. Solving for k, we have $-5700k = \ln\frac{1}{2} \Rightarrow k = \frac{-\ln\frac{1}{2}}{5700} \approx 0.0001216$.

(b) From the given information, we have $0.91 = e^{-kt}$, where t is the age of the shroud. Solving for t, we have $t = \frac{-\ln 0.91}{k} \approx 775.6$ years.

8.6 Solutions

1. Let $D(t)$ be the quantity of dead leaves (in grams) per square centimeter. Then $\frac{dD}{dt} = 3 - 0.75D$, where t is in years. Again, we factor out -0.75 and then separate variables.

$$\begin{aligned} \frac{dD}{dt} &= -0.75(D-4) \\ \int \frac{dD}{D-4} &= \int -0.75\,dt \\ \ln|D-4| &= -0.75t + C \\ |D-4| &= e^{-0.75t+C} = e^{-0.75t}e^{C} \\ D &= 4 + Ae^{-0.75t}, \text{ where } A = \pm e^{C}. \end{aligned}$$

If initially the ground is clear, the solution looks like:

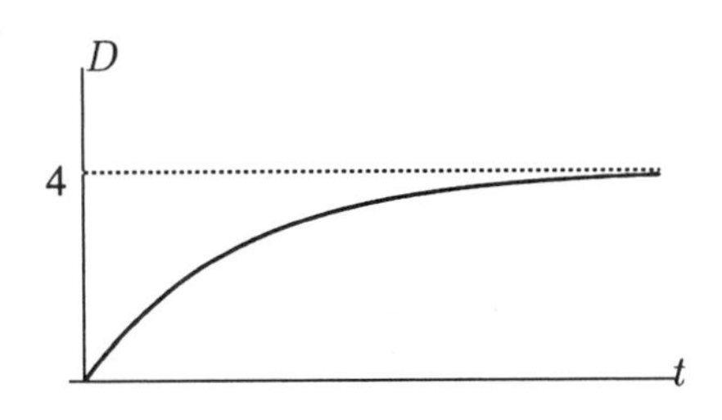

The equilibrium level is 4 grams per square centimeter, regardless of the initial condition.

3. (a)
$$\frac{dT}{dt} = -k(T - A),$$
where $A = 10$ is the outside temperature.

(b) Integrating on both sides,
$$\int \frac{dT}{T - A} = -\int k\, dt.$$
Then $\ln|T - A| = -kt + C$, $T = A + Be^{-kt}$. So $T = A + (T_0 - A)e^{-kt}$, where $T_0 = 68$ is the initial temperature. Thus
$$T = 10 + 58e^{-kt}.$$
Letting 1:00 p.m. be $t = 0$, then at 10:00 p.m. ($t = 9$), we have
$$\begin{aligned} 57 &= 10 + (68 - 10)e^{-9k} \\ \frac{47}{58} &= e^{-9k} \\ \ln\frac{47}{58} &= -9k \\ k &= -\frac{1}{9}\ln\frac{47}{58} \approx 0.0234. \end{aligned}$$
At 7:00 a.m. the next morning ($t = 18$) we have
$$\begin{aligned} T &\approx 10 + (68 - 10)e^{18(-0.0234)} \\ &= 10 + 58(0.66) \\ &\approx 48°\text{F} \end{aligned}$$
Your pipes won't freeze.

(c) We assumed that the temperature outside the house stayed constant at 10°F. This is probably incorrect because the temperature was most likely warmer during the day (between 1 p.m. and 10 p.m.) and colder after (between 10 p.m. and 7 a.m.). Thus, when the temperature in the house dropped from 68°F to 57°F between 1 p.m. and 10 p.m., the outside temperature was probably higher than 10°F, which changes our calculation of the value of the constant k. The house temperature will most certainly be lower than 48°F at 7 a.m., but not by much—not enough to freeze.

5. (a) Use the fact that
$$\left(\begin{array}{c}\text{Rate at which}\\ \text{balance is increasing}\end{array}\right) = \left(\begin{array}{c}\text{Rate interest}\\ \text{is accrued}\end{array}\right) - \left(\begin{array}{c}\text{Rate payments}\\ \text{are made}\end{array}\right).$$
Thus $\frac{dB}{dt} = 0.05B - 12000$.

(b) We solve the equation by separation of variables. First, however, we factor out a 0.05 on the right hand side of the equation to make the work easier.
$\frac{dB}{dt} = 0.05(B - 240000) \Rightarrow \frac{dB}{B-240000} = 0.05\,dt \Rightarrow \int \frac{dB}{B-240000} = \int 0.05\,dt$, so
$\ln|B - 240000| = 0.05t + C \Rightarrow |B - 240000| = e^{0.05t+C} = e^{0.05t}e^{C}$, so
$B - 240000 = Ae^{0.05t}$, where $A = \pm e^{C}$.
If the initial balance is B_0, then $B_0 - 240000 = Ae^0 = A$, thus $B - 240000 = (B_0 - 240000)e^{0.05t}$, so $B = (B_0 - 240000)e^{0.05t} + 240000$.

(c) To find the initial balance such that the account has a 0 balance after 20 years, we solve $0 = (B_0 - 240000)e^{(0.05)20} + 240000 = (B_0 - 240000)e^{1} + 240000$,

$$B_0 = 240000 - \frac{240000}{e} \approx \$151708.93.$$

7. (a) $\frac{dp}{dt} = k(p - p_0)$, where k is the proportionality constant of the Evans Price Adjustment model. Notice that $k < 0$, since if $p > p_0$ then $\frac{dp}{dt}$ should be negative, and if $p < p_0$ then $\frac{dp}{dt}$ should be positive.

(b) Separating variables, we have $\frac{dp}{p-p_0} = k\,dt$. Solving, we find $p = p_0 + (p - I)e^{kt}$, where I is the initial price.

(c)

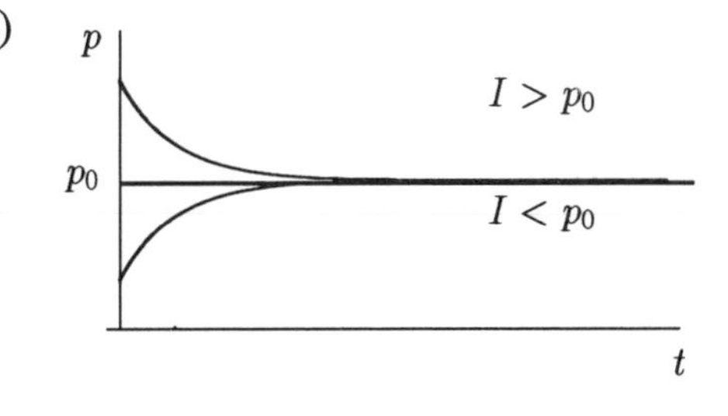

(d) As $t \to \infty, p \to p_0$. We see this in the solution in (b), since as $t \to \infty, e^{kt} \to 0$. (Remember $k < 0$!) In other words, as $t \to \infty$, p approaches the equilibrium price p_0.

9. (a) The rate at which salt enters the pool is

$$(10 \text{ grams/liter}) \times (60 \text{ liters/minute}) = (600 \text{ grams/minute})$$

The rate at which salt leaves the pool depends on the concentration of salt in the pool. At time t, the concentration is $\frac{S(t)}{2\times10^6 \text{ liters}}$, where $S(t)$ is measured in grams. Thus the rate at which salt leaves the pool is

$$\frac{S(t) \text{ grams}}{2 \times 10^6 \text{ liters}} \times \frac{60 \text{ liters}}{\text{minute}} = \frac{3S(t) \text{ grams}}{10^5 \text{ minutes}}.$$

Thus

$$\frac{dS}{dt} = 600 - \frac{3S}{100000}.$$

(b) $\frac{dS}{dt} = -\frac{3}{100000}(S - 20000000)$
$\int \frac{dS}{S-20000000} = \int -\frac{3}{100000}\,dt$
$\ln|S - 20000000| = -\frac{3}{100000}t + C$
$S = 20000000 - Ae^{-\frac{3}{100000}t}$
Since $S = 0$ at $t = 0$, $A = 20000000$. Thus $S(t) = 20000000 - 20000000e^{-\frac{3}{100000}t}$.

(c) As $t \to \infty$, $e^{-\frac{3}{100000}t} \to 0$, so $S(t) \to 20000000$ grams. The concentration approaches 10 grams/liter. Note that this makes sense; we'd expect the concentration of salt in the pool to become closer and closer to the concentration of salt being poured into the pool as $t \to \infty$.

10. (a) $\frac{dM}{dt} = IM$, where I is the interest rate at time t. In this case, however, I is not constant, but depends on t. Using the beginning of 1975 as $t = 0$, and measuring time in years, we have $I = 0.50 + 0.25t$.
Thus our differential equation is $\frac{dM}{dt} = (0.50 + 0.25t)M$.

(b)

$$\begin{aligned}
\frac{dM}{dt} &= (0.50 + 0.25t)M \\
\int \frac{dM}{M} &= \int (0.50 + 0.25t)\,dt \\
\ln|M| &= 0.50t + 0.125t^2 + C \\
M &= Ae^{0.50t+0.125t^2}.
\end{aligned}$$

Since the initial deposit is 100,000 cruzieros, we also know that $M = 100,000$ when $t = 0$, and so we have $A = 100,000$. Thus $M = 100,000e^{0.50t+0.125t^2}$.

(c) On January 1, 1980, we have $t = 5$. Thus $M = 100,000e^{0.50(5)+0.125(5)^2} \approx 27,727,000$ cruzieros.

11. (a) If the interest rate had been constant, the equation relating money and time would have been $M = 100,000e^{rt}$ where r is the constant interest rate. Using Problem 10(c), we see that for the amount of money to be equal after 5 years we need

$$\begin{aligned}
100,000e^{0.50(5)+0.125(5)^2} &= 100,000e^{5r} \\
5r &= 0.50(5) + 0.125(5^2) \\
r &= 1.125 = 112.5\%
\end{aligned}$$

(b)

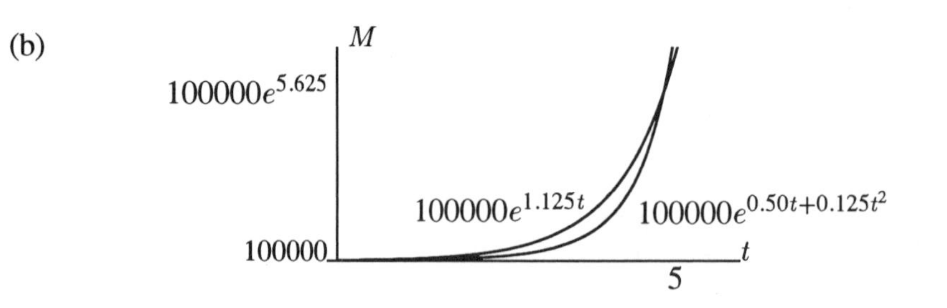

8.7 Solutions

1. $P = (6.6 \times 10^6)e^{0.002t}$.

3. Rewriting the equation as $\frac{1}{P}\frac{dP}{dt} = \frac{(100-P)}{1000}$, we see that this is a logistic equation. Before looking at its solution, we explain intuitively why there must always be at least 100 individuals. Since the population begins at 200, $\frac{dP}{dt}$ is initially negative, so the population decreases. It continues to do so while $P > 100$. If the population ever reached 100, however, then $\frac{dP}{dt}$ would be 0. This means the population would stop changing – so if the population ever decreased to 100, that's where it would stay.

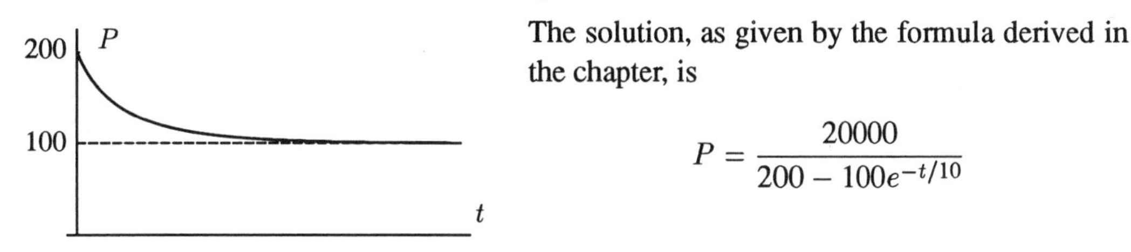

The solution, as given by the formula derived in the chapter, is

$$P = \frac{20000}{200 - 100e^{-t/10}}$$

5. (a) Let I be the number of informed people. Then this model predicts that $\frac{dI}{dt} = k(M - I)$ for some positive constant k. Solving this, we find the solution is

$$I = M - Me^{-kt}.$$

Notice that $\frac{dI}{dt}$ is largest when I is smallest, so the information spreads fastest in the beginning, at $t = 0$.

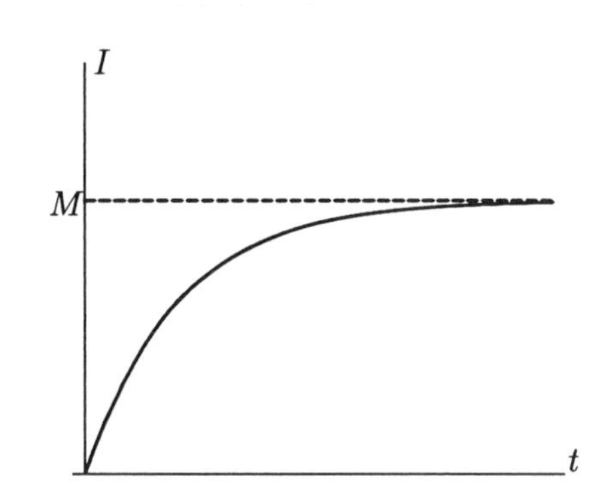

(b) In this case, the model suggests that $\frac{dI}{dt} = kI(M - I)$ for some positive constant k. From the result derived in the chapter, we know the solution to this is

$$I = \frac{MI_0}{I_0 + (M - I_0)e^{-kMt}}.$$

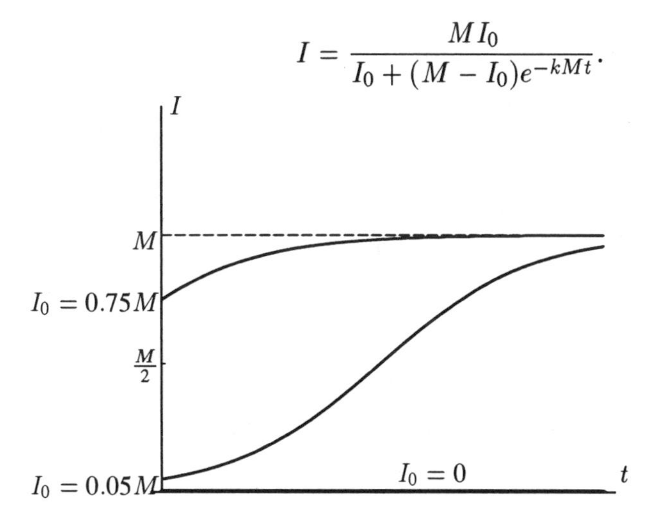

i. If $I_0 = 0$. Then $I = 0$ for all t. In other words, if nobody knows something, it doesn't spread by word of mouth!

ii. If $I_0 = 0.05M$, then $\frac{dI}{dt}$ is increasing up to $I = \frac{M}{2}$. Thus, the information is spreading fastest at $I = \frac{M}{2}$.

iii. If $I_0 = 0.75M$, then since $\frac{dI}{dt}$ is always decreasing for $I > \frac{M}{2}$, $\frac{dI}{dt}$ is largest when $t = 0$.

7. (a) Here we have, where t = years since 1800:

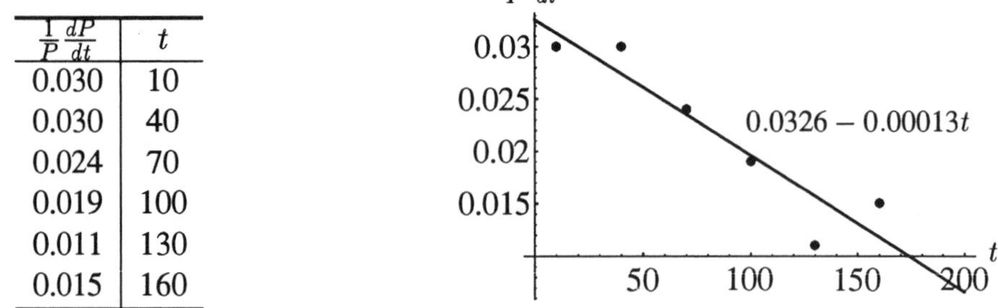

$\frac{1}{P}\frac{dP}{dt}$	t
0.030	10
0.030	40
0.024	70
0.019	100
0.011	130
0.015	160

Graphing the data and fitting a line, we get $\frac{1}{P}\frac{dP}{dt} = 0.0326 - 0.00013t$ as our guess.

(b) $\frac{dP}{dt}$ will be positive and P will increase until $0.0326 = 0.00013t$, i.e. until $t \approx 250$ or about the year 2050.

(c) $\frac{dP}{P} = (0.0326 - 0.00013t)\,dt$.
$\int \frac{dP}{P} = \int (0.0326 - 0.00013t)\,dt$.
$P = Ae^{0.0326t - 0.000065t^2}$
Using the fact that $P = 5.3$ when $t = 0$, we get $P = 5.3e^{0.0326t - 0.000065t^2}$

8.8 Solutions

1. This is an example of a predator-prey relationship. Normally, we would expect the worm population, without any predator after it, to increase at a rate proportional to the population size. In other words, we would expect $\frac{dx}{dt} = ax$. However, since there are predators (robins), $\frac{dx}{dt}$ won't be that big. We must lessen $\frac{dx}{dt}$. It makes sense to lessen it by an amount proportional to the interaction between robins and worms. We can think of this interaction as proportional to the size of the product of the populations. Thus, $\frac{dx}{dt} = ax - bxy$, or $\frac{1}{x}\frac{dx}{dt} = a - by$. a thus represents the normal growth rate for worms, and b is related to the rate at which worms are killed by robins. Similarly, we set $\frac{1}{y}\frac{dy}{dt} = -c + kx$. Here, $-c$ is the natural growth rate of the robin population without food. Notice the sign is negative, since without food the robin population would decline. k is related to how the robin population grows given its own size and the worm food supply; it is given a positive sign since the food supply will help the robin population grow.

3. There is symmetry across the line $y = x$. Indeed, since $\frac{dy}{dx} = \frac{y(x-1)}{x(1-y)}$, if we switch x and y we get $\frac{dx}{dy} = \frac{x(y-1)}{y(1-x)}$, so $\frac{dy}{dx} = \frac{y(1-x)}{x(y-1)}$. Since switching x and y changes nothing, the slope field must be symmetric across the line $y = x$. The slope field shows that the solution curves are either spirals or closed curves. Since there is symmetry about the line $y = x$, the solutions must in fact be closed curves.

5. Sketching the trajectory through the point $(2, 2)$ on the slope field given shows that the maximum robin population is about 2500, and the minimum robin population is about 500. When the robin population is at its maximum, the worm population is about 1,000,000.

7. It will work somewhat; the maximum number the robins reach will increase. However, the minimum number the robins reach will decrease as well. (See graph of slope field.) In the long term, the robin-worm populations will again fall into a cycle.

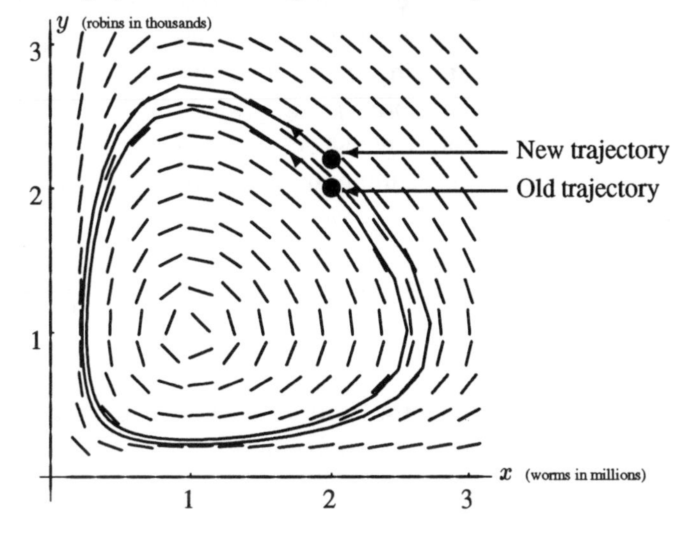

Note that if too many robins are added the minimum number may get so small the model may fail since a small number of robins are more susceptible to disaster.

9. If alone, the x and y populations each grow exponentially, because the equations become $\dfrac{dx}{dt} = 0.01x$ and $\dfrac{dy}{dt} = 0.2y$. For each population, the presence of the other decreases their growth rate. The two populations are therefore competitors—they may be eating each other's food, for instance.

11. (a) Equilibrium points are where $\frac{dx}{dt} = 0$ and $\frac{dy}{dt} = 0$.
$15x - 3xy = 0$ gives $3x(5 - y) = 0$, so $x = 0$ or $y = 5$
$-14y + 7xy = 0$ gives $7y(-2 + x) = 0$, so $x = 2$ or $y = 0$.
The solutions are thus (0,0) and (2,5).

(b) At $x = 2$, $y = 0$ we have $\frac{dy}{dt} = 0$ but $\frac{dx}{dt} = 15(2) - 3(2)(0) = 30 \neq 0$. Thus $x = 2$, $y = 0$ is not an equilibrium point.

13. (a) If B were not present, then we'd have $A' = 2A$, so company A's net worth would grow exponentially. Similarly, if A were not present, B would grow exponentially. The two companies restrain each other's growth, probably by competing for the market.

(b) To find equilibrium points, find the solutions of the pair of equations

$$\begin{aligned} A' &= 2A - AB = 0 \\ B' &= B - AB = 0 \end{aligned}$$

The first equation has solutions $A = 0$ or $B = 2$. The second has solutions $B = 0$ or $A = 1$. Thus the equilibrium points are (0,0) and (1,2).

(c) In the long run, one of the companies will go out of business. Two of the trajectories in the figure below go towards the A axis; they represent A surviving and B going out of business. The trajectories going towards the B axis represent A going out of business. Notice both the equilibrium points are unstable.

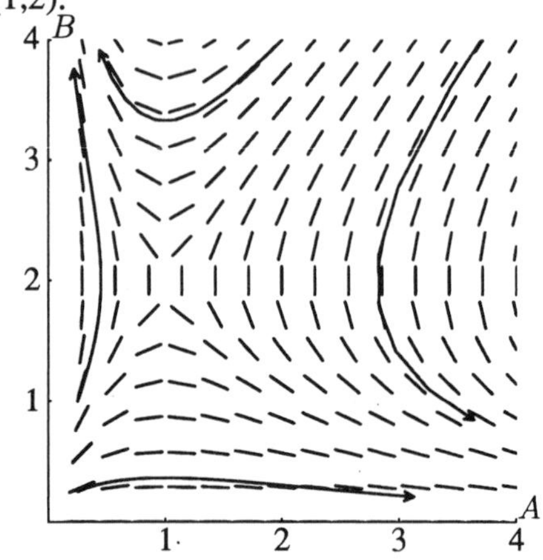

15. (a) Both X and Y would grow freely without each other, and their interaction inhibits their growth. Thus, X and Y are competitors.

(b) Equilibrium points are where $4y - 2xy = 0$ and $x - 2xy = 0$. This occurs at (0,0), $(2, \frac{1}{2})$.

(c)

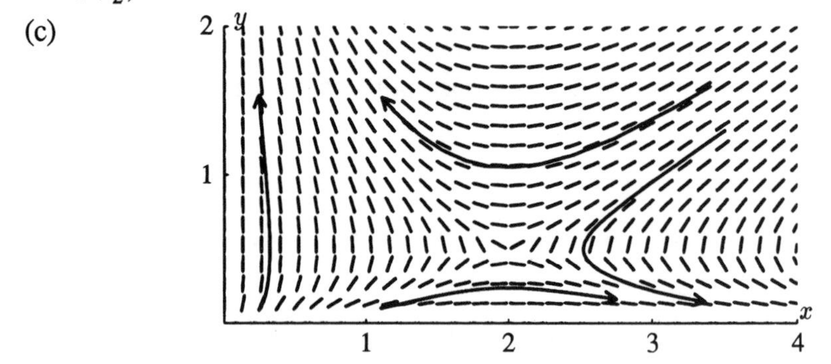

17. (a) $\dfrac{\frac{dy}{dt}}{\frac{dx}{dt}} = \dfrac{dy}{dx} = \dfrac{-3y - xy}{-2x - xy} = \dfrac{y(x+3)}{x(y+2)}$. Thus

$$\left(\frac{y+2}{y}\right) dy = \left(\frac{x+3}{x}\right) dx \Rightarrow \int (1 + \frac{2}{y})\, dy = \int (1 + \frac{3}{x})\, dx$$

. So, $y + 2\ln|y| = x + 3\ln|x| + C$. Since x and y are non-negative, $y + 2\ln y = x + 3\ln x + C$. This is as far as we can go with this equation – we cannot solve for y in terms of x, for example. We can, however, put it in the form $e^{y+2\ln y} = e^{x+3\ln x+C}$, or $y^2e^y = Ax^3e^x$.

(b) An equilibrium state satisfies $\frac{dx}{dt} = -2x - xy = 0$ and $\frac{dy}{dt} = -3y - xy = 0$. Solving the first equation, we have $-x(y+2) = 0$, so $x = 0$ or $y = -2$. The second equation has solutions $y = 0$ or $x = -3$. Since $x, y \geq 0$, the only equilibrium point is (0,0).

(c) We can use either of our forms for the solution. Looking at $y^2e^y = Ax^3e^x$, we see that if x and y are very small positive numbers, then $e^x \approx e^y \approx 1$. Thus, $y^2 \approx Ax^3$, or $\frac{y^2}{x^3} \approx A$, a constant. Looking at $y + 2\ln y = x + 3\ln x + C$, we note that if x and y are small, then they are negligible compared to $\ln y$ and $\ln x$. Thus, $2\ln y \approx 3\ln x + C$, giving $\ln y^2 - \ln x^3 \approx C$, so $\ln\frac{y^2}{x^3} \approx C$ and therefore $\frac{y^2}{x^3} \approx e^C$, a constant.

(d) If $x(0) = 4$ and $y(0) = 8$, then $8 + 2\ln 8 = 4 + 3\ln 4 + C$. Note that $2\ln 8 = 3\ln 4 = \ln 64$, giving $4 = C$. So the phase trajectory is $y + 2\ln y = x + 3\ln x + 4$. (Or equivalently, $y^2e^y = e^4x^3e^x = x^3e^{x+4}$).

(e) If the concentrations are equal, then $y + 2\ln y = y + 3\ln y + 4$, giving $-\ln y = 4$ or $y = e^{-4}$. Thus, they are equal when $y = x = e^{-4} \approx 0.0183$.

(f) Using part (c), we have that if x is small, $\frac{y^2}{x^3} \approx e^4$. Since $x = e^{-10}$ is certainly small, $\frac{y^2}{e^{-30}} \approx e^4$, and $y \approx e^{-13}$.

8.9 Solutions

1. If $y = 2\cos t + 3\sin t$, then $y' = -2\sin t + 3\cos t$ and $y'' = -2\cos t - 3\sin t$. Thus, $y'' + y = 0$.

3. $y = A\cos\alpha t$
$y' = -\alpha A\sin\alpha t$
$y'' = -\alpha^2 A\cos\alpha t$
If $y'' + 5y = 0$, then $-\alpha^2 A\cos\alpha t + 5A\cos\alpha t = 0$, so $A(5 - \alpha^2)\cos\alpha t = 0$. This is true for all t if $A = 0$, or if $\alpha = \pm\sqrt{5}$.
We also have the initial condition: $y'(1) = -\alpha A\sin\alpha = 3$. Notice that this equation will not work if $A = 0$. If $\alpha = \sqrt{5}$, then
$A = -\frac{3}{\sqrt{5}\sin\sqrt{5}} \approx -1.705$.
Similarly, if $\alpha = -\sqrt{5}$, we find that $A \approx -1.705$. Thus, the possible values are $A = -\frac{3}{\sqrt{5}\sin\sqrt{5}} \approx -1.705$ and $\alpha = \pm\sqrt{5}$.

5. At $t = 0$, we find $y = 0$. Since $-1 \leq \sin 3t \leq 1$, y ranges from -0.5 to 0.5, so at $t = 0$ it is starting in the middle. Since $y' = -1.5\cos 3t$, we see $y' = -1.5$ when $t = 0$, so the mass is moving downward.

7. (a) We recall that the chapter stated that the general solution to the equation $\frac{d^2s}{dt^2} + \omega^2 s = 0$ is $s(t) = C_1\cos\omega t + C_2\sin\omega t$. Using this, we find that the general solution of $y'' = -4y$ is $y(t) = C_1\cos 2t + C_2\sin 2t$.

   (b) Notice that $y'(t) = -2C_1\sin 2t + 2C_2\cos 2t$.

   i.

$$\begin{aligned} y(0) &= 5 = C_1\cos 0 + C_2\sin 0 = C_1, \text{ so } C_1 = 5. \\ y'(0) &= 0 = -2C_1\sin 0 + 2C_2\cos 0 = C_2, \text{ so } C_2 = 0. \end{aligned}$$

   Thus, $y(t) = 5\cos 2t$ is the solution.

   ii.

$$\begin{aligned} y(0) &= 0 = C_1\cos 0 + C_2\sin 0 = C_1, \text{ so } C_1 = 0. \\ y'(0) &= 10 = -2C_1\sin 0 + 2C_2\cos 0 = C_2, \text{ so } C_2 = 5. \end{aligned}$$

   Thus, $y(t) = 5\sin 2t$ is the solution.

   iii. As in part i), $y(0) = 5$. So, $C_1 = 5$.
   $y'(0) = 5$. So, $C_2 = \frac{5}{2}$.
   Thus, $y(t) = 5\cos 2t + \frac{5}{2}\sin 2t$ is the solution.

   (c)

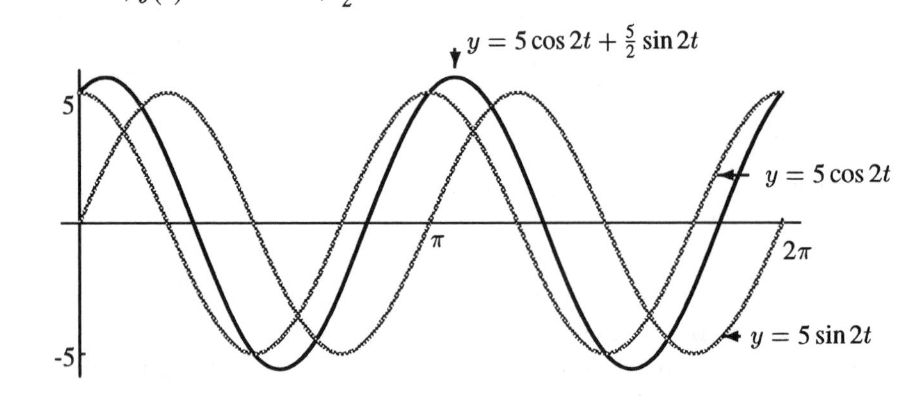

   Notice that the third function (like the other two) is just a single oscillation.

9. All the differential equations have solutions of the form $s(t) = C_1\sin\omega t + C_2\cos\omega t$. Since for all of them, $s'(0) = 0$, we have $s'(0) = 0 = C_1\omega\cos 0 - C_2\omega\sin 0 = 0$, giving $C_1\omega = 0$. Thus, either $C_1 = 0$ or $\omega = 0$. If $\omega = 0$, then $s(t)$ is a constant function, and since the equations represent oscillating springs, we don't want $s(t)$ to be a constant function. Thus,

$C_1 = 0$, so all four equations have solutions of the form $s(t) = C\cos\omega t$.
i) $s'' + 4s = 0$, so $\omega = \sqrt{4} = 2$. $s(0) = C\cos 0 = C = 5$. Thus, $s(t) = 5\cos 2t$.
ii) $s'' + \frac{1}{4}s = 0$, so $\omega = \sqrt{\frac{1}{4}} = \frac{1}{2}$. $s(0) = C\cos 0 = C = 10$. Thus, $s(t) = 10\cos\frac{1}{2}t$.
iii) $s'' + 6s = 0$, so $\omega = \sqrt{6}$. $s(0) = C = 4$, Thus, $s(t) = 4\cos\sqrt{6}t$.
iv) $s'' + \frac{1}{6}s = 0$, so $\omega = \sqrt{\frac{1}{6}}$. $s(0) = C = 20$. Thus, $s(t) = 20\cos\sqrt{\frac{1}{6}}t$.

(a) Spring iii) has the shortest period, $\frac{2\pi}{\sqrt{6}}$. (Other periods are $\pi, 4\pi, 2\pi\sqrt{6}$)

(b) Spring iv) has the largest amplitude, 20.

(c) Spring iv) has the longest period, $2\pi\sqrt{6}$.

(d) Spring i) has the largest maximum velocity. We can see this by looking at $v(t) = s'(t) = -C\omega\sin\omega t$. The velocity is just a sine function, so we look for the derivative with the biggest amplitude, which will have the greatest value. The velocity function for Spring i) has amplitude 10, the largest of the four springs. (The other velocity amplitudes are $10 \cdot \frac{1}{2} = 5, 4\sqrt{6} \approx 9.8, \frac{20}{\sqrt{6}} \approx 8.2$)

11. (a) If x_0 is increased, the amplitude of the function x is increased, but the period remains the same. In other words, the pendulum will start higher, but the time to swing back and forth will stay the same.

(b) If l is increased, the period of the function x is increased. (Remember, the period of $x_0\cos\sqrt{\frac{g}{l}}t$ is $\frac{2\pi}{\sqrt{\frac{l}{g}}}$.) In other words, it will take longer for the pendulum to swing back and forth.

13. The amplitude is $A = \sqrt{7^2 + 24^2} = \sqrt{625} = 25$.
The phase shift, φ, is given by $\tan\varphi = \frac{24}{7}$, so $\varphi = \arctan\frac{24}{7} \approx 1.287$ or $\varphi \approx -1.855$.
Since $C_1 = 24 > 0$, we want $\varphi = 1.287$.

so the solution is $25\sin(\omega t + 1.287)$.

15. (a) $36\dfrac{d^2I}{dt^2} + \dfrac{I}{9} = 0$ so $\frac{d^2I}{dt^2} = -\frac{I}{324}$.
Thus,

$$\begin{aligned} I &= C_1\cos\frac{1}{18}t + C_2\sin\frac{1}{18}t. \\ I(0) &= 0 = C_1\cos 0 + C_2\sin 0 = C_1, \\ \text{so } C_1 &= 0. \end{aligned}$$

So, $I = C_2 \sin \frac{1}{18}t$, and

$$\begin{aligned} I' &= \frac{1}{18}C_2 \cos \frac{1}{18}t. \\ I'(0) &= 2 = \frac{1}{18}C_2 \cos\left(\frac{1}{18}\cdot 0\right) = \frac{1}{18}C_2, \\ \text{so } C_2 &= 36. \end{aligned}$$

Therefore, $I = 36 \sin \frac{1}{18}t$.

(b) As in part (a), $I = C_1 \cos \frac{1}{18}t + C_2 \sin \frac{1}{18}t$.
According to the initial conditions:

$$\begin{aligned} I(0) &= 6 = C_1 \cos 0 + C_2 \sin 0 = C_1, \\ \text{so } C_1 &= 6. \end{aligned}$$

So $I = 6 \cos \frac{1}{18}t + C_2 \sin \frac{1}{18}t$.
Thus,

$$\begin{aligned} I' &= -\frac{1}{3} \sin \frac{1}{18}t + \frac{1}{18}C_2 \cos \frac{1}{18}t. \\ I'(0) &= 0 = -\frac{1}{3} \sin\left(\frac{1}{18}\cdot 0\right) + \frac{1}{18}C_2 \cos\left(\frac{1}{18}\cdot 0\right) = \frac{1}{18}C_2, \\ \text{so } C_2 &= 0. \end{aligned}$$

Therefore, $I = 6 \cos \frac{1}{18}t$.

17. We know that the general formula for I will be of the form:

$$I = C_1 \cos \omega t + C_2 \sin \omega t.$$

Thus, as $t \to \infty$, the current does not approach a limit. Instead, the current varies sinusoidally. A positive current means charge moves forward along the circuit, from the capacitor to the inductor. A negative current means charge moves backwards in the circuit, from the inductor back to the capacitor. The charge sloshes back and forth, from capacitor to inductor, just as our idealized spring bounces up and down, forever.

8.10 Solutions

1.

s	v	$ds = v\Delta t$	$dv = -s\Delta t$
0	1	0.5	0
0.5	1	0.5	−0.25
1	0.75	0.375	−0.5

3. Solution depends on software. Here are formulas for an Excel spreadsheet:

	A	B	C	D	E	F
1	i	ti	yi	vi	$\frac{dy}{dt}$	$\frac{dv}{dt}$
2	1	0.00	2.00	0.00	=D2	= −2*(D2−C2)
3	=A2	=B2+0.1	=C2+E2*0.1	=D2+F2*0.1	=D3	= −2*(D3−C3)

8.11 Solutions

1. The characteristic equation is $r^2 + 4r + 3 = 0$, so $r = -1$ or -3.
   Therefore $y(t) = C_1e^{-t} + C_2e^{-3t}$.

3. The characteristic equation is $r^2 + 4r + 5 = 0$, so $r = -2 \pm i$.
   Therefore $y(t) = C_1e^{-2t}\cos t + C_2e^{-2t}\sin t$.

5. The characteristic equation is $r^2 + 7 = 0$, so $r = \pm\sqrt{7}i$.
   Therefore $s(t) = C_1\cos\sqrt{7}t + C_2\sin\sqrt{7}t$.

7. The characteristic equation is $r^2 + 4r + 8 = 0$, so $r = -2 \pm 2i$.
   Therefore $x(t) = C_1e^{-2t}\cos 2t + C_2e^{-2t}\sin 2t$.

9. The characteristic equation is $r^2 + 6r + 5 = 0$, so $r = -1$ or -5.
   Therefore $y(t) = C_1e^{-t} + C_2e^{-5t}$.
   $y'(t) = -C_1e^{-t} - 5C_2e^{-5t}$
   $y'(0) = 0 = -C_1 - 5C_2$
   $y(0) = 1 = C_1 + C_2$
   Therefore $C_2 = -1/4$, $C_1 = 5/4$ and $y(t) = \frac{5}{4}e^{-t} - \frac{1}{4}e^{-5t}$.

11. The characteristic equation is $r^2 + 6r + 10 = 0$, so $r = -3 \pm i$.
    Therefore $y(t) = C_1e^{-3t}\cos t + C_2e^{-3t}\sin t$.
    $y'(t) = C_1[e^{-3t}(-\sin t) + (-3e^{-3t})\cos t] + C_2[e^{-3t}\cos t + (-3e^{-3t})\sin t]$
    $y'(0) = 2 = -3C_1 + C_2$
    $y(0) = 0 = C_1$
    Therefore $C_1 = 0, C_2 = 2$ and $y(t) = 2e^{-3t}\sin t$.

13. The characteristic equation is $r^2 + 2r + 2 = 0$, so $r = -1 \pm i$.
    Therefore $p(t) = C_1e^{-t}\cos t + C_2e^{-t}\sin t$.
    $p(0) = 0 = C_1$ so $p(t) = C_2e^{-t}\sin t$
    $p(\pi/2) = 20 = C_2e^{-\pi/2}\sin\frac{\pi}{2}$ so $C_2 = 20e^{\pi/2}$
    Therefore $p(t) = 20e^{\frac{\pi}{2}}e^{-t}\sin t = 20e^{\frac{\pi}{2}-t}\sin t$.

15. i) $x'' + 4x = 0$ is the equation of a spring and so goes with (d).

    ii) $x'' - 4x + 0$ has characteristic equation $D^2 - 4$ and so is a heavily damped oscillator ($b^2 - 4ac = 16$), so it goes with (c).

    iii) $x'' - 0.2x' + 1.01x = 0$ has characteristic equation $D^2 - 0.2D + 1.01$ so $b^2 - 4ac = 0.04 - 4.04 = -4$ so this is a damped oscillator, with $D = 0.1 \pm i$. So the solutions are

    $$C_1e^{(0.1+i)t} + C_2e^{(0.1-i)t} = e^{0.1t}(A\sin t + B\cos t)$$

    Notice that $e^{0.1t}$ increases as t does, so this goes with (a).

    iv) $x'' + 0.2x' + 1.01x$ has characteristic equation $D^2 + 0.2D + 1.01$ so $b^2 - 4ac = -4$. Again this is a damped harmonic oscillator. We have $D = -0.1 \pm i$ and so the solution is $x = e^{-0.1t}(A\sin t + B\cos t)$, which goes with (b).

17. The restoring force is given by $F_{\text{spring}} = -ks$, so we look for the smallest coefficient of s. Spring (iv) exerts the smallest restoring force.

19. All of these differential equations have solutions of the form $C_1e^{\alpha t}\cos\beta t + C_2e^{\alpha t}\sin\beta t$. The spring with the longest period has the smallest β. Since $i\beta$ is the complex part of the roots of the characteristic equation, $\beta = \frac{1}{2}(\sqrt{4c - b^2})$. Thus spring (iii) has the longest period.

21. Recall that $s'' + bs' + c = 0$ is overdamped if the discriminant $b^2 - 4c > 0$, critically damped if $b^2 - 4c = 0$, and underdamped if $b^2 - 4c < 0$. Since $b^2 - 4c = 16 - 4c$, the circuit is overdamped if $c < 4$, critically damped if $c = 4$, and underdamped if $c > 4$.

23. Recall that $s'' + bs' + cs = 0$ is overdamped if the discriminant $b^2 - 4c > 0$, critically damped if $b^2 - 4c = 0$, and underdamped if $b^2 - 4c < 0$. Since $b^2 - 4c = 36 - 4c$, the solution is overdamped if $c < 9$, critically damped if $c = 9$, and underdamped if $c > 9$.

25. The characteristic equation is $r^2 + r - 2 = 0$, so $r = 1$ or -2. Therefore $z(t) = C_1 e^t + C_2 e^{-2t}$. Since $e^t \to \infty$ as $t \to \infty$, we must have $C_1 = 0$. Therefore $z(t) = C_2 e^{-2t}$. Furthermore, $z(0) = 3 = C_2$, so $z(t) = 3e^{-2t}$.

26. The differential equation is $I'' + 2I' + \frac{1}{4}I = 0$, so the characteristic equation is $r^2 + 2r + \frac{1}{4} = 0$. This has roots $\dfrac{-2 \pm \sqrt{3}}{2} = -1 \pm \dfrac{\sqrt{3}}{2}$. Thus, the general solution is

$$\begin{aligned} I(t) &= C_1 e^{(-1+\frac{\sqrt{3}}{2})t} + C_2 e^{(-1-\frac{\sqrt{3}}{2})t}. \\ \text{Also, } I'(t) &= C_1\left(-1+\frac{\sqrt{3}}{2}\right)e^{(-1+\frac{\sqrt{3}}{2})t} + C_2\left(-1-\frac{\sqrt{3}}{2}\right)e^{(-1-\frac{\sqrt{3}}{2})t}. \end{aligned}$$

We have

(a)

$$\begin{aligned} I(0) &= C_1 + C_2 = 0 \\ \text{and } I'(0) &= \left(-1+\frac{\sqrt{3}}{2}\right)C_1 + \left(-1-\frac{\sqrt{3}}{2}\right)C_2 = 2. \end{aligned}$$

Using the formula for $I(t)$, we have $C_1 = -C_2$. Using the formula for $I'(t)$, we have:

$$\begin{aligned} 2 &= \left(-1+\frac{\sqrt{3}}{2}\right)(-C_2) + \left(-1-\frac{\sqrt{3}}{2}\right)C_2 = -\sqrt{3}C_2 \\ \text{so,} \quad C_2 &= -\frac{2}{\sqrt{3}}. \end{aligned}$$

Thus, $C_1 = \dfrac{2}{\sqrt{3}}$, and $I(t) = \dfrac{2}{\sqrt{3}}\left(e^{(-1+\frac{\sqrt{3}}{2})t} - e^{(-1-\frac{\sqrt{3}}{2})t}\right)$.

(b) We have

$$\begin{aligned} I(0) &= C_1 + C_2 = 2 \\ \text{and} \quad I'(0) &= \left(-1+\frac{\sqrt{3}}{2}\right)C_1 + \left(-1-\frac{\sqrt{3}}{2}\right)C_2 = 0. \end{aligned}$$

Using the first equation, we have $C_1 = 2 - C_2$. Thus,

$$\begin{aligned}\left(-1+\frac{\sqrt{3}}{2}\right)(2-C_2) \;+\; \left(-1-\frac{\sqrt{3}}{2}\right)C_2 = 0\\ -\sqrt{3}C_2 &= 2-\sqrt{3}\\ C_2 &= -\frac{2-\sqrt{3}}{\sqrt{3}}\\ \text{and } C_1 &= 2-C_2 = \frac{2+\sqrt{3}}{\sqrt{3}}.\end{aligned}$$

Thus, $I(t) = \frac{1}{\sqrt{3}}\left((2+\sqrt{3})e^{(-1+\frac{\sqrt{3}}{2})t} - (2-\sqrt{3})e^{(-1-\frac{\sqrt{3}}{2})t}\right)$.

27. In this case, the differential equation describing current is $I''+I'+\frac{1}{4}I = 0$, so the characteristic equation is $r^2+r+\frac{1}{4} = 0$. This equation has one root, $r = -\frac{1}{2}$, so the equation for current is

$$\begin{aligned}I(t) &= (C_1+C_2t)e^{-\frac{1}{2}t}\\ \text{Also, } I'(t) &= -\frac{1}{2}(C_1+C_2t)e^{-\frac{1}{2}t}+C_2e^{-\frac{1}{2}t}\\ &= \left(C_2-\frac{C_1}{2}-\frac{C_2t}{2}\right)e^{-\frac{1}{2}t}.\end{aligned}$$

(a) We have

$$\begin{aligned}I(0) &= C_1 = 0,\\ I'(0) &= C_2-\frac{C_1}{2} = 2.\end{aligned}$$

Thus, $C_1 = 0$, $C_2 = 2$, and

$$I(t) = 2te^{-\frac{1}{2}t}.$$

(b) We have

$$\begin{aligned}I(0) &= C_1 = 2,\\ I'(0) &= C_2-\frac{C_1}{2} = 0.\end{aligned}$$

Thus, $C_1 = 2$, $C_2 = 1$, and

$$I(t) = (2+t)e^{-\frac{1}{2}t}.$$

(c) The resistance was decreased by exactly the amount to switch the circuit from the overdamped case to the critically damped case. Comparing the solutions of parts (a) and (b) in Problems 26, we find that in the critically damped case the current goes to 0 much faster as $t \to \infty$.

29. The differential equation for the current in a circuit, given a resistance R, a capacitance C, and and inductance L, is

$$LI'' + RI' + \frac{I}{C} = 0.$$

The corresponding characteristic equation is $Lr^2 + Rr + \frac{1}{C} = 0$. This equation has roots

$$r = -\frac{R}{2L} \pm \frac{\sqrt{R^2 - \frac{4L}{C}}}{2L}.$$

(a) If $R^2 - \frac{4L}{C} < 0$, the solution is

$$I(t) = e^{-\frac{R}{2L}t}(A\sin\omega t + B\cos\omega t) \text{ for some } A \text{ and } B,$$

where $\omega = \frac{\sqrt{R^2 - \frac{4L}{C}}}{2L}$. As $t \to \infty$, $I(t)$ clearly goes to 0.

(b) If $R^2 - \frac{4L}{C} = 0$, the solution is

$$I(t) = e^{-\frac{R}{t}}(A + Bt) \text{ for some } A \text{ and } B.$$

Again, as $t \to \infty$, the current goes to 0.

(c) If $R^2 - \frac{4L}{C} > 0$, the solution is

$$I(t) = Ae^{r_1 t} + Be^{r_2 t} \text{ for some } A \text{ and } B,$$

where

$$r_1 = -\frac{R}{2L} + \frac{\sqrt{R^2 - \frac{4L}{C}}}{2L}, \quad \text{and} \quad r_2 = -\frac{R}{2L} - \frac{\sqrt{R^2 - \frac{4L}{C}}}{2L}.$$

Notice that r_2 is clearly negative. r_1 is also negative since

$$\begin{aligned} \frac{\sqrt{R^2 - \frac{4L}{C}}}{2L} &< \frac{\sqrt{R^2}}{2L} \quad (L \text{ and } C \text{ are positive}) \\ &= \frac{R}{2L}. \end{aligned}$$

Since r_1 and r_2 are negative, again $I(t) \to 0$, as $t \to \infty$.

Thus, for any circuit with a resistor, a capacitor and an inductor, $I(t) \to 0$ as $t \to \infty$. Compare this with Problem 17 in Section 8.9, where we showed that in a circuit with just a capacitor and inductor, the current varied along a sine curve.

8.12 Solutions

1. (1,0)

3. $(-2,0)$

5. $(\frac{5\sqrt{3}}{2}, -\frac{5}{2})$

7. $(0,-3)$

9. $2e^{\frac{i\pi}{2}}$

11. $\sqrt{2}e^{\frac{i\pi}{4}}$

13. $0e^{i\theta}$, for any θ.

15. $\sqrt{10}e^{i\theta}$, where $\theta = \arctan(-3) \approx -1.249 + \pi = 1.893$ is an angle in the second quadrant.

17. $-3-4i$

19. $-5+12i$

21. $\frac{1}{4} - \frac{9i}{8}$

23. $\cos\frac{2\pi}{3} + i\sin\frac{2\pi}{3} = -\frac{1}{2} + i\frac{\sqrt{3}}{2}$

25. $5^3(\cos\frac{3\pi}{2} + i\sin\frac{3\pi}{2}) = -125i$

27. One value of $\sqrt{i}$ is $\sqrt{e^{i\frac{\pi}{2}}} = (e^{i\frac{\pi}{2}})^{\frac{1}{2}} = e^{i\frac{\pi}{4}} = \cos\frac{\pi}{4} + i\sin\frac{\pi}{4} = \frac{\sqrt{2}}{2} + i\frac{\sqrt{2}}{2}$

29. One value of $\sqrt[3]{i}$ is $\sqrt[3]{e^{i\frac{\pi}{2}}} = (e^{i\frac{\pi}{2}})^{\frac{1}{3}} = e^{i\frac{\pi}{6}} = \cos\frac{\pi}{6} + i\sin\frac{\pi}{6} = \frac{\sqrt{3}}{2} + \frac{i}{2}$

31. $(1+i)^{100} = (\sqrt{2}e^{i\frac{\pi}{4}})^{100} = (2^{\frac{1}{2}})^{100}(e^{i\frac{\pi}{4}})^{100} = 2^{50}\cdot e^{i\cdot 25\pi} = 2^{50}\cos 25\pi + i2^{50}\sin 25\pi = -2^{50}$

33. One value of $(-4+4i)^{2/3}$ is $[\sqrt{32}e^{i\frac{3\pi}{4}}]^{2/3} = (\sqrt{32})^{2/3}e^{i\frac{\pi}{2}} = 2^{\frac{10}{3}}\cos\frac{\pi}{2} + i2^{\frac{10}{3}}\sin\frac{\pi}{2} = 8i\sqrt[3]{2}$

35. One value of $(\sqrt{3}+i)^{-1/2}$ is
$(2e^{i\frac{\pi}{6}})^{-1/2} = \frac{1}{\sqrt{2}}e^{i(-\frac{\pi}{12})} = \frac{1}{\sqrt{2}}\cos(-\frac{\pi}{12}) + i\frac{1}{\sqrt{2}}\sin(-\frac{\pi}{12}) \approx 0.683 - 0.183i$

37. (a)

$$z_1z_2 = (-3-i\sqrt{3})(-1+i\sqrt{3}) = 3 + (\sqrt{3})^2 + i(\sqrt{3} - 3\sqrt{3}) = 6 - i2\sqrt{3}.$$

$$\frac{z_1}{z_2} = \frac{-3-i\sqrt{3}}{-1+i\sqrt{3}}\cdot\frac{-1-i\sqrt{3}}{-1-i\sqrt{3}} = \frac{3-(\sqrt{3})^2 + i(\sqrt{3}+3\sqrt{3})}{(-1)^2+(\sqrt{3})^2} = \frac{i\cdot 4\sqrt{3}}{4} = i\sqrt{3}.$$

(b) We find (r_1, θ_1) corresponding to $z_1 = -3 - i\sqrt{3}$.
$r_1 = \sqrt{(-3)^2 + (\sqrt{3})^2} = \sqrt{12} = 2\sqrt{3}.$
$\tan\theta_1 = \frac{-\sqrt{3}}{-3} = \frac{\sqrt{3}}{3}$, so $\theta_1 = \frac{7\pi}{6}$.
Thus $-3 - i\sqrt{3} = r_1e^{i\theta_1} = 2\sqrt{3}\,e^{i\frac{7\pi}{6}}$.

We find (r_2, θ_2) corresponding to $z_2 = -1 + i\sqrt{3}$.
$r_2 = \sqrt{(-1)^2 + (\sqrt{3})^2} = 2;$
$\tan\theta_2 = \frac{\sqrt{3}}{-1} = -\sqrt{3}$, so $\theta_2 = \frac{2\pi}{3}$.
Thus, $-1 + i\sqrt{3} = r_2e^{i\theta_2} = 2e^{i\frac{2\pi}{3}}$.

We now calculate z_1z_2 and $\frac{z_1}{z_2}$.

$$\begin{aligned} z_1z_2 &= \left(2\sqrt{3}e^{i\frac{7\pi}{6}}\right)\left(2e^{i\frac{2\pi}{3}}\right) = 4\sqrt{3}e^{i(\frac{7\pi}{6}+\frac{2\pi}{3})} = 4\sqrt{3}e^{i\frac{11\pi}{6}} \\ &= 4\sqrt{3}\left[\cos\frac{11\pi}{6} + i\sin\frac{11\pi}{6}\right] = 4\sqrt{3}\left[\frac{\sqrt{3}}{2} - i\frac{1}{2}\right] = 6 - i2\sqrt{3}. \end{aligned}$$

$$\begin{aligned}\frac{z_1}{z_2} &= \frac{2\sqrt{3}e^{i\frac{7\pi}{6}}}{2e^{i\frac{2\pi}{3}}} = \sqrt{3}e^{i(\frac{7\pi}{6}-\frac{2\pi}{3})} = \sqrt{3}e^{i\frac{\pi}{2}} \\ &= \sqrt{3}\left(\cos\frac{\pi}{2} + i\sin\frac{\pi}{2}\right) = i\sqrt{3}.\end{aligned}$$

These agrees with the values found in (a).

39. For each pair of Cartesian coordinates, there is more than one pair of polar coordinates for that point. For example, if $(x, y) = (1, 0)$ then $(r, \theta) = (1, 0)$, $(r, \theta) = (1, 2\pi)$, and $(r, \theta) = (1, 4\pi)$ all represent the same point.

41. True, since $(x - iy)(x + iy) = x^2 + y^2$ is real.

43. False. Let $f(x) = x$. Then $f(i) = i$ but $f(\bar{i}) = \bar{i} = -i$.

45. False, since $(1 + 2i)^2 = -3 + 4i$.

47. Using Euler's formula, we have:

$$e^{i(2\theta)} = \cos 2\theta + i\sin 2\theta$$

On the other hand,

$$e^{i(2\theta)} = (e^{i\theta})^2 = (\cos\theta + i\sin\theta)^2 = (\cos^2\theta - \sin^2\theta) + i(2\cos\theta\sin\theta)$$

Equating imaginary parts, we find

$$\sin 2\theta = 2\sin\theta\cos\theta.$$

49. $\frac{d}{d\theta}(e^{i\theta}) = ie^{i\theta} = i(\cos\theta + i\sin\theta) = -\sin\theta + i\cos\theta$

Since in addition $\frac{d}{d\theta}(e^{i\theta}) = \frac{d}{d\theta}(\cos\theta + i\sin\theta) = \frac{d}{d\theta}(\cos\theta) + i\frac{d}{d\theta}(\sin\theta)$, by equating imaginary parts, we conclude that $\frac{d}{d\theta}\sin\theta = \cos\theta$.

8.13 Answers to Miscellaneous Exercises for Chapter 8

1. $\frac{dP}{dt} = 0.03P + 400$ so $\int \frac{dP}{P+\frac{40000}{3}} = \int 0.03dt$.
$\ln|P + \frac{40000}{3}| = 0.03t + C$ giving $P = Ae^{0.03t} - \frac{40000}{3}$. Since $P(0) = 0, A = \frac{40000}{3}$, therefore $P = \frac{40000}{3}(e^{0.03t} - 1)$.

3. $\frac{df}{dx} = \sqrt{xf(x)}$ gives $\int \frac{df}{\sqrt{f(x)}} = \int \sqrt{x}\,dx \Rightarrow 2\sqrt{f(x)} = \frac{2}{3}x^{\frac{3}{2}} + C$. Since $f(1) = 1$, we have $2 = \frac{2}{3} + C$ so $C = \frac{4}{3}$. Thus, $2\sqrt{f(x)} = \frac{2}{3}x^{\frac{3}{2}} + \frac{4}{3}$, so $f(x) = (\frac{1}{3}x^{\frac{3}{2}} + \frac{2}{3})^2$.
[Note: this is only defined for $x \geq 0$]

5. $\frac{dy}{dx} = e^{x-y}$ giving $\int e^y\,dy = \int e^x\,dx$ so $e^y = e^x + C$. Since $y(0) = 1$, we have $e^1 = e^0 + C$ so $C = e - 1$. Thus, $e^y = e^x + e - 1$, so $y = \ln(e^x + e - 1)$.
[Note: $e^x + e - 1 > 0$ always.]

7. $2\sin x - y^2\frac{dy}{dx} = 0$ giving $2\sin x = y^2\frac{dy}{dx}$. $\int 2\sin x\,dx = \int y^2\,dy$ so $-2\cos x = \frac{y^3}{3} + C$. Since $y(0) = 3$ we have $-2 = 9 + C$, so $C = -11$. Thus, $-2\cos x = \frac{y^3}{3} - 11$ giving $y = \sqrt[3]{33 - 6\cos x}$.

9. $\frac{dy}{dx} + xy^2 = 0$ means $\frac{dy}{dx} = -xy^2$, so $\int \frac{dy}{y^2} = \int -x\,dx$ giving $-\frac{1}{y} = -\frac{x^2}{2} + C$. Since $y(1) = 1$ we have $-1 = -\frac{1}{2} + C$ so $C = -\frac{1}{2}$. Thus, $-\frac{1}{y} = -\frac{x^2}{2} - \frac{1}{2}$ giving $y = \frac{2}{x^2+1}$.

11. $\frac{dy}{dx} = \frac{0.2y(18+0.1x)}{x(100+0.5y)}$ giving $\int \frac{(100+0.5y)}{0.2y}\,dy = \int \frac{18+0.1x}{x}\,dx$, so

$$\int \left(\frac{500}{y} + \frac{5}{2}\right)\,dy = \int \left(\frac{18}{x} + \frac{1}{10}\right)\,dx.$$

Therefore, $500\ln|y| + \frac{5}{2}y = 18\ln|x| + \frac{1}{10}x + C$. Since the curve passes through (10,10), $500\ln 10 + 25 = 18\ln 10 + 1 + C$, so $C = 482\ln 10 + 24$. Thus, the solution is $500\ln|y| + \frac{5}{2}y = 18\ln|x| + \frac{1}{10}x + 482\ln 10 + 24$. We cannot solve for y in terms of x, so we leave the answer in this form.

13. Using separation of variables and the integral tables, you can show that solutions are of the form $t = \frac{1}{4}(\ln|w - 7| - \ln|w - 3|) + C$, or $w = \frac{4}{1-Ae^{4t}} + 3$. The equilibrium values, where $\frac{dw}{dt} = 0$, are $w = 3$ and $w = 7$. Graphs of the solutions can also be sketched directly from the graph of $\frac{dw}{dt}$ against w.

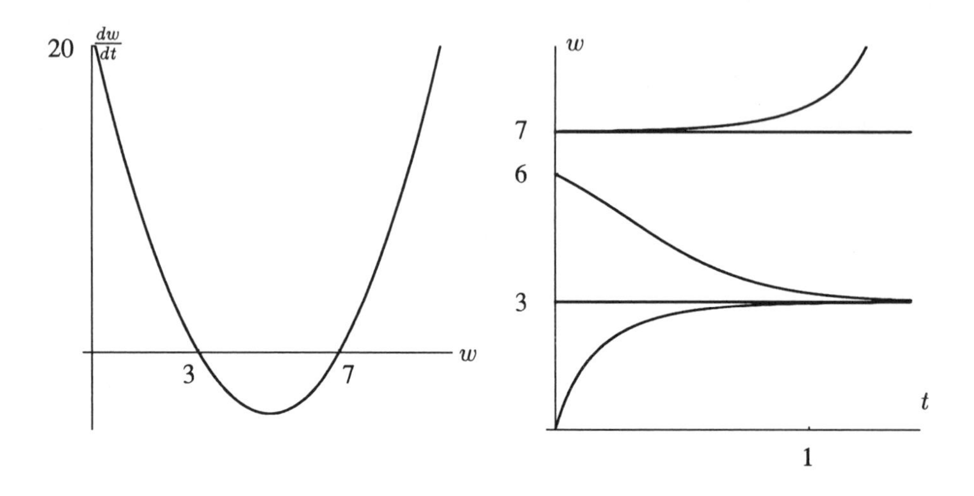

15. (a) 1 step: $\Delta y = \frac{1}{(\cos x)(\cos y)}\Delta x = \frac{1}{(\cos 0)(\cos 0)}\frac{1}{2} = \frac{1}{2}$.
Thus, using 1 step, we get $(\frac{1}{2}, \frac{1}{2})$ as our approximation.

(b) 2 steps: $\Delta x = \frac{1}{4}$.

x	y	$\Delta y = \frac{1}{(\cos x)(\cos y)}\Delta x$
0	0	0.25
0.25	0.25	0.266
0.5	0.516	

Thus, using 2 steps, we get (0.5,0.516) as our approximation.

(c) 4 steps: $\Delta x = \frac{1}{8}$

x	y	$\Delta y = \frac{1}{(\cos x)(\cos y)}\Delta x$
0	0	0.125
0.125	0.125	0.127
0.25	0.252	0.133
0.375	0.385	0.145
0.5	0.530	

Thus, using 4 steps, we get (0.5,0.530) as our approximation.

(d) $\frac{dy}{dx} = \frac{1}{(\cos x)(\cos y)} \Rightarrow \int \cos y\, dy = \int \frac{dx}{\cos x} \Rightarrow \sin y = \frac{1}{2}\ln\left|\frac{(\sin x)+1}{(\sin x)-1}\right| + C.$

Our curve passes through (0,0), so, $0 = 0+C$, and $C = 0$. Therefore $y = \arcsin\left(\frac{1}{2}\ln\left|\frac{(\sin x)+1}{(\sin x)-1}\right|\right)$.

When $x = \frac{1}{2}$, $y \approx 0.549$. Our answers in (a)-(c) are all underestimates. In each case, the error is about $\frac{1}{n+1}$, where n is the number of steps. We expect the error to be approximately proportional to $\frac{1}{n}$, so this seems reasonable.

17. For the equation $9z'' + z = 0$, the characteristic equation is

$$9r^2 + 1 = 0$$

If we write this in the form $r^2 + br + c = 0$, we have that $r^2 + 1/9 = 0$ and

$$b^2 - 4c = 0 - (4)(1/9) = -4/9 < 0$$

This indicates underdamped motion and since the roots of the characteristic equation are $r = \pm\frac{1}{3}i$, the general equation is

$$y(t) = C_1 \cos\left(\frac{1}{3}t\right) + C_2 \sin\left(\frac{1}{3}t\right)$$

19. For the equation $x'' + 2x' + 10x = 0$, the characteristic equation is

$$r^2 + 2r + 10 = 0$$

We have that

$$b^2 - 4c = 2^2 - 4(10) = -36 < 0$$

This indicates underdamped motion and since the roots of the characteristic equation are $r = -1 \pm 3i$, the general solution is

$$y(t) = C_1 e^{-t} \cos 3t + C_2 e^{-t} \sin 3t$$

21. Recall that $s'' + bs' + cs = 0$ is overdamped if the discriminant $b^2 - 4c > 0$, critically damped if $b^2 - 4c = 0$, and underdamped if $b^2 - 4c < 0$. This has discriminant $b^2 - 4c = b^2 + 64$. Since $b^2 + 64$ is always positive, the solution is always overdamped.

23. (a) Since the amount leaving the blood is proportional to the quantity in the blood,

$$\frac{dQ}{dt} = -kQ \quad \text{for some positive constant } k.$$

Thus $Q = Q_0 e^{-kt}$, where Q_0 is the initial quantity in the bloodstream. Only 20% is left in the blood after 3 hours. Thus $0.20 = e^{-3k}$, so $k = \frac{\ln 0.20}{-3} \approx 0.5365$. Therefore $Q = Q_0 e^{-0.5365t}$.

(b) Since 20% is left after 3 hours, after 6 hours only 20% of that 20% will be left. Thus after 6 hours only 4% will be left, so if the patient is given 100 mg, only 4 mg will be left 6 hours later.

25. (a) $\frac{dT}{dt} = -k(T - A)$, where $A = 68$ is the temperature of room.

(b) $\int \frac{dT}{T - A} = -\int k\,dt, \ln|T-A| = -kt+C, T = A+Be^{-kt}$, so $T = A+(T_0-A)e^{-kt}$, where $T_0 = 90.3$ is the initiall temperature. Thus

$$T = 68 + (90.3 - 68)e^{-kt}.$$

(c) Letting 9 a.m. be $t = 0$ (with initial temperature of 90.3°F), then at 10 a.m., $t = 1$, so

$$\begin{aligned} 89.0 &= 68 + (90.3 - 68)e^{-k} \\ 21 &= 22.3e^{-k} \\ k &= -\ln\frac{21}{22.3} \approx 0.06. \end{aligned}$$

We want to know when T was equal to 98.6°F, the temperature of a live body, so

$$\begin{aligned} 98.6 &= 68 + (90.3 - 68)e^{(-0.06)t} \\ \ln\frac{30.6}{22.3} &= -0.06t \\ t &= \left(-\frac{1}{0.06}\right)\ln\frac{30.6}{22.3} \\ t &\approx -5.27. \end{aligned}$$

The professor was killed approximately $5\frac{1}{4}$ hours prior to 9 a.m., at 3:45 a.m.

26. (a) For this situation,

$$\begin{pmatrix}\text{money added}\\ \text{to account}\end{pmatrix} = \begin{pmatrix}\text{money added}\\ \text{via interest}\end{pmatrix} + \begin{pmatrix}\text{money}\\ \text{deposited}\end{pmatrix}$$

Translating this into an equation yields

$$\frac{df}{dt} = 0.1f + 1200$$

(b) Solving this equation via separation of variables gives

$$\begin{aligned} \frac{df}{dt} &= 0.1f + 1200 \\ &= (0.1)(f + 12000) \end{aligned}$$

So

$$\int \frac{df}{f + 12000} = \int 0.1\,dt$$

and

$$\ln|f + 12000| = 0.1t + C$$

solving for f,

$$|f + 12000| = e^{(0.1)t+C} = e^{C}e^{(0.1)t}$$

or

$$f = Ae^{0.1t} - 12000, \text{ (where } A = e^{c})$$

We may find A using the initial condition $f(0) = 0$

$$A - 12000 = 0 \quad \text{or} \quad A = 12000$$

So the solution is

$$f(t) = 12000(e^{0.1t} - 1)$$

(c) After 5 years, the balance is

$$\begin{aligned} f(5) &= 12000(e^{(0.1)(5)} - 1) \\ &= 7784.66 \end{aligned}$$

27. (a) The balance in the account at the beginning of the month is given by the following sum

$$\begin{pmatrix}\text{balance in}\\ \text{account}\end{pmatrix} = \begin{pmatrix}\text{previous month's}\\ \text{balance}\end{pmatrix} + \begin{pmatrix}\text{interest on}\\ \text{previous month's balance}\end{pmatrix} + \begin{pmatrix}\text{monthly deposit}\\ \text{of \$100}\end{pmatrix}$$

Denote month i's balance by B_i. Assuming the interest is compounded continuously, we have

$$\begin{pmatrix}\text{previous month's}\\ \text{balance}\end{pmatrix} + \begin{pmatrix}\text{interest on previous}\\ \text{month's balance}\end{pmatrix} = B_{i-1}e^{0.1/12}.$$

Since the interest rate is $10\% = 0.1$ per year, interest is $\frac{0.1}{12}$ per month. So at month i, the balance is

$$B_i = B_{i-1}e^{\frac{0.1}{12}} + 100$$

Explicitly, we have for the five years (60 months) in equation:

$$\begin{aligned} B_0 &= 0 \\ B_1 &= B_0 e^{\frac{0.1}{12}} + 100 \\ B_2 &= B_1 e^{\frac{0.1}{12}} + 100 \\ B_3 &= B_2 e^{\frac{0.1}{12}} + 100 \\ &\vdots \\ B_{60} &= B_{59} e^{\frac{0.1}{12}} + 100 \end{aligned}$$

In other words,

$$
\begin{aligned}
B_1 &= 100 \\
B_2 &= 100e^{\frac{0.1}{12}} + 100 \\
B_3 &= (100e^{\frac{0.1}{12}} + 100)e^{\frac{0.1}{12}} + 100 \\
&= 100e^{\frac{(0.1)2}{12}} + 100e^{\frac{0.1}{12}} + 100 \\
B_4 &= 100e^{\frac{(0.1)3}{12}} + 100e^{\frac{(0.1)2}{12}} + 100e^{\frac{(0.1)}{12}} + 100 \\
\vdots \quad & \quad \vdots \\
B_{60} &= 100e^{\frac{(0.1)59}{12}} + 100e^{\frac{(0.1)58}{12}} + \cdots + 100e^{\frac{(0.1)1}{12}} + 100 \\
B_{60} &= \sum_{k=0}^{59} 100e^{\frac{(0.1)k}{12}}
\end{aligned}
$$

(b) The sum $B_{60} = \sum_{k=0}^{59} 100e^{\frac{(0.1)k}{12}}$ can be written as $S = \sum_{k=0}^{59} 1200e^{\frac{(0.1)k}{12}}(\frac{1}{12})$ which is the left Riemann sum for $\int_0^5 1200e^{(0.1)t}dt$, with $\Delta t = \frac{1}{12}$ and $N = 60$. If we make the substitution $t = 5 - u$, we get

$$\int_0^5 1200e^{(0.1)t}\,dt = -\int_5^0 1200e^{(0.1)(5-u)}\,du = \int_0^5 1200e^{(0.1)(5-u)}\,du$$

Evaluating the sum on a calculator gives $S = 7752.26$.

(c) Since the answer for Problem 26(c) should be an approximation of the integral, we expect that the number optained in part (b) of this problem should be close to the number obtained in Problem 26(c). Note that the integrand of part (b), $1200e^{0.1(5-t)}$, is a decreasing function, so we expect that a right-hand-sum would underestimate the integral. Comparison of the answers obtained in parts (b) and Problem 26 confirm this.

29. (a) The insects grow exponentially with no birds around (the equation becomes $\frac{dx}{dt} = 3x$); the birds die out exponentially with no insects to feed on ($\frac{dy}{dt} = -10y$). The interaction increases the birds' growth rate (the $+\ 0.001xy$ term is positive), but decreases the insects' (the $-\ 0.02xy$ term is negative). This is as you would expect: having the insects around helps the birds; having birds around hurts the insects.

(b) $(3 - 0.02y)x = 0$
$-(10 - 0.001x)y = 0$
Solutions are $(0, 0)$ and $(10{,}000, 150)$

(c) $\frac{dy}{dx} = \frac{y(-10+0.001x)}{x(3-0.02y)}$.
Solving $\int \frac{-10+0.001x}{x}dx = \int \frac{3-0.02y}{y}dy$ yields $3\ln y - 0.02y = -10\ln x + 0.001x + C$. Using the initial point $A = (10{,}000, 160)$, we have $3\ln 160 - 0.02(160) = -10\ln 10{,}000 + 0.001(10{,}000) + C$. Thus $C \approx 94.13$ and $3\ln y - 0.02y = -10\ln x + 0.001x + 94.1$

(d) One can verify that the equation is satisfied by points B, C, D, as given, by plugging them in on a calculator.

(e)

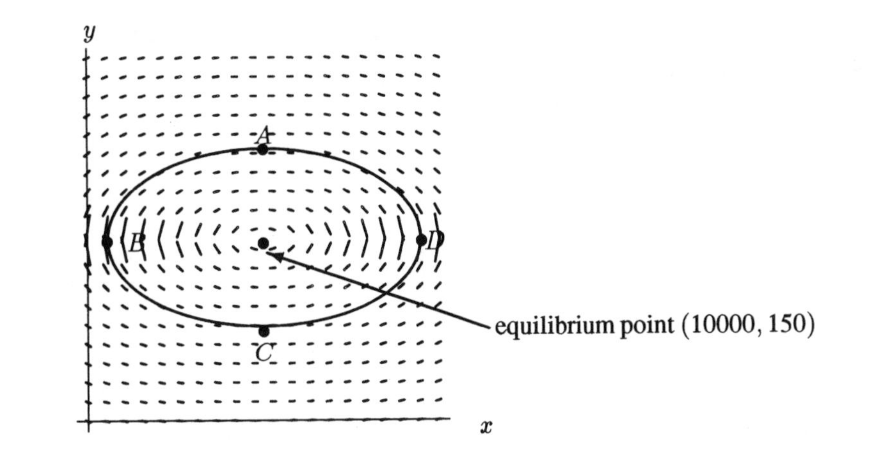

(f) Consider point A. $\frac{dy}{dt} = 0$, and $\frac{dx}{dt} = 3(10,000) - 0.02(10,000)(160) = -2000 < 0$ so x is decreasing at point A. Hence rotation is counterclockwise in the phase plane, and the order of traversal is $A \longrightarrow B \longrightarrow C \longrightarrow D$.

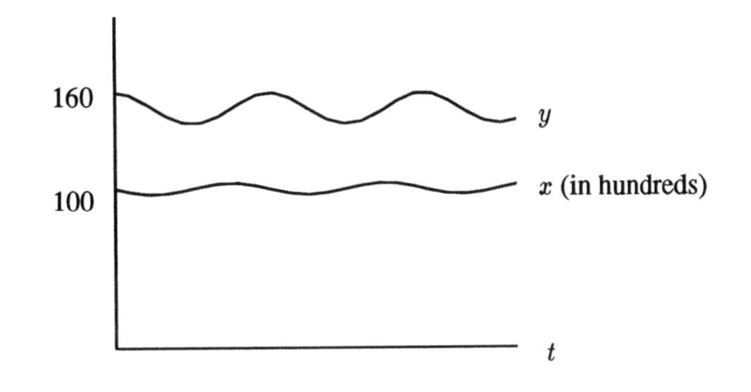

(h) At points A and C, $\frac{dy}{dx} = 0$, and at B and D, $\frac{dx}{dy} = 0$, so as you may have already guessed, these points are extrema: y is maximized at A, minimized at C; x is maximized at D, minimized at B.

Chapter 9

9.1 Solutions

1. Let $f(x)=\cos x$. Then $f(0)=\cos(0)=1$, and

$$\begin{aligned} f'(x) &= -\sin x & f'(0) &= 0\\ f''(x) &= -\cos x & f''(0) &= -1\\ f'''(x) &= \sin x & f'''(0) &= 0\\ f^{(4)}(x) &= \cos x & f^{(4)}(0) &= 1\\ f^{(5)}(x) &= -\sin x & f^{(5)}(0) &= 0\\ f^{(6)}(x) &= -\cos x & f^{(6)}(0) &= -1. \end{aligned}$$

Thus,

$$\begin{aligned} P_2(x) &= 1-\frac{x^2}{2!};\\ P_4(x) &= 1-\frac{x^2}{2!}+\frac{x^4}{4!};\\ P_6(x) &= 1-\frac{x^2}{2!}+\frac{x^4}{4!}-\frac{x^6}{6!}. \end{aligned}$$

2. Let $f(x)=\sqrt{1+x}=(1+x)^{1/2}$. Then $f(0)=1$, and

$$\begin{aligned} f'(x) &= \tfrac{1}{2}(1+x)^{-1/2} & f'(0) &= \tfrac{1}{2}\\ f''(x) &= -\tfrac{1}{4}(1+x)^{-3/2} & f''(0) &= -\tfrac{1}{4}\\ f'''(x) &= \tfrac{3}{8}(1+x)^{-5/2} & f'''(0) &= \tfrac{3}{8}\\ f^{(4)}(x) &= -\tfrac{15}{16}(1+x)^{-7/2} & f^{(4)}(0) &= -\tfrac{15}{16} \end{aligned}$$

Thus,

$$\begin{aligned} P_2(x) &= 1+\frac{1}{2}x-\frac{1}{8}x^2;\\ P_3(x) &= 1+\frac{1}{2}x-\frac{1}{8}x^2+\frac{1}{16}x^3;\\ P_4(x) &= 1+\frac{1}{2}x-\frac{1}{8}x^2+\frac{1}{16}x^3-\frac{5}{128}x^4. \end{aligned}$$

3. Let $f(x)=\dfrac{1}{1-x}=(1-x)^{-1}$. Then $f(0)=1$.

$$\begin{aligned} f'(x) &= 1!(1-x)^{-2} & f'(0) &= 1!\\ f''(x) &= 2!(1-x)^{-3} & f''(0) &= 2!\\ f'''(x) &= 3!(1-x)^{-4} & f'''(0) &= 3!\\ f^{(4)}(x) &= 4!(1-x)^{-5} & f^{(4)}(0) &= 4!\\ f^{(5)}(x) &= 5!(1-x)^{-6} & f^{(5)}(0) &= 5!\\ f^{(6)}(x) &= 6!(1-x)^{-7} & f^{(6)}(0) &= 6!\\ f^{(7)}(x) &= 7!(1-x)^{-8} & f^{(7)}(0) &= 7! \end{aligned}$$

$$\begin{aligned} P_3(x) &= 1 + x + x^2 + x^3; \\ P_5(x) &= 1 + x + x^2 + x^3 + x^4 + x^5; \\ P_7(x) &= 1 + x + x^2 + x^3 + x^4 + x^5 + x^6 + x^7. \end{aligned}$$

4. Let $\dfrac{1}{1+x} = (1+x)^{-1}$. Then $f(0) = 1$.

$$\begin{aligned} f'(x) &= -1!(1+x)^{-2} & f'(0) &= -1 \\ f''(x) &= 2!(1+x)^{-3} & f''(0) &= 2! \\ f'''(x) &= -3!(1+x)^{-4} & f'''(0) &= -3! \\ f^{(4)}(x) &= 4!(1+x)^{-5} & f^{(4)}(0) &= 4! \\ f^{(5)}(x) &= -5!(1+x)^{-6} & f^{(5)}(0) &= -5! \\ f^{(6)}(x) &= 6!(1+x)^{-7} & f^{(6)}(0) &= 6! \\ f^{(7)}(x) &= -7!(1+x)^{-8} & f^{(7)}(0) &= -7! \\ f^{(8)}(x) &= 8!(1+x)^{-9} & f^{(8)}(0) &= 8! \end{aligned}$$

$$\begin{aligned} P_4(x) &= 1 - x + x^2 - x^3 + x^4; \\ P_6(x) &= 1 - x + x^2 - x^3 + x^4 - x^5 + x^6; \\ P_8(x) &= 1 - x + x^2 - x^3 + x^4 - x^5 + x^6 - x^7 + x^8. \end{aligned}$$

5. Let $f(x) = \tan x$. So $f(0) = \tan 0 = 0$, and

$$\begin{aligned} f'(x) &= \frac{1}{\cos^2 x} & f'(0) &= 1, \\ f''(x) &= \frac{2\sin x}{\cos^3 x} & f''(0) &= 0, \\ f'''(x) &= \frac{2}{\cos^2 x} + \frac{6\sin^2 x}{\cos^4 x} & f'''(0) &= 2, \\ f^{(4)}(x) &= \frac{16\sin x}{\cos^3 x} + \frac{24\sin^3 x}{\cos^5 x} & f^{(4)}(0) &= 0, \\ f^{(5)}(x) &= \frac{16}{\cos^2 x} + \frac{120\sin^2 x}{\cos^4 x} + \frac{120\sin^4 x}{\cos^6 x} & f^{(5)}(0) &= 16. \end{aligned}$$

Thus,

$$\begin{aligned} P_3(x) &= x + \frac{x^3}{3}, \\ P_4(x) &= x + \frac{x^3}{3}, \\ P_5(x) &= x + \frac{x^3}{3} + \frac{2x^5}{15}. \end{aligned}$$

7. Let $f(x) = \ln(1+x)$. Then $f(0) = \ln 1 = 0$, and

$$\begin{aligned}
f'(x) &= (1+x)^{-1} & f'(0) &= 1,\\
f''(x) &= (-1)(1+x)^{-2} & f''(0) &= -1,\\
f'''(x) &= 2(1+x)^{-3} & f'''(0) &= 2,\\
f^{(4)}(x) &= -3!(1+x)^{-4} & f^{(4)}(0) &= -3!,\\
f^{(5)}(x) &= 4!(1+x)^{-5} & f^{(5)}(0) &= 4!,\\
f^{(6)}(x) &= -5!(1+x)^{-6} & f^{(6)}(0) &= -5!,\\
f^{(7)}(x) &= 6!(1+x)^{-7} & f^{(7)}(0) &= 6!,\\
f^{(8)}(x) &= -7!(1+x)^{-8} & f^{(8)}(0) &= -7!,\\
f^{(9)}(x) &= 8!(1+x)^{-9} & f^{(9)}(0) &= 8!
\end{aligned}$$

So,

$$\begin{aligned}
P_5(x) &= x - \frac{x^2}{2} + \frac{x^3}{3} - \frac{x^4}{4} + \frac{x^5}{5},\\
P_7(x) &= x - \frac{x^2}{2} + \frac{x^3}{3} - \frac{x^4}{4} + \frac{x^5}{5} - \frac{x^6}{6} + \frac{x^7}{7},\\
P_9(x) &= x - \frac{x^2}{2} + \frac{x^3}{3} - \frac{x^4}{4} + \frac{x^5}{5} - \frac{x^6}{6} + \frac{x^7}{7} - \frac{x^8}{8} + \frac{x^9}{9}.
\end{aligned}$$

8. Let $f(x) = \sqrt[3]{1-x} = (1-x)^{1/3}$. Then $f(0) = 1$, and

$$\begin{aligned}
f'(x) &= -\tfrac{1}{3}(1-x)^{-2/3} & f'(0) &= -\tfrac{1}{3},\\
f''(x) &= -\tfrac{2}{3^2}(1-x)^{-5/3} & f''(0) &= -\tfrac{2}{3^2},\\
f'''(x) &= -\tfrac{10}{3^3}(1-x)^{-8/3} & f'''(0) &= -\tfrac{10}{3^3},\\
f^{(4)}(x) &= -\tfrac{80}{3^4}(1-x)^{-11/3} & f^{(4)}(0) &= -\tfrac{80}{3^4}.
\end{aligned}$$

Then,

$$\begin{aligned}
P_2(x) &= 1 - \frac{1}{3}x - \frac{1}{2!}\frac{2}{3^2}x^2\\
&= 1 - \frac{1}{3}x - \frac{1}{9}x^2,\\
P_3(x) &= P_2(x) - \frac{1}{3!}\left(\frac{10}{3^3}\right)x^3\\
&= 1 - \frac{1}{3}x - \frac{1}{9}x^2 - \frac{5}{81}x^3,\\
P_4(x) &= P_3(x) - \frac{1}{4!}\frac{80}{3^4}x^4\\
&= 1 - \frac{1}{3}x - \frac{1}{9}x^2 - \frac{5}{81}x^3 - \frac{10}{243}x^4.
\end{aligned}$$

9. Let $f(x) = \dfrac{1}{\sqrt{1+x}} = (1+x)^{-1/2}$. Then $f(0) = 1$

$$\begin{array}{rclcrcl} f'(x) &=& -\frac{1}{2}(1+x)^{-3/2} & & f'(0) &=& -\frac{1}{2}, \\ f''(x) &=& \frac{3}{2^2}(1+x)^{-5/2} & & f''(0) &=& \frac{3}{2^2}, \\ f'''(x) &=& -\frac{3\cdot 5}{2^3}(1+x)^{-7/2} & & f'''(0) &=& -\frac{3\cdot 5}{2^3}, \\ f^{(4)}(x) &=& \frac{3\cdot 5\cdot 7}{2^4}(1+x)^{-9/2} & & f^{(4)}(0) &=& \frac{3\cdot 5\cdot 7}{2^4}. \end{array}$$

Then,

$$\begin{aligned} P_2(x) &= 1 - \frac{1}{2}x + \frac{1}{2!}\frac{3}{2^2}x^2 \\ &= 1 - \frac{1}{2}x + \frac{3}{8}x^2, \\ P_3(x) &= P_2(x) - \frac{1}{3!}\frac{3\cdot 5}{2^3}x^3 \\ &= 1 - \frac{1}{2}x + \frac{3}{8}x^2 - \frac{5}{16}x^3, \\ P_4(x) &= P_3(x) + \frac{1}{4!}\frac{3\cdot 5\cdot 7}{2^4}x^4 \\ &= 1 - \frac{1}{2}x + \frac{3}{8}x^2 - \frac{5}{16}x^3 + \frac{35}{128}x^4. \end{aligned}$$

10. Let $f(x) = (1+x)^{\alpha}$.

(a) Suppose that $\alpha = 0$. Then $f(x) = 1$ and $f^{(k)}(x) = 0$ for any $k \geq 1$. Thus $P_2(x) = P_3(x) = P_4(x) = 1$.

(b) If $\alpha = 1$ then $f(x) = 1 + x$, so

$$\begin{array}{rcll} f(0) &=& 1, & \\ f'(x) &=& 1, & \\ f^{(k)}(x) &=& 0 & \quad k \geq 2. \end{array}$$

Thus $P_2(x) = P_3(x) = P_4(x) = 1 + x$.

(c) For $\alpha \neq 1, \alpha \neq 0$:

$$\begin{array}{rclrcl} f(x) &=& (1+x)^{\alpha} & f(0) &=& 1 \\ f'(x) &=& \alpha(1+x)^{\alpha-1} & f'(0) &=& \alpha \\ f''(x) &=& \alpha(\alpha-1)(1+x)^{\alpha-2} & f''(0) &=& \alpha(\alpha-1) \\ f'''(x) &=& \alpha(\alpha-1)(\alpha-2)(1+x)^{\alpha-3} & f'''(0) &=& \alpha(\alpha-1)(\alpha-2) \\ f^{(4)}(x) &=& \alpha(\alpha-1)(\alpha-2)(\alpha-3)(1+x)^{\alpha-4} & f^{(4)}(0) &=& \alpha(\alpha-1)(\alpha-2)(\alpha-3) \end{array}$$

$$\begin{aligned} P_2(x) &= 1 + \alpha x + \frac{\alpha(\alpha-1)}{2}x^2, \\ P_3(x) &= 1 + \alpha x + \frac{\alpha(\alpha-1)}{2}x^2 + \frac{\alpha(\alpha-1)(\alpha-2)}{6}x^3, \\ P_4(x) &= 1 + \alpha x + \frac{\alpha(\alpha-1)}{2}x^2 + \frac{\alpha(\alpha-1)(\alpha-2)}{6}x^3 + \frac{\alpha(\alpha-1)(\alpha-2)(\alpha-3)}{24}x^4. \end{aligned}$$

11. For Problem 2, substitute $\alpha = \frac{1}{2}$ in the result of Problem 10:

$$\begin{aligned}(1+x)^{\frac{1}{2}} &= 1+\frac{1}{2}x+\frac{\frac{1}{2}(\frac{1}{2}-1)}{2}x^2+\frac{\frac{1}{2}(\frac{1}{2}-1)(\frac{1}{2}-2)}{6}x^3\\ &\quad +\frac{\frac{1}{2}(\frac{1}{2}-1)(\frac{1}{2}-2)(\frac{1}{2}-3)}{24}x^4\\ &= 1+\frac{1}{2}x-\frac{1}{8}x^2+\frac{1}{16}x^3-\frac{5}{128}x^4,\end{aligned}$$

as before.
For Problem 3, substitute $\alpha = -1$ and replace x by $-x$ in Problem 10, giving $(1-x)^{-1} = \frac{1}{1-x}.$
For Problem 4, substitute $\alpha = -1$ in Problem 10, giving $(1+x)^{-1} = \frac{1}{1+x}.$
For Problem 8, let $\alpha = 1/3$ and replace x by $-x$ in Problem 10, giving: $(1-x)^{1/3} = \sqrt[3]{1-x}.$

13. Let $f(x) = \cos x$. $f(\frac{\pi}{2}) = 0$.

$$\begin{aligned} f'(x) &= -\sin x & f'(\tfrac{\pi}{2}) &= -1,\\ f''(x) &= -\cos x & f''(\tfrac{\pi}{2}) &= 0,\\ f'''(x) &= \sin x & f'''(\tfrac{\pi}{2}) &= 1,\\ f^{(4)}(x) &= \cos x & f^{(4)}(\tfrac{\pi}{2}) &= 0.\end{aligned}$$

So,

$$\begin{aligned}P_4(x) &= 0-\left(x-\frac{\pi}{2}\right)+0+\frac{1}{3!}\left(x-\frac{\pi}{2}\right)^3\\ &= -\left(x-\frac{\pi}{2}\right)+\frac{1}{3!}\left(x-\frac{\pi}{2}\right)^3.\end{aligned}$$

15. Let $f(x) = \cos x$. Then $\cos\frac{\pi}{4} = \sin\frac{\pi}{4} = \frac{\sqrt{2}}{2}$.
Then $f'(x) = -\sin x$, $f''(x) = -\cos x$, and $f'''(x) = \sin x$, so the Taylor polynomial for $\cos x$ of degree three about $x = \pi/4$ is

$$\begin{aligned}P_3(x) &= \cos\frac{\pi}{4}+\left(-\sin\frac{\pi}{4}\right)\left(x-\frac{\pi}{4}\right)+\frac{-\cos\frac{\pi}{4}}{2!}\left(x-\frac{\pi}{4}\right)^2+\frac{\sin\frac{\pi}{4}}{3!}\left(x-\frac{\pi}{4}\right)^3\\ &= \frac{\sqrt{2}}{2}\left(1-\left(x-\frac{\pi}{4}\right)-\frac{1}{2}\left(x-\frac{\pi}{4}\right)^2+\frac{1}{6}\left(x-\frac{\pi}{4}\right)^3\right).\end{aligned}$$

17. Let $f(x) = \sqrt{1-x} = (1-x)^{1/2}$. Then $f'(x) = -\frac{1}{2}(1-x)^{-1/2}$, $f''(x) = -\frac{1}{4}(1-x)^{-3/2}$, $f'''(x) = -\frac{3}{8}(1-x)^{-5/2}$. So $f(0) = 1$, $f'(0) = -\frac{1}{2}$, $f''(0) = -\frac{1}{4}$, $f'''(0) = -\frac{3}{8}$, and

$$\begin{aligned}P_3(x) &= 1-\frac{1}{2}x-\frac{1}{4}\frac{1}{2!}x^2-\frac{3}{8}\frac{1}{3!}x^3\\ &= 1-\frac{1}{2}x-\frac{1}{8}x^2-\frac{1}{16}x^3.\end{aligned}$$

19. Let $f(x)=\dfrac{1}{1+x}=(1+x)^{-1}$. Then $f'(x)=-(1+x)^{-2}$, $f''(x)=2(1+x)^{-3}$, $f'''(x)=-6(1+x)^{-4}$, $f^{(4)}(x)=24(1+x)^{-5}$. So $f(2)=\frac{1}{3}$, $f'(2)=-\frac{1}{3^2}$, $f''(2)=\frac{2}{3^3}$, $f'''(2)=-\frac{6}{3^4}$, and $f^{(4)}(2)=\frac{24}{3^5}$. Therefore,

$$\begin{aligned} P_4(x) &= \frac{1}{3}-\frac{1}{3^2}(x-2)+\frac{2}{3^3}\frac{1}{2!}(x-2)^2-\frac{6}{3^4}\frac{1}{3!}(x-2)^3+\frac{24}{3^5}\frac{1}{4!}(x-2)^4 \\ &= \frac{1}{3}\left(1-\frac{1}{3}(x-2)+\frac{1}{3^2}(x-2)^2-\frac{1}{3^3}(x-2)^3+\frac{1}{3^4}(x-2)^4\right). \end{aligned}$$

21. This is the same as Example 7, page 632 except we need two more terms:

$$\begin{aligned} f^{(5)}(x) &= 24x^{-5} & f^{(5)}(1) &= 24, \\ f^{(6)}(x) &= -120x^{-6} & f^{(6)}(1) &= -120. \end{aligned}$$

So,

$$\begin{aligned} P_6(x) &= P_4(x)+\frac{24}{5!}(x-1)^5+\frac{-120}{6!}(x-1)^6 \\ &= (x-1)-\frac{(x-1)^2}{2}+\frac{(x-1)^3}{3}-\frac{(x-1)^4}{4}+\frac{(x-1)^5}{5}-\frac{(x-1)^6}{6}. \end{aligned}$$

23. Since $P_2(x)$ is the second degree Taylor polynomial for $f(x)$ about $x=0$, $P_2(0)=f(0)$, which says $a=f(0)$; $\left.\dfrac{d}{dx}P_2(x)\right|_{x=0}=f'(0)$, so $b=f'(0)$; and $\left.\dfrac{d^2}{dx^2}P_2(x)\right|_{x=0}=f''(0)$, then $2c=f''(0)$.
As we can see now, a is the y-intercept of $f(x)$, b is the slope of the tangent line to $f(x)$ at $x=0$ and c tells us the concavity of $f(x)$ near $x=0$.
So $c<0$ since f is concave down; $b>0$ since f is increasing; $a>0$ since $f(0)>0$.

25. Since $P_2(x)$ is the second degree Taylor polynomial for $f(x)$ about $x=0$, $P_2(0)=f(0)$, which says $a=f(0)$; $\left.\dfrac{d}{dx}P_2(x)\right|_{x=0}=f'(0)$, so $b=f'(0)$; and $\left.\dfrac{d^2}{dx^2}P_2(x)\right|_{x=0}=f''(0)$, then $2c=f''(0)$.
As we can see now, a is the y-intercept of $f(x)$, b is the slope of the tangent line to $f(x)$ at $x=0$ and c tells us the concavity of $f(x)$ near $x=0$.
So $a<0$, $b>0$ and $c>0$.

27.

$$\lim_{x\to 0}\frac{\sin x}{x}=\lim_{x\to 0}\frac{x-\frac{x^3}{3!}}{x}=\lim_{x\to 0}\left(1-\frac{x^2}{3!}\right)=1.$$

29. For $f(h) = e^h$, $P_4(h) = 1 + h + \frac{h^2}{2} + \frac{h^3}{3!} + \frac{h^4}{4!}$. So,

(a)

$$\begin{aligned}\lim_{h\to 0}\frac{e^h - 1 - h}{h^2} &= \lim_{h\to 0}\frac{e^h - 1 - h}{h^2} \\ &= \lim_{h\to 0}\frac{\frac{h^2}{2} + \frac{h^3}{3!} + \frac{h^4}{4!}}{h^2} = \lim_{h\to 0}\left(\frac{1}{2} + \frac{h}{3!} + \frac{h^2}{4!}\right) \\ &= \frac{1}{2}.\end{aligned}$$

(b)

$$\begin{aligned}\lim_{h\to 0}\frac{e^h - 1 - h - \frac{h^2}{2}}{h^3} &= \lim_{h\to 0}\frac{P_4(h) - 1 - h - \frac{h^2}{2}}{h^3} \\ &= \lim_{h\to 0}\frac{\frac{h^3}{3!} + \frac{h^4}{4!}}{h^3} = \lim_{h\to 0}\left(\frac{1}{3!} + \frac{h}{4!}\right) \\ &= \frac{1}{3!} = \frac{1}{6}.\end{aligned}$$

Using Taylor polynomials of higher degree would not have changed the results since the terms with higher powers of h all go to zero as $h \to 0$.

31.

$$\begin{aligned}f(x) &= 4x^2 - 7x + 2 & f(0) &= 2 \\ f'(x) &= 8x - 7 & f'(0) &= -7 \\ f''(x) &= 8 & f''(0) &= 8,\end{aligned}$$

so $P_2(x) = 2 + (-7)x + \frac{8}{2}x^2 = 4x^2 - 7x + 2$. We notice that $f(x) = P_2(x)$ in this case.

33. (a) We'll make the following conjecture:
"If $f(x)$ is a polynomial of degree n, i.e.

$$f(x) = a_0 + a_1x + a_2x^2 + \cdots + a_{n-1}x^{n-1} + a_nx^n,$$

then $P_n(x)$, the nth degree Taylor polynomial for $f(x)$ about $x = 0$ is $f(x)$ itself."

(b) All we need to do is to calculate $P_n(x)$, the nth degree Taylor polynomial for f about $x = 0$ and see if it is the same as $f(x)$.

$$\begin{aligned}C_0 &= f(0) = a_0; \\ C_1 &= f'(0) = \left(a_1 + 2a_2x + \cdots + na_nx^{n-1}\right)\Big|_{x=0} \\ &= a_1; \\ C_2 &= f''(0) = \left(2a_2 + 3\cdot 2a_3x + \cdots + n(n-1)a_nx^{n-2}\right)\Big|_{x=0} \\ &= 2!a_2.\end{aligned}$$

If we continue doing this, we'll see in general

$$C_k = f^{(k)}(0) = k!a_k, \qquad k = 1, 2, 3, \cdots, n.$$

So, $a_k = \dfrac{C_k}{k!}$, $k = 1, 2, 3, \cdots, n$. Therefore,

$$\begin{aligned} P_n(x) &= C_0 + \frac{C_1}{1!}x + \frac{C_2}{2!}x^2 + \cdots + \frac{C_n}{n!}x^n \\ &= a_0 + a_1 x + a_2 x^2 + \cdots + a_n x^n \\ &= f(x). \end{aligned}$$

35. Let $f(x)$ be a function that has derivatives up to order n at $x = a$. Let

$$P_n(x) = C_0 + C_1(x - a) + \cdots + C_n(x - a)^n$$

be the polynomial of degree n that approximates $f(x)$ about $x = a$. We require that $P_n(x)$ and all of its first n derivatives agree with those of the function $f(x)$ at $x = a$, i.e. we want

$$\begin{aligned} f(a) &= P_n(a), \\ f'(a) &= P_n'(a), \\ f''(a) &= P_n''(a), \\ &\vdots \\ f^{(n)}(a) &= P_n^{(n)}(a). \end{aligned}$$

When we substitute $x = a$ in $P_n(x)$, all the terms except the first drop out, so

$$f(a) = C_0.$$

Now differentiate $P_n(x)$:

$$P_n'(x) = C_1 + 2C_2(x - a) + 3C_3(x - a)^2 + \cdots + nC_n(x - a)^{n-1}.$$

Substituting $x = a$ again, which yields

$$f'(a) = P_n'(a) = C_1.$$

Differentiate $P_n'(x)$:

$$P_n''(x) = 2C_2 + 3 \cdot 2C_3(x - a) + \cdots + n(n - 1)C_n(x - a)^{n-2}$$

and substitute $x = a$ again:

$$f''(a) = P_n''(a) = 2C_2.$$

Differentiating and substituting again gives

$$f'''(a) = P_n'''(a) = 3 \cdot 2C_3.$$

Similarly,

$$f^{(k)}(a) = P_n{}^{(k)}(a) = k!C_k.$$

So, $C_0 = f(a)$, $C_1 = f'(a)$, $C_2 = \frac{f''(a)}{2!}$, $C_3 = \frac{f'''(a)}{3!}$, and so on.

If we adopt the convention that $f^{(0)}(a) = f(a)$ and $0! = 1$, then

$$C_k = \frac{f^{(k)}(a)}{k!}, \; k = 0, 1, 2, \cdots, n.$$

Therefore,

$$\begin{aligned} f(x) \approx P_n(x) &= C_0 + C_1(x-a) + C_2(x-a)^2 \cdots + C_n(x-a)^n \\ &= f(a) + f'(a)(x-a) + \frac{f''(a)}{2!}(x-a)^2 + \cdots + \frac{f^{(n)}(a)}{n!}(x-a)^n. \end{aligned}$$

9.2 Solutions

1.

$$\begin{aligned} f(x) &= \tfrac{1}{1-x} = (1-x)^{-1} & f(0) &= 1 \\ f'(x) &= -(1-x)^{-2}(-1) = (1-x)^{-2} & f'(0) &= 1 \\ f''(x) &= -2(1-x)^{-3}(-1) = 2(1-x)^{-3} & f''(0) &= 2 \\ f'''(x) &= -6(1-x)^{-4}(-1) = 6(1-x)^{-4} & f'''(0) &= 6 \end{aligned}$$

$$\begin{aligned} f(x) = \frac{1}{1-x} &= 1 + 1\cdot x + \frac{2x^2}{2!} + \frac{6x^3}{3!} + \cdots \\ &= 1 + x + x^2 + x^3 + \cdots \end{aligned}$$

3.

$$\begin{aligned} f(z) &= \arctan z & f(0) &= 0 \\ f'(z) &= \tfrac{1}{1+z^2} = (1+z^2)^{-1} & f'(0) &= 1 \\ f''(z) &= -(1+z^2)^{-2}(2z) & f''(0) &= 0 \\ f'''(z) &= -2(1+z^2)^{-2} + 2z\,(2(1+z^2)^{-3})\,2z & f'''(0) &= -2 \\ &= -2(1+z^2)^{-2} + 8z^2(1+z^2)^{-3} & & \\ f^{(4)}(z) &= 4(1+z^2)^{-3}(2z) + 16z(1+z^2)^{-3} + 8z^2(-3)(1+z^2)^{-4}(2z) & f^4(0) &= 0 \end{aligned}$$

$$\begin{aligned} f(z) = \arctan z &= 0 + 1\cdot z + \frac{0\cdot z^2}{2!} + \frac{(-2)z^3}{3!} + \frac{0\cdot z^4}{4!} + \cdots \\ &= z - \frac{2z^3}{3!} \cdots \end{aligned}$$

5.

$$\begin{aligned} f(x) &= \tfrac{1}{\sqrt{1+x}} = (1+x)^{-\frac{1}{2}} & f(0) &= 1 \\ f'(x) &= -\tfrac{1}{2}(1+x)^{-\frac{3}{2}} & f'(0) &= -\tfrac{1}{2} \\ f''(x) &= \tfrac{3}{4}(1+x)^{-\frac{5}{2}} & f''(0) &= \tfrac{3}{4} \\ f'''(x) &= -\tfrac{15}{8}(1+x)^{-\frac{7}{2}} & f'''(0) &= -\tfrac{15}{8} \end{aligned}$$

$$\begin{aligned} f(x) = \frac{1}{\sqrt{1+x}} &= 1 + (-\frac{1}{2})x + \frac{(\frac{3}{4})x^2}{2!} + \frac{(-\frac{15}{8})x^3}{3!} + \cdots \\ &= 1 - \frac{x}{2} + \frac{3x^2}{8} - \frac{5x^3}{16} + \cdots \end{aligned}$$

7.

$$\begin{aligned} f(x) &= \tfrac{1}{x} & f(1) &= 1 \\ f'(x) &= -\tfrac{1}{x^2} & f'(1) &= -1 \\ f''(x) &= \tfrac{2}{x^3} & f''(1) &= 2 \\ f'''(x) &= -\tfrac{6}{x^4} & f'''(1) &= -6 \end{aligned}$$

$$\frac{1}{x} = 1 - (x-1) + \frac{2(x-1)^2}{2!} - \frac{6(x-1)^3}{3!} + \cdots$$

$$\frac{1}{x} = 1 - (x-1) + (x-1)^2 - (x-1)^3 + \cdots$$

9. Again using the derivatives found in Problem 7, we have

$$f(2) = \frac{1}{2}, \quad f'(2) = -\frac{1}{4}, \quad f''(2) = \frac{1}{4}, \quad f'''(2) = -\frac{3}{8}.$$

$$\frac{1}{x} = \frac{1}{2} - \frac{x-2}{4} + \frac{(x-2)^2}{4 \cdot 2!} - \frac{3(x-2)^3}{8 \cdot 3!} + \cdots$$

$$\frac{1}{x} = \frac{1}{2} - \frac{(x-2)}{4} + \frac{(x-2)^2}{8} - \frac{(x-2)^3}{16} + \cdots$$

11.

$$\begin{aligned} f(\theta) &= \cos\theta & f(\tfrac{\pi}{4}) &= \tfrac{\sqrt{2}}{2} \\ f'(\theta) &= -\sin\theta & f'(\tfrac{\pi}{4}) &= -\tfrac{\sqrt{2}}{2} \\ f''(\theta) &= -\cos\theta & f''(\tfrac{\pi}{4}) &= -\tfrac{\sqrt{2}}{2} \\ f'''(\theta) &= \sin\theta & f'''(\tfrac{\pi}{4}) &= \tfrac{\sqrt{2}}{2} \end{aligned}$$

$$\begin{aligned} \cos\theta &= \frac{\sqrt{2}}{2} - \frac{\sqrt{2}}{2}(\theta - \frac{\pi}{4}) - \frac{\sqrt{2}}{2}\frac{(\theta - \frac{\pi}{4})^2}{2!} + \frac{\sqrt{2}}{2}\frac{(\theta - \frac{\pi}{4})^3}{3!} \cdots \\ &= \frac{\sqrt{2}}{2} - \frac{\sqrt{2}}{2}\left(\theta - \frac{\pi}{4}\right) - \frac{\sqrt{2}}{4}\left(\theta - \frac{\pi}{4}\right)^2 + \frac{\sqrt{2}}{12}\left(\theta - \frac{\pi}{4}\right)^3 \cdots \end{aligned}$$

13.
$$\begin{aligned} f(x) &= \tan x & f(\tfrac{\pi}{4}) &= 1 \\ f'(x) &= \frac{1}{\cos^2 x} & f'(\tfrac{\pi}{4}) &= 2 \\ f''(x) &= \frac{-2(-\sin x)}{\cos^3 x} = \frac{2\sin x}{\cos^3 x} & f''(\tfrac{\pi}{4}) &= 4 \\ f'''(x) &= \frac{-6\sin x(-\sin x)}{\cos^4 x} + \frac{2}{\cos^2 x} & f'''(\tfrac{\pi}{4}) &= 16 \end{aligned}$$

$$\tan x = 1 + 2\left(x - \frac{\pi}{4}\right) + 4\frac{(x-\frac{\pi}{4})^2}{2!} + 16\frac{(x-\frac{\pi}{4})^3}{3!} \cdots$$

$$\tan x = 1 + 2\left(x - \frac{\pi}{4}\right) + 2\left(x - \frac{\pi}{4}\right)^2 + \frac{8}{3}\left(x - \frac{\pi}{4}\right)^3 \cdots$$

15. The derivatives for $\tan\alpha$ are given in Problem 13. Then $f(0) = 0$, $f'(0) = 1$, $f''(0) = 0$, and $f'''(0) = 2$. So $\tan\alpha \approx \alpha + \frac{\alpha^3}{3} \cdots$

$$\begin{aligned} \lim_{\alpha\to 0} \frac{\tan\alpha}{\alpha} &= \lim_{\alpha\to 0} \frac{\alpha + \frac{\alpha^3}{3} + \cdots}{\alpha} \\ &= \lim_{\alpha\to 0}\left(1 + \frac{\alpha^2}{3} + \cdots\right) \\ &= 1. \end{aligned}$$

16. Using Binomial series with $\alpha = 1/2$, we have $\sqrt{1+x} = (1+x)^{1/2} = 1 + \frac{x}{2} - \frac{x^2}{8} + \cdots$. So,

$$\begin{aligned} \lim_{x\to 0} \frac{\sqrt{1+x} - 1}{x} &= \lim_{x\to 0} \frac{\left(1 + \frac{x}{2} - \frac{x^2}{8} + \cdots\right) - 1}{x} \\ &= \lim_{x\to 0}\left(\frac{1}{2} - \frac{x}{8} + \cdots\right) \\ &= \frac{1}{2}. \end{aligned}$$

17. Using the same expansion as in Problem 16, we have

$$\begin{aligned} \lim_{h\to 0} \frac{h}{\sqrt{1+h} - 1} &= \lim_{h\to 0} \frac{h}{\left(1 + \frac{h}{2} - \frac{h^2}{8} + \cdots\right) - 1} \\ &= \lim_{h\to 0} \frac{1}{\frac{1}{2} - \frac{h}{8} + \cdots} \\ &= 2. \end{aligned}$$

19. Expand $\cos\theta$ at $\theta = \pi/2$ first.

$$\begin{aligned}
\cos\theta|_{\theta=\frac{\pi}{2}} &= 0 \\
\left.\frac{d\cos\theta}{d\theta}\right|_{\theta=\frac{\pi}{2}} &= -\sin\theta|_{\theta=\frac{\pi}{2}} = -1 \\
\left.\frac{d^2\cos\theta}{d\theta^2}\right|_{\theta=\frac{\pi}{2}} &= -\cos\theta|_{\theta=\frac{\pi}{2}} = 0 \\
\left.\frac{d^3\cos\theta}{d\theta^3}\right|_{\theta=\frac{\pi}{2}} &= \sin\theta|_{\theta=\frac{\pi}{2}} = 1.
\end{aligned}$$

So, $\cos\theta = -\left(\theta - \frac{\pi}{2}\right) + \frac{1}{3!}\left(\theta - \frac{\pi}{2}\right)^3 - \cdots$ Thus,

$$\begin{aligned}
\lim_{\theta\to\frac{\pi}{2}} \frac{\cos\theta}{\theta - \frac{\pi}{2}} &= \lim_{\theta\to\frac{\pi}{2}} \frac{-(\theta - \frac{\pi}{2}) + \frac{1}{3!}(\theta - \frac{\pi}{2})^3 - \cdots}{\theta - \frac{\pi}{2}} \\
&= \lim_{\theta\to\frac{\pi}{2}} \left(-1 + \frac{1}{3!}\left(\theta - \frac{\pi}{2}\right)^2 - \cdots\right) \\
&= -1.
\end{aligned}$$

21. (a)

$$\begin{aligned}
f(x) &= \sin x^2 \\
f'(x) &= (\cos x^2)2x \\
f''(x) &= (-\sin x^2)4x^2 + (\cos x^2)2 \\
f'''(x) &= (-\cos x^2)8x^3 + (-\sin x^2)8x + (-\sin x^2)4x \\
&= (-\cos x^2)8x^3 + (-\sin x^2)12x \\
f^{(4)}(x) &= (\sin x^2)16x^4 + (-\cos x^2)24x^2 + (-\cos x^2)24x^2 + (-\sin x^2)12 \\
&= (\sin x^2)16x^4 + (-\cos x^2)48x^2 + (-\sin x^2)12 \\
f^{(5)}(x) &= (\cos x^2)32x^5 + (\sin x^2)64x^3 + (\sin x^2)96x^3 + (-\cos x^2)96x + (-\cos x^2)24x \\
&= (\cos x^2)32x^5 + (\sin x^2)160x^3 + (-\cos x^2)120x \\
f^{(6)}(x) &= (-\sin x^2)64x^6 + (\cos x^2)160x^4 + (\cos x^2)320x^4 + (\sin x^2)480x^2 \\
&\quad +(\sin x^2)240x^2 + (-\cos x^2)120 \\
&= (-\sin x^2)64x^6 + (\cos x^2)480x^4 + (\sin x^2)720x^2 + (-\cos x^2)120
\end{aligned}$$

So,

$$\begin{aligned}
f(0) &= 0 & \qquad f^{(4)}(0) &= 0 \\
f'(0) &= 0 & f^{(5)}(0) &= 0 \\
f''(0) &= 2 & f^{(6)}(0) &= -120 \\
f'''(0) &= 0 & &
\end{aligned}$$

Thus

$$\begin{aligned} f(x) &= \sin x^2 \\ &= \frac{2}{2!}x^2 - \frac{120}{6!}x^6 + \cdots \\ &= x^2 - \frac{1}{3!}x^6 + \cdots \end{aligned}$$

As we can see, the amount of calculation in order to find the higher derivatives of $\sin x^2$ increases very rapidly. In fact, the next non-zero term in the Taylor expansion of $\sin x^2$ is the 10th derivative term, which really requires a lot of work to get.

(b)

$$\sin x = x - \frac{1}{3!}x^3 + \frac{1}{5!}x^5 - \cdots$$

The first couple of coefficients of the above expansion are the same as those in the previous part. If we substitute x^2 for x in the Taylor expansion of $\sin x$ we should get the Taylor expansion of $\sin x^2$.

$$\begin{aligned} \sin x^2 &= x^2 - \frac{1}{3!}(x^2)^3 + \frac{1}{5!}(x^2)^5 - \cdots \\ &= x^2 - \frac{1}{3!}x^6 + \frac{1}{5!}x^{10} - \cdots \end{aligned}$$

23. By looking at the graph we can see that the Taylor polynomials are reasonable approximations for the function $f(x) = \sqrt{1+x}$ between $x = -0.25$ and $x = 0.25$. Thus a good guess is that the interval of convergence is $-0.25 < x < 0.25$.

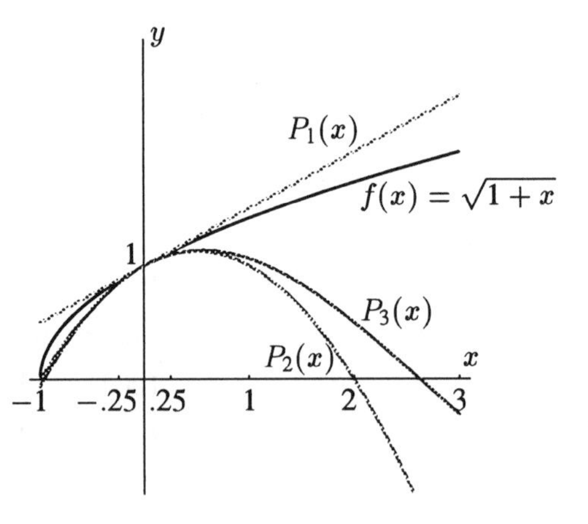

25. $C_1 = f'(0)/1!$, so $f'(0) = 1!C_1 = 1 \cdot 1 = 1$.

Similarly, $f''(0) = 2!C_2 = 2! \cdot \frac{1}{2} = 1$;

$f'''(0) = 3!C_3 = 3! \cdot \frac{1}{3} = 2! = 2$;

$f^{(10)}(0) = 10!C_{10} = 10! \cdot \frac{1}{10} = \frac{10!}{10} = 9! = 362880$.

9.3 Solutions

1. Substitute $y = -x$ into $e^y = 1 + y + \frac{y^2}{2!} + \frac{y^3}{3!} \cdots$ giving

$$\begin{aligned} e^{-x} &= 1 + (-x) + \frac{(-x)^2}{2!} + \frac{(-x)^3}{3!} \cdots \\ &= 1 - x + \frac{x^2}{2!} - \frac{x^3}{3!} \cdots \end{aligned}$$

3. Substitute $x = \theta^2$ into series for $\cos x$:

$$\begin{aligned} \cos(\theta^2) &= 1 - \frac{(\theta^2)^2}{2!} + \frac{(\theta^2)^4}{4!} - \frac{(\theta^2)^6}{6!} \cdots \\ &= 1 - \frac{\theta^4}{2!} + \frac{\theta^8}{4!} - \frac{\theta^{12}}{6!} \cdots \end{aligned}$$

5. Substituting $x = -2y$ into $\ln(1 + x) = x - \frac{x^2}{2} + \frac{x^3}{3} - \frac{x^4}{4} \cdots$ gives

$$\begin{aligned} \ln(1 - 2y) &= (-2y) - \frac{(-2y)^2}{2} + \frac{(-2y)^3}{3} - \frac{(-2y)^4}{4} \cdots \\ &= -2y - 2y^2 - \frac{8}{3}y^3 - 4y^4 \cdots \end{aligned}$$

7. Since $\frac{d}{dx}(\arcsin x) = \frac{1}{\sqrt{1-x^2}} = 1 + \frac{1}{2}x^2 + \frac{3}{8}x^4 + \frac{5}{16}x^6 \cdots$, integrating gives

$$\arcsin x = c + x + \frac{1}{6}x^3 + \frac{3}{40}x^5 + \frac{5}{112}x^7 \cdots$$

Since $\arcsin 0 = 0$, $c = 0$.

9.

$$\begin{aligned} \phi^3 \cos(\phi^2) &= \phi^3 \left(1 - \frac{(\phi^2)^2}{2!} + \frac{(\phi^2)^4}{4!} - \frac{(\phi^2)^6}{6!} \cdots\right) \\ &= \phi^3 - \frac{\phi^7}{2!} + \frac{\phi^{11}}{4!} - \frac{\phi^{15}}{6!} \cdots \end{aligned}$$

11.

$$e^t \cos t = \left(1 + t + \frac{t^2}{2!} + \frac{t^3}{3!} + \frac{t^4}{4!} \cdots\right)\left(1 - \frac{t^2}{2!} + \frac{t^4}{4!} - \frac{t^6}{6!} \cdots\right)$$

Multiplying out and collecting terms gives

$$\begin{aligned} e^t \cos t &= 1 + t + \left(\frac{t^2}{2!} - \frac{t^2}{2!}\right) + \left(\frac{t^3}{3!} - \frac{t^3}{2!}\right) + \left(\frac{t^4}{4!} + \frac{t^4}{4!} - \frac{t^4}{(2!)^2}\right) \cdots \\ &= 1 + t - \frac{t^3}{3} - \frac{t^4}{6} \cdots \end{aligned}$$

13.

$$\begin{aligned} \frac{1}{2+x} &= \frac{1}{2(1+\frac{x}{2})} = \frac{1}{2}\left(1 + \frac{x}{2}\right)^{-1} \\ &= \frac{1}{2}\left(1 - \frac{x}{2} + \left(\frac{x}{2}\right)^2 - \left(\frac{x}{2}\right)^3 \cdots\right) \\ &= \frac{1}{2} - \frac{1}{4}x + \frac{1}{8}x^2 - \frac{1}{16}x^3 \cdots \end{aligned}$$

15. From the series for $\ln(1+y)$,

$$\ln(1+y) = y - \frac{y^2}{2} + \frac{y^3}{3} - \frac{y^4}{4} \cdots,$$

we get

$$\ln(1+y^2) = y^2 - \frac{y^4}{2} + \frac{y^6}{3} - \frac{y^8}{4} \cdots$$

The Taylor series for $\sin y$ is

$$\sin y = y - \frac{y^3}{3!} + \frac{y^5}{5!} - \frac{y^7}{7!} \cdots$$

So

$$\sin y^2 = y^2 - \frac{y^6}{3!} + \frac{y^{10}}{5!} - \frac{y^{14}}{7!} \cdots$$

The Taylor series for $\cos y$ is

$$\cos y = 1 - \frac{y^2}{2!} + \frac{y^4}{4!} - \frac{y^6}{6!} \cdots$$

So

$$1 - \cos y = \frac{y^2}{2!} - \frac{y^4}{4!} + \frac{y^6}{6!} \cdots$$

Near $y = 0$, we can drop terms beyond the fourth degree in each expression:

$$\begin{aligned} \ln(1 + y^2) &\approx y^2 - \frac{y^4}{4!} \\ \sin y^2 &\approx y^2 \\ 1 - \cos y &\approx \frac{y^2}{2!} - \frac{y^4}{4!} \end{aligned}$$

(Note: These functions are all even, so what holds for negative y will hold for positive y.) Clearly $1 - \cos y$ is smallest, because the y^2 term has a factor of $\frac{1}{2}$. Thus, for small y,

$$\frac{y^2}{2} - \frac{y^4}{4!} < y^2 - \frac{y^4}{4!} < y^2$$

so

$$1 - \cos y < \ln(1 + y^2) < \sin(y^2)$$

17. (a) $e^{-x^2} = 1 - x^2 + \frac{x^4}{2!} - \frac{x^6}{3!} \cdots$

$\frac{1}{1 + x^2} = 1 - x^2 + x^4 - x^6 \cdots$

Notice that the first two terms are the same in both series.

(b) $\frac{1}{1 + x^2}$ is greater.

(c) Even, because the only terms involved are of even degree.

(d) The coefficients for e^{-x^2} become extremely small for higher powers of x, and we can "counteract" the effect of these powers for large values of x. The series for $\frac{1}{1+x^2}$ has no such coefficients.

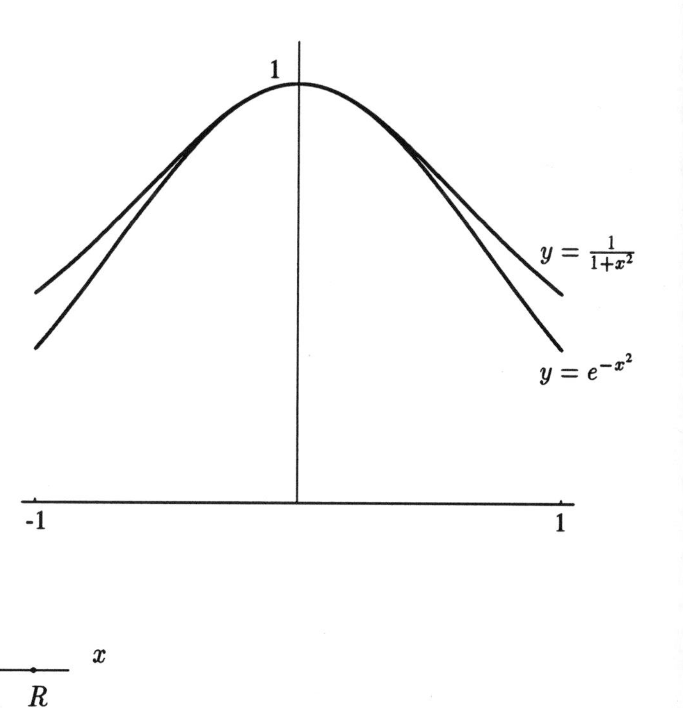

19.

Q Q x

-1 1 R

$$\begin{aligned} E &= kQ\left(\frac{1}{(R-1)^2} - \frac{1}{(R+1)^2}\right) \\ &= \frac{kQ}{R^2}\left(\frac{1}{(1-\frac{1}{R})^2} - \frac{1}{(1+\frac{1}{R})^2}\right) \end{aligned}$$

Since $|\frac{1}{R}| < 1$, we can expand the two terms using the Binomial Expansion:

$$\begin{aligned} \frac{1}{(1-\frac{1}{R})^2} &= \left(1-\frac{1}{R}\right)^{-2} = 1 - 2\left(-\frac{1}{R}\right) + (-2)(-3)\frac{(-\frac{1}{R})^2}{2!} + (-2)(-3)(-4)\frac{(-\frac{1}{R})^3}{3!}\cdots \\ \frac{1}{(1+\frac{1}{R})^2} &= \left(1+\frac{1}{R}\right)^{-2} = 1 - 2\left(\frac{1}{R}\right) + (-2)(-3)\frac{(\frac{1}{R})^2}{2!} + (-2)(-3)(-4)\frac{(\frac{1}{R})^3}{3!}\cdots \end{aligned}$$

Substituting, we get:

$$\begin{aligned} E &= \frac{kQ}{R^2}\left[1 + \frac{2}{R} + \frac{3}{R^2} + \frac{4}{R^3}\cdots - \left(1 - \frac{2}{R} + \frac{3}{R^2} - \frac{4}{R^3}\cdots\right)\right] \\ &\approx \frac{kQ}{R^2}\left(\frac{4}{R} + \frac{8}{R^3}\right) \end{aligned}$$

using only the first two non-zero terms.

21. This time we are interested in how a function behaves at large values in its domain. Therefore, we don't want to expand $V = 2\pi\sigma(\sqrt{R^2+a^2} - R)$ about $R = 0$. We want to find a variable which becomes small as R gets large. Since $R > a$

$$V = R2\pi\sigma\left(\sqrt{1+\frac{a^2}{R^2}} - 1\right).$$

We can now expand a series in terms of $(\frac{a}{R})^2$. This may seem strange, but suspend your disbelief. The Taylor series for $\sqrt{1+\frac{a^2}{R^2}}$ is

$$1 + \frac{1}{2}\frac{a^2}{R^2} + \frac{(1/2)(-1/2)}{2}\left(\frac{a^2}{R^2}\right)^2 + \cdots$$

So $V = R2\pi\sigma\left(1 + \frac{1}{2}\frac{a^2}{R^2} - \frac{1}{8}\left(\frac{a^2}{R^2}\right)^2 + \cdots - 1\right)$. For large R, we can drop the $-\frac{1}{8}\frac{a^4}{R^4}$ term and terms of higher power, so

$$V \approx \frac{\pi\sigma a^2}{R}.$$

Notice that what we really did by expanding around $(\frac{a}{R})^2 = 0$ was expanding around $R = \infty$. We then get a series that is valid for large R instead of small R.

23. (a) At $r = a$, the force between the atoms is 0.

(b) There will be an attractive force, pulling them back together.

(c) There will be a repulsive force, pushing the atoms apart.

(d) $F = F(a) + F'(a)(r-a) + F''(a)\frac{(r-a)^2}{2!} \cdots$

(e) $F(a) = 0$, so discarding all but the first non-zero terms, we get $F = F'(a)(r-a)$. $F'(a)$ is a negative number, so for r slightly greater than a, the force is negative (attractive). For r slightly less than a, the force is positive (repulsive).

9.4 Solutions

1. (a) The Taylor polynomial of degree 0 about $t = 0$ for $f(t) = e^t$ is simply $f(0) = 1$. Since $e^t \geq 1$ on $[0, 0.5]$, the approximation is an underestimate.

   (b) Using the error formula,

$$|E_n| = \left|\frac{f^{(n+1)}(c)(b-a)^{n+1}}{(n+1)!}\right| \text{ with } n = 0,$$

   we get

$$|E_0| = |f'(c)(b-a)|$$

   We need to find the value of c such that $f'(c)$ is a maximum, for $0 \leq c \leq 0.5$. Since $f'(t) = e^t$ is increasing on $[0, 0.5]$, $f'(c)$ will be a maximum when $c = 0.5$. So

$$\begin{aligned} |E_0| &\leq |f'(0.5)(0.5-0)| \\ &= 0.5f'(0.5) \\ &= 0.5e^{0.5} \\ &\approx 0.83. \end{aligned}$$

   (Note: By looking at a graph of $f(t)$ and its 0-th order approximation, it is easy to see that the greatest error occurs when $t = 0.5$, and the error is $e^{0.5} - 1 \approx 0.65 < 0.83$. So our error approximation works.)

3. (a) θ is the first degree approximation of $f(\theta) = \sin\theta$. $P_1(\theta) = \theta$ is an overestimate for $0 < \theta \leq 1$, and is an underestimate for $-1 \leq \theta < 0$. (This can be seen easily from a graph.)

   (b) Using the first order error formula,

$$|E_1| = \left|\frac{f^{(2)}(c)(b-a)^2}{2}\right| = \left|\frac{f^{(2)}(c)1^2}{2}\right| = \frac{|f^{(2)}(c)|}{2}.$$

   For what value of c on $[-1, 1]$ is $|f^{(2)}(c)|$ a maximum? Well, $f^{(2)}(\theta) = -\sin\theta$. $|-\sin\theta| = |\sin\theta|$ has maximum $|\sin 1|$ and $|\sin(-1)| = \sin 1$. So

$$|E_1| \leq \frac{\sin 1}{2} \approx 0.42$$

   This error estimate is very large and not particularly useful.

5. Let $f(x) = \sqrt{1+x}$. We will use a Taylor polynomial with $x = 1$ to approximate $\sqrt{2}$. The error in the Taylor approximation of degree three for $f(x) = \sqrt{2}$ about $x = 0$ is:

$$|E_3| = \left|\frac{f^{(4)}(c)(1-0)^4}{4!}\right|$$

for some c between 0 and 1. Now, $f^{(4)}(c) = -\frac{15}{16}(1+c)^{-\frac{7}{2}} = \frac{-15}{16(1+c)^{\frac{7}{2}}}$. Since $1 \leq (1+c)^{\frac{7}{2}}$ for c between 0 and 1, $|f^{(4)}(c)| = \frac{15}{16(1+c)^{\frac{7}{2}}} \leq \frac{15}{16}$ for c between 0 and 1. Thus,

$$|E_3| \leq \frac{15}{16 \cdot 4!} < 0.04$$

7. Let $f(x) = (1-x)^{\frac{1}{3}}$ and use $x = 0.5$, so that, $f(0.5) = (0.5)^{\frac{1}{3}}$. The error in the Taylor approximation of degree 3 for $f(0.5) = 0.5^{\frac{1}{3}}$ about $x = 0$ is:

$$|E_3| = \left|\frac{f^{(4)}(c)(0.5-0)^4}{4!}\right|$$

for some c between 0 and $\frac{1}{2}$. Now, $f^{(4)}(c) = -\frac{80}{81}(1-c)^{-\frac{11}{3}}$. By looking at the graph of $(1-x)^{-\frac{11}{3}}$, we see that $|f^{(4)}(c)|$ is maximized for c between 0 and 0.5 when $c = 0.5$. Thus, $|E_3| \leq \frac{80}{81}2^{\frac{11}{3}}(0.5)^4\frac{1}{4!} \approx 0.033$.

9. Let $f(x) = (1+x)^{-\frac{1}{2}} = \frac{1}{\sqrt{1+x}}$ with $x = 2$. The error in the Taylor approximation of degree three for $f(2) = \frac{1}{\sqrt{3}}$ about $x = 0$ is:

$$|E_3| = \left|\frac{f^{(4)}(c)(2-0)^4}{4!}\right|$$

for some c between 0 and 2. Since $f^{(4)}(c) = \frac{105}{16}(1+c)^{-\frac{9}{2}}$, we see that if c is between 0 and 2, $|f^{(4)}c)| \leq \frac{105}{16}$. Thus,

$$|E_3| \leq \frac{105}{16}16\frac{1}{4!} = \frac{105}{24}.$$

Again, this is not a very helpful bound on the error, but that is to be expected as the Taylor series does not converge at $x = 2$. (At $x = 2$, we are outside the interval of convergence.)

10. The maximum possible error for the nth degree Taylor polynomial about $x = 0$ approximating $\cos x$ is $|E_n| = \left|\frac{\cos^{(n+1)}(c)x^{n+1}}{(n+1)!}\right|$, where c is some value between 0 and x. In particular, if x is between 0 and 1, $|E_n| \leq \frac{|\cos^{(n+1)}(c)|}{(n+1)!}$. Now the derivatives of $\cos x$ are simply $\cos x, \sin x, -\cos x$, and $-\sin x$. The largest value these ever take is 1, so $|\cos^{(n+1)}(c)| \leq 1$, and thus $|E_n| \leq \frac{1}{(n+1)!}$. The same argument works for $\sin x$.

11. By the results of Problem 10, if we approximate cos 1 using the nth degree polynomial, the error is at most $\frac{1}{(n+1)!}$. For the answer to be correct to four decimal places, the error must be less than 0.00005. Thus, the first n such that $\frac{1}{(n+1)!} \leq 0.00005$ will work. In particular, when $n = 7$, $\frac{1}{8!} = \frac{1}{40370} \leq 0.00005$, so the 7th degree Taylor polynomial will give the desired result. For six decimal places, we need $\frac{1}{(n+1)!} \leq 0.0000005$. Since $n = 9$ works, the 9th degree Taylor polynomial is sufficient.

13. (a) Let $f(x) = \ln x$. If we wish to approximate $\ln x$, the error in the nth degree Taylor polynomial approximation about $x = 1$ is:

$$|E_n| = \left|\frac{f^{(n+1)}(c)(x-1)^{n+1}}{(n+1)!}\right|$$

for some c between 1 and x. To bound $|E_n|$, we notice that the nth derivative of $f(x) = \ln x$ is just $f^{(n)}(x) = \frac{(-1)^{n+1}(n-1)!}{x^n}$. Thus, replacing n by $n+1$, gives $f^{(n+1)}(c) = \frac{(-1)^{n+2}n!}{c^{n+1}}$, where c is between 1 and x. Since $x > 1$, then $|f^{(n+1)}(c)| \leq n!$, and $|E_n| \leq \frac{(x-1)^{n+1}}{n+1}$.

(b) If $1 < x < 2$, then $0 \leq |E_n| \leq \frac{(x-1)^{n+1}}{n+1} \leq \frac{1}{n+1}$. Since $\lim_{n\to\infty} \frac{1}{n+1} = 0, |E_n| \to 0$ as $n \to \infty$. If $x > 2$, then $|E_n| \leq \frac{(x-1)^{n+1}}{n+1}$. Let $z = x - 1 > 1$. Then $|E_n| \leq \frac{z^{n+1}}{n+1}$. But $\frac{z^{n+1}}{n+1} \to \infty$ as $n \to \infty$, since the numerator grows exponentially with n, while the denominator only grows linearly with n. Thus, as $n \to \infty$, this bound on $|E_n|$ also goes to ∞.

9.5 Solutions

1. (a) i. The graph of $y = \sin x + \frac{1}{3}\sin 3x$ looks like

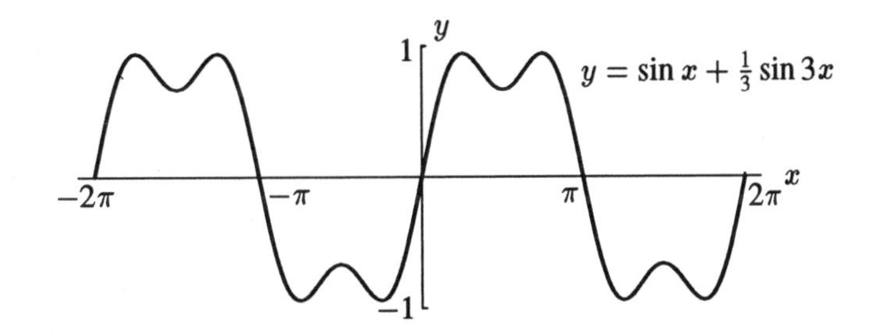

ii. The graph of $y = \sin x + \frac{1}{3}\sin 3x + \frac{1}{5}\sin 5x$ looks like

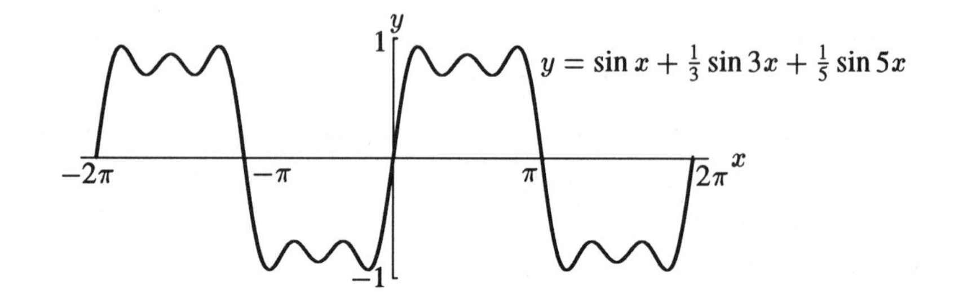

(b) Following the pattern, we add the term $\frac{1}{7}\sin 7x$.

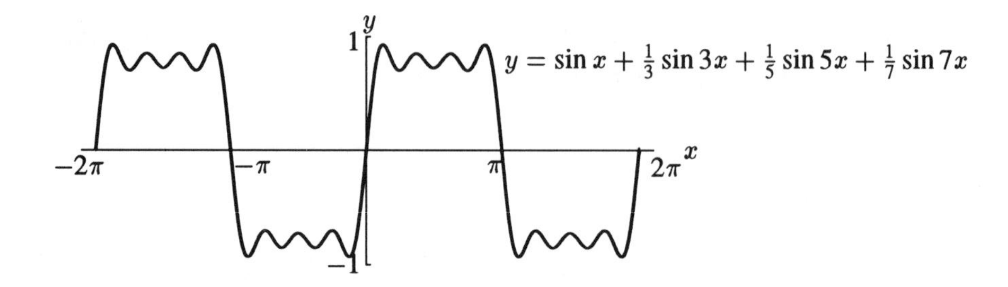

(c) The equation is

$$f(x) = \begin{cases} 1 & -2\pi \leq x < -\pi \\ -1 & -\pi \leq x < 0 \\ 1 & 0 \leq x < \pi \\ -1 & \pi \leq x < 2\pi \end{cases}$$

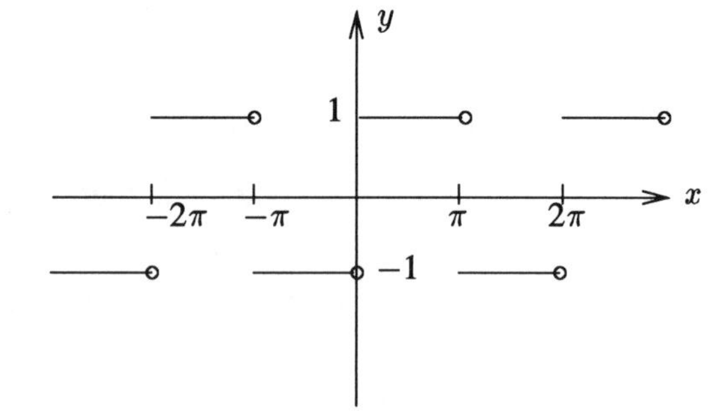

The square wave function is not continuous at $x = 0, \pm\pi, \pm 2\pi, \ldots$

3. First,

$$a_0 = \frac{1}{2\pi}\int_{-\pi}^{\pi} f(x)\,dx = \frac{1}{2\pi}\left[\int_{-\pi}^{0} -x\,dx + \int_{0}^{\pi} x\,dx\right] = \frac{1}{2\pi}\left[-\frac{x^2}{2}\bigg|_{-\pi}^{0} + \frac{x^2}{2}\bigg|_{0}^{\pi}\right] = \frac{\pi}{2}.$$

To find the a_i's, we use the integral table. For $n \geq 1$,

$$\begin{aligned} a_n = \frac{1}{\pi}\int_{-\pi}^{\pi} f(x)\cos(nx)\,dx &= \frac{1}{\pi}\left[\int_{-\pi}^{0} -x\cos(nx)\,dx + \int_{0}^{\pi} x\cos(nx)\,dx\right] \\ &= \frac{1}{\pi}\left[\left(-\frac{x}{n}\sin(nx) - \frac{1}{n^2}\cos(nx)\right)\bigg|_{-\pi}^{0}\right. \\ &\qquad \left. + \left(\frac{x}{n}\sin(nx) + \frac{1}{n^2}\cos(nx)\right)\bigg|_{0}^{\pi}\right] \end{aligned}$$

$$= \frac{1}{\pi}\left(-\frac{1}{n^2}+\frac{1}{n^2}\cos(-n\pi)+\frac{1}{n^2}\cos(n\pi)-\frac{1}{n^2}\right)$$
$$= \frac{2}{\pi n^2}(\cos n\pi - 1)$$

Thus, $a_1 = -\frac{4}{\pi}$, $a_2 = 0$, and $a_3 = -\frac{4}{9\pi}$.
To find the b_i's, note that $f(x)$ is even, so for $n \geq 1$, $f(x)\sin(nx)$ is odd. Thus, $\int_{-\pi}^{\pi} f(x)\sin(nx) = 0$, so all the b_i's are 0.
Thus $F_1 = F_2 = \frac{\pi}{2} - \frac{4}{\pi}\cos x$, $F_3 = \frac{\pi}{2} - \frac{4}{\pi}\cos x - \frac{4}{9\pi}\cos 3x$.

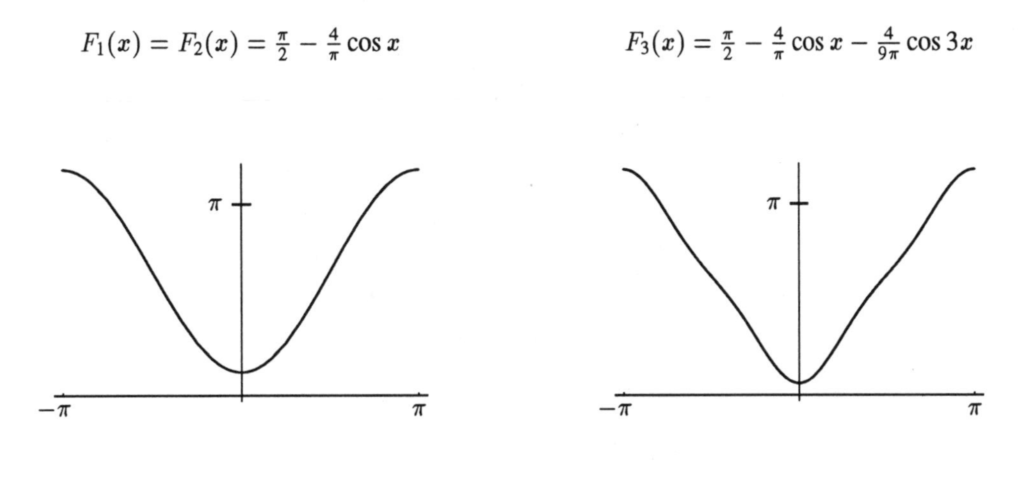

$F_1(x) = F_2(x) = \frac{\pi}{2} - \frac{4}{\pi}\cos x$
$F_3(x) = \frac{\pi}{2} - \frac{4}{\pi}\cos x - \frac{4}{9\pi}\cos 3x$
π
$-\pi$
π
π
$-\pi$
π

5. First, we find a_0.

$$a_0 = \frac{1}{2\pi}\int_{-\pi}^{\pi} x^2\,dx = \frac{1}{2\pi}\left(\frac{x^3}{3}\Big|_{-\pi}^{\pi}\right) = \frac{\pi^2}{3}.$$

To find a_n, $n \geq 1$, we use the integral table.

$$\begin{aligned} a_n = \frac{1}{\pi}\int_{-\pi}^{\pi} x^2\cos nx\,dx &= \frac{1}{\pi}\left[\frac{x^2}{n}\sin(nx)+\frac{2x}{n^2}\cos(nx)-\frac{2}{n^3}\sin(nx)\right]\Bigg|_{-\pi}^{\pi} \\ &= \frac{1}{\pi}\left[\frac{2\pi}{n^2}\cos(n\pi)+\frac{2\pi}{n^2}\cos(-n\pi)\right] \\ &= \frac{4}{n^2}\cos(n\pi). \end{aligned}$$

Again, $\cos(n\pi) = (-1)^n$ for all integers n, so $a_n = (-1)^n \frac{4}{n^2}$. Note that

$$b_n = \frac{1}{\pi} \int_{-\pi}^{\pi} x^2 \sin nx \, dx.$$

x^2 is an even function, and $\sin nx$ is odd, so $x^2 \sin nx$ is odd. Thus $\int_{-\pi}^{\pi} x^2 \sin nx \, dx = 0$, and $b_n = 0$ for all n.
We deduce that the nth Fourier polynomial for f (where $n \geq 1$) is

$$F_n(x) = \frac{\pi^2}{3} + \sum_{i=1}^{n} (-1)^i \frac{4}{i^2} \cos(ix).$$

In particular, we have the following graphs:

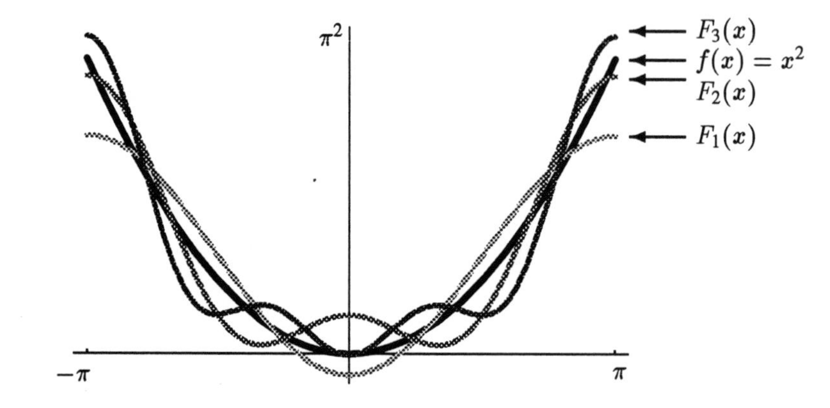

7. Let r_k and s_k be the Fourier coefficients of $Af + Bg$. Then

$$\begin{aligned} r_0 &= \frac{1}{2\pi} \int_{-\pi}^{\pi} \Big[Af(x) + Bg(x) \Big] \, dx \\ &= A\Big[\frac{1}{2\pi} \int_{-\pi}^{\pi} f(x)\, dx\Big] + B\Big[\frac{1}{2\pi} \int_{-\pi}^{\pi} g(x)\, dx\Big] \\ &= Aa_0 + Bc_0. \end{aligned}$$

Similarly,

$$\begin{aligned} r_k &= \frac{1}{\pi} \int_{-\pi}^{\pi} \Big[Af(x) + Bg(x) \Big] \cos(kx) \, dx \\ &= A\Big[\frac{1}{\pi} \int_{-\pi}^{\pi} f(x) \cos(kx)\, dx\Big] + B\Big[\frac{1}{\pi} \int_{-\pi}^{\pi} g(x) \cos(kx)\, dx\Big] \\ &= Aa_k + Bc_k. \end{aligned}$$

And finally,

$$\begin{aligned}
s_k &= \frac{1}{\pi}\int_{-\pi}^{\pi}\Big[Af(x)+Bg(x)\Big]\sin(kx)\,dx \\
&= A\Big[\frac{1}{\pi}\int_{-\pi}^{\pi} f(x)\sin(kx)\,dx\Big] + B\Big[\frac{1}{\pi}\int_{-\pi}^{\pi} g(x)\sin(kx)\,dx\Big] \\
&= Ac_k + Bd_k.
\end{aligned}$$

9. We have $f(x) = x, 0 \le x < 1$. Let $t = 2\pi x - \pi$. Notice that as x varies from 0 to 1, t varies from $-\pi$ to π. Thus if we rewrite the function in terms of t, we can find the Fourier series in terms of t in the usual way. To do this, let $g(t) = f(x) = x = \frac{t+\pi}{2\pi}$ on $-\pi \le t < \pi$. We now find the fourth degree Fourier polynomial for g.

$$a_o = \frac{1}{2\pi}\int_{-\pi}^{\pi} g(t)dt = \frac{1}{2\pi}\int_{-\pi}^{\pi}\frac{t+\pi}{2\pi}dt = \frac{1}{(2\pi)^2}(\frac{t^2}{2}+\pi t)|_{-\pi}^{\pi} = \frac{1}{2}$$

Notice, a_0 is the average value of both f and g. For $n \ge 1$

$$\begin{aligned}
a_n &= \frac{1}{\pi}\int_{-\pi}^{\pi}\frac{t+\pi}{2\pi}\cos(nt)dt = \frac{1}{2\pi^2}\int_{-\pi}^{\pi}(t\cos(nt)+\pi\cos(nt))dt \\
&= \frac{1}{2\pi^2}\left[\frac{t}{n}\sin(nt)+\frac{1}{n^2}\cos(nt)+\frac{\pi}{n}\sin(nt)\right]\Big|_{-\pi}^{\pi} \\
&= 0
\end{aligned}$$

$$\begin{aligned}
b_n &= \frac{1}{\pi}\int_{-\pi}^{\pi}\frac{t+\pi}{2\pi}\sin(nt)dt = \frac{1}{2\pi^2}\int_{-\pi}^{\pi}(t\sin(nt)+\pi\sin(nt))dt \\
&= \frac{1}{2\pi^2}\left[-\frac{t}{n}\cos(nt)+\frac{1}{n^2}\sin(nt)-\frac{\pi}{n}\cos(nt)\right]\Big|_{-\pi}^{\pi} \\
&= \frac{1}{2\pi^2}(-\frac{4\pi}{n}\cos(\pi n)) = -\frac{2}{\pi n}\cos(\pi n) = \frac{2}{\pi n}(-1)^{n+1}
\end{aligned}$$

(Note: We get the integrals for a_n and b_n using the integral table)
Thus, the Fourier polynomial of degree 4 for g is:

$$G_4(t) = \frac{1}{2} + \frac{2}{\pi}\sin t - \frac{1}{\pi}\sin 2t + \frac{2}{3\pi}\sin 3t - \frac{1}{2\pi}\sin 4t$$

Now, since $g(t) = f(x)$, the Fourier polynomial of degree 4 for f can be found by replacing t in terms of x again. Thus,

$$F_4(x) = \frac{1}{2} + \frac{2}{\pi}\sin(2\pi x-\pi) - \frac{1}{\pi}\sin(4\pi x-2\pi) + \frac{2}{3\pi}\sin(6\pi x-3\pi) - \frac{1}{2\pi}\sin(8\pi x-4\pi).$$

Now, using the fact that $\sin(x-\pi) = -\sin x$ and $\sin(x-2\pi) = \sin x$, etc., we have:

$$F_4(x) = \frac{1}{2} - \frac{2}{\pi}\sin(2\pi x) - \frac{1}{\pi}\sin(4\pi x) - \frac{2}{3\pi}\sin(6\pi x) - \frac{1}{2\pi}\sin(8\pi x).$$

1

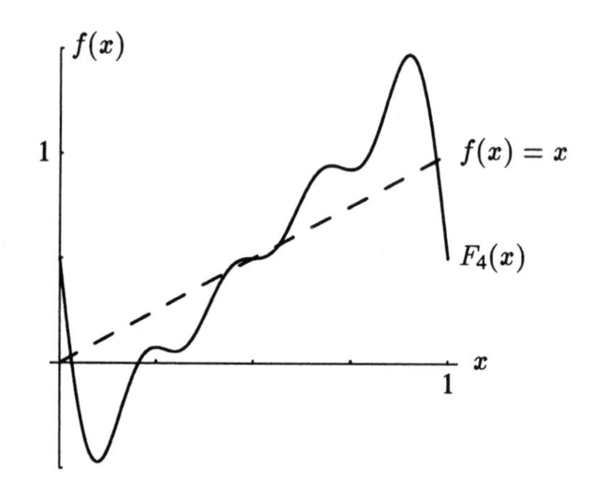

11. By formula II-12 of the integral table,

$$\begin{aligned}\int_{-\pi}^{\pi} \sin kx \cos mx \, dx &= \frac{1}{m^2-k^2}\Big(m\sin(kx)\sin(mx)+k\cos(kx)\cos(mx)\Big)\Big|_{-\pi}^{\pi} \\ &= \frac{1}{m^2-k^2}\Big[m\sin(k\pi)\sin(m\pi)+k\cos(k\pi)\cos(m\pi) \\ &\qquad -m\sin(-k\pi)\sin(-m\pi)-k\cos(-k\pi)\cos(-m\pi)\Big]\end{aligned}$$

Since k and m are positive integers, $\sin(k\pi) = \sin(m\pi) = \sin(-k\pi) = \sin(-m\pi) = 0$. Also, $\cos(k\pi) = \cos(-k\pi)$ since cos is even. Thus this expression reduces to 0. [Note: one could also note that $\sin kx \cos mx$ is odd, so $\int_{-\pi}^{\pi} \sin kx \cos mx \, dx$ must be 0.]

13. Let us make the substitution $u = mx$, $dx = \frac{1}{m}du$.
Then,

$$\int_{-\pi}^{\pi} \cos^2 mx \, dx = \frac{1}{m}\int_{u=-m\pi}^{u=m\pi} \cos^2 u \, du$$

By Formula IV-18 of the integral table,

$$\begin{aligned} &= \frac{1}{m}\left[\frac{1}{2}\cos u \sin u\right]\Big|_{-m\pi}^{m\pi} + \frac{1}{m}\frac{1}{2}\int_{-m\pi}^{m\pi} 1 \, du \\ &= 0 + \frac{1}{2m}u\Big|_{-m\pi}^{m\pi} \\ &= \frac{1}{2m}u\Big|_{-m\pi}^{m\pi} \\ &= \frac{1}{2m}(2m\pi) = \pi. \end{aligned}$$

15. The easiest way to do this is to use Problem 13.

$$\begin{aligned}\int_{-\pi}^{\pi} \sin^2 mx\,dx = \int_{-\pi}^{\pi} (1-\cos^2 mx)\,dx &= \int_{-\pi}^{\pi} dx - \int_{-\pi}^{\pi} \cos^2 mx\,dx \\ &= 2\pi - \pi \quad \text{using Problem 13} \\ &= \pi.\end{aligned}$$

9.6 Answers to Miscellaneous Exercises for Chapter 9

1. Substituting $y = t^2$ in $\sin y = y - \frac{y^3}{3!} + \frac{y^5}{5!} - \frac{y^7}{7!} \cdots$ gives $\sin t^2 = t^2 - \frac{t^6}{3!} + \frac{t^{10}}{5!} - \frac{t^{14}}{7!} \cdots$

3. Substituting $y = -4z^2$ into $\frac{1}{1+y} = 1 - y + y^2 - y^3 \cdots$ gives $\frac{1}{1-4z^2} = 1 + 4z^2 + 16z^4 + 64z^6 \cdots$

5. $\frac{a}{a+b} = \frac{a}{a(1+\frac{b}{a})} = \left(1+\frac{b}{a}\right)^{-1} = 1 - \frac{b}{a} + \left(\frac{b}{a}\right)^2 - \left(\frac{b}{a}\right)^3 \cdots$

7. $\sin x \approx -\frac{1}{\sqrt{2}} + \frac{1}{\sqrt{2}}\left(x + \frac{\pi}{4}\right) + \frac{1}{2\sqrt{2}}\left(x + \frac{\pi}{4}\right)^2$

9. $\ln x \approx \ln 2 + \frac{1}{2}(x-2) - \frac{1}{8}(x-2)^2$

11. (a) The series for $\frac{\sin 2\theta}{\theta}$ is

$$\begin{aligned}\frac{\sin 2\theta}{\theta} &= \frac{1}{\theta}\left(2\theta - \frac{(2\theta)^3}{3!} + \frac{(2\theta)^5}{5!} \cdots\right) \\ &= 2 - \frac{4\theta^2}{3} + \frac{4\theta^4}{15} \cdots\end{aligned}$$

so $\lim_{\theta \to 0} \frac{\sin 2\theta}{\theta} = 2$.

(b) Near $\theta = 0$, we make the approximation

$$\frac{\sin 2\theta}{\theta} \approx 2 - \frac{4}{3}\theta^2$$

so the parabola is $y = 2 - \frac{4}{3}\theta^2$.

13. (a) $f(t) = te^t$.
Use the Taylor expansion for e^t :

$$\begin{aligned} f(t) &= t\left(1+t+\frac{t^2}{2!}+\frac{t^3}{3!}\cdots\right) \\ &= t+t^2+\frac{t^3}{2!}+\frac{t^4}{3!}\cdots \end{aligned}$$

(b)

$$\begin{aligned} \int_0^x f(t)\,dt = \int_0^x te^t\,dt &= \int_0^x \left(t+t^2+\frac{t^3}{2!}+\frac{t^4}{3!}\cdots\right)dt \\ &= \frac{t^2}{2}+\frac{t^3}{3}+\frac{t^4}{4\cdot 2!}+\frac{t^5}{5\cdot 3!}\cdots\Big|_0^x \\ &= \frac{x^2}{2}+\frac{x^3}{3}+\frac{x^4}{4\cdot 2!}+\frac{x^3}{5\cdot 3!}\cdots \end{aligned}$$

(c) Substitute $x = 1$:

$$\int_0^1 te^t\,dt = \frac{1}{2}+\frac{1}{3}+\frac{1}{4\cdot 2!}+\frac{1}{5\cdot 3!}\cdots$$

In the integral above, to integrate by parts, let $u = t,\ dv = e^t\,dt$, so $du = dt,\ v = e^t$.

$$\int_0^1 te^t\,dt = te^t\Big|_0^1 - \int_0^1 e^t\,dt = e-(e-1) = 1$$

$$\text{Hence } \frac{1}{2}+\frac{1}{3}+\frac{1}{4\cdot 2!}+\frac{1}{5\cdot 3!}\cdots = 1.$$

15. (a) Since the expression under the square root sign, $1-\frac{v^2}{c^2}$ must be positive in order to give a real value of m (like we would expect.) We have

$$\begin{aligned} 1-\frac{v^2}{c^2} &> 0 \\ \frac{v^2}{c^2} &< 1 \\ v^2 &< c^2 \\ \text{or } -c < v &< c. \end{aligned}$$

In other words, the object can never travel faster that the speed of light.

(b)

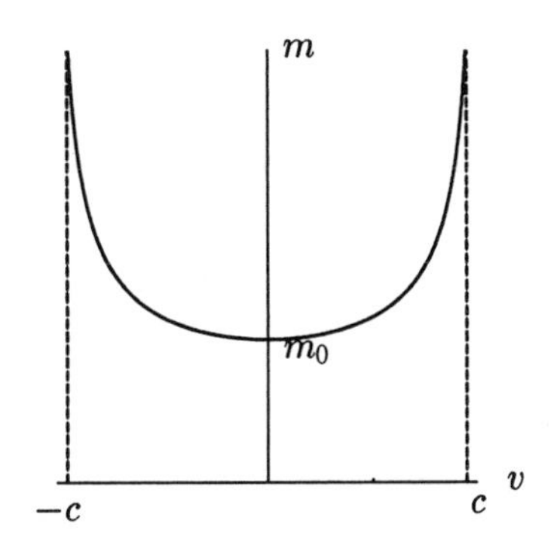

(c) Notice that $m = m_0\left(1 - \frac{v^2}{c^2}\right)^{-1/2}$. If we substitute $u = -\frac{v^2}{c^2}$, we get $m = m_0(1 + u)^{-1/2}$ and we can use the binomial expansion to get:

$$\begin{aligned} m &= m_0\left(1 - \frac{1}{2}u + \frac{(-1/2)(-3/2)}{2!}u^2 \cdots\right) \\ &= m_0\left(1 + \frac{1}{2}\frac{v^2}{c^2} + \frac{3}{8}\frac{v^4}{c^4} \cdots\right) \end{aligned}$$

(d) We would expect this series to converge only for values of the original function that exists, namely when $|v| < c$.

17. (a) To find when V takes on its minimum values, set $\frac{dV}{dr} = 0$. So

$$\begin{aligned} -V_0\frac{d}{dr}\left(2\left(\frac{r_0}{r}\right)^6 - \left(\frac{r_0}{r}\right)^{12}\right) &= 0 \\ -V_0\left(-12r_0^6r^{-7} + 12r_0^{12}r^{-13}\right) &= 0 \\ 12r_0^6r^{-7} &= 12r_0^{12}r^{-13} \\ r_0^6 &= r^6 \\ r &= r_0 \end{aligned}$$

Thus, $V = -V_0(2(1)^6 - (1)^{12}) = -V_0$.
(Note: We discard the negative root $-r_0$ since negative separations are really the same as the positive separations.)

(b)

$$\begin{aligned} V(r) &= -V_0(2(\frac{r_0}{r})^6 - (\frac{r_0}{r})^{12}) \\ V'(r) &= -V_0(-12r_0^6r^{-7} + 12r_0^{12}r^{-13}) \\ V''(r) &= -V_0(84r_0^6r^{-8} - 156r_0^{12}r^{-14}) \end{aligned}$$

$$\begin{aligned} V(r_0) &= -V_0 \\ V'(r_0) &= 0 \\ V''(r_0) &= 72V_0r_0^{-2} \end{aligned}$$

The Taylor series is thus:

$$V(r) = -V_0 + 72V_0 r_0^{-2} \cdot (r - r_0)^2 \cdot \frac{1}{2} \cdots$$

(c) The difference between V and its minimum value $-V_0$ is

$$V - (-V_0) = 36V_0 \frac{(r - r_0)^2}{r_0^2},$$

which is proportional to $(r - r_0)^2$.

(d) From part (a) we know that $\frac{dV}{dr} = 0$ when $r = r_0$, hence $F = 0$ when $r = r_0$. Since $V(r) = -V_0 \left(1 - 36\frac{(r - r_0)^2}{r_0^2}\right)$ to the second order,

$$F = -\frac{dV}{dr} = 72 \cdot \frac{r - r_0}{r_0^2}(-V_0) = -72V_0 \frac{r - r_0}{r_o^2}.$$

So, F is proportional to $(r - r_0)$.

19.

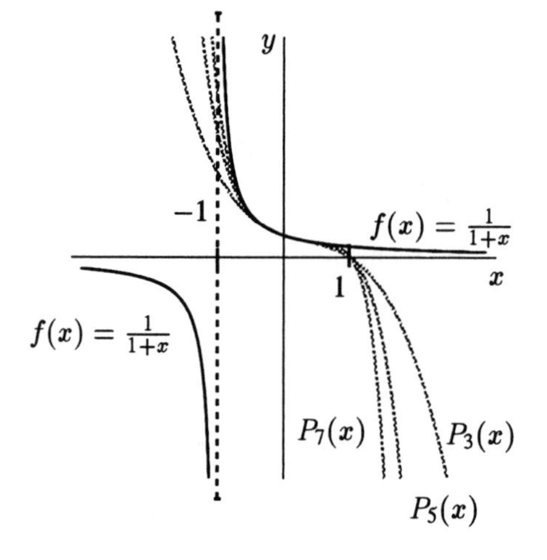

The graph suggests that the Taylor polynomials converge to $f(x) = \dfrac{1}{1+x}$ on the interval $(-1, 1)$.

21. (a) Since $g^{(k)}(0)$ exists for all $k \geq 0$, and $g'(0) = 0$ because g has a critical point at $x = 0$. For $n \geq 2$,

$$g(x) \approx P_n(x) = g(0) + \frac{g''(0)}{2!}x^2 + \frac{g'''(0)}{3!}x^3 + \cdots + \frac{g^{(n)}(0)}{n!}x^n.$$

(b) The Second Derivative test says that if $g''(0) > 0$, then 0 is a local minimum and if $g''(0) < 0$, 0 is a local maximum.

(c) Let $n = 2$. Then $P_2(x) = g(0) + \frac{g''(0)}{2!}x^2$. So, for x near 0,

$$g(x) - g(0) \approx \frac{g''(0)}{2!}x^2.$$

If $g''(0) > 0$, then $g(x) - g(0) \geq 0$, as long as x stays near 0. In other words, there exists a small interval around $x = 0$ such that for any x in this interval $g(x) \geq g(0)$. So $g(0)$ is a local minimum.
The case when $g''(0) < 0$ is treated similarly and we have that $g(0)$ is a local maximum.

23. (a) Since $4 \arctan 1 = \pi$, we approximate π by approximating $4 \arctan x$ by Taylor polynomials with $x = 1$. Let $f(x) = 4 \arctan x$. We find the Taylor polynomial of f about $x = 0$.

$$\begin{aligned} f(0) &= 0 & & \\ f'(x) &= \tfrac{4}{1+x^2} & f'(0) &= 4 \\ f''(x) &= -\tfrac{8x}{(1+x^2)^2} & f''(0) &= 0 \\ f'''(x) &= -\tfrac{8}{(1+x^2)^2} + \tfrac{32x^2}{(1+x^2)^3} & f'''(0) &= -8 \end{aligned}$$

Thus, the third degree Taylor polynomial for f is $F_3(x) = \frac{4x}{1!} - \frac{8}{3!}x^3 = 4x - \frac{4}{3}x^3$. In particular, $F_3(1) = 4 - \frac{4}{3} = \frac{8}{3} \approx 2.67$.

(b) We now approximate π by looking at $g(x) = 2 \arcsin x$ about $x = 0$ and substituting $x = 1$.

$$\begin{aligned} g(0) &= 0 & & \\ g'(x) &= \tfrac{2}{\sqrt{1-x^2}} & g'(0) &= 2 \\ g''(x) &= \tfrac{2x}{(1-x^2)^{\frac{3}{2}}} & g''(0) &= 0 \\ g'''(x) &= \tfrac{2}{(1-x^2)^{\frac{3}{2}}} + \tfrac{6x^2}{(1-x^2)^{\frac{5}{2}}} & g'''(0) &= 2 \end{aligned}$$

Thus, the third degree Taylor polynomial for g is

$$G_3(x) = \frac{2x}{1!} + \frac{2x^3}{3!} = 2x + \frac{1}{3}x^3.$$

In particular, $G_3(1) = \frac{7}{3} \approx 2.33$.

(c) The maximum error is:

$$|E_n| = \left| \frac{f^{(4)}(c)(1-0)^{n+1}}{(n+1)!} \right| = \frac{|f^{(4)}(c)|}{(n+1)!}$$

for some c between 0 and 1. Since

$$f^{(4)}(x) = -\frac{192x^3}{(1+x^2)^4} + \frac{96x}{(1+x^2)^3},$$

now use a graphing calculator to see that the maximum value for $|f^{(4)}(c)|$, where c is between 0 and 1, is about 18.6. Thus, $E_n \leq \frac{18.67}{4!} \approx 0.78$. (Notice that $\pi \approx 3.14$ is within 0.78 of 2.67.)

(d) Recall that

$$|E_n| = \left|\frac{g^{(n+1)}(c)(b-a)^{n+1}}{(n+1)!}\right|,$$

where $g(x) = \arcsin x$. Notice that the derivatives of $\arcsin x$ contain terms of the form $(1-x^2)^{-a}$, for some positive a. In fact, the more derivatives you take, the bigger a gets. The problem is that $(1-x^2)^{-a}$ is unbounded for $0 \leq x < 1$. That is, as x gets close to 1, $(1-x^2)^{-a}$ approaches ∞. Thus, we cannot get a bound on the derivative $f^{(n+1)}(c)$, which means that we cannot get a bound on $|E_n|$. Thus, using $\arctan x$, where we can get a bound for the derivatives, seems like a better idea, because we can get a bound on the error of the approximation.

25. Let us begin by finding the Fourier coefficients for $f(x)$. Since f is odd, $\int_{-\pi}^{\pi} f(x)\,dx = 0$ and $\int_{-\pi}^{\pi} f(x)\cos nx\,dx = 0$. Thus $a_i = 0$ for all $i \geq 0$. On the other hand,

$$\begin{aligned}
b_i = \frac{1}{\pi}\int_{-\pi}^{\pi} f(x)\sin nx\,dx &= \frac{1}{\pi}\left[\int_{-\pi}^{0} -\sin(nx)\,dx + \int_{0}^{\pi}\sin(nx)\,dx\right] \\
&= \frac{1}{\pi}\left[\frac{1}{n}\cos(nx)\Big|_{-\pi}^{0} - \frac{1}{n}\cos(nx)\Big|_{0}^{\pi}\right] \\
&= \frac{1}{n\pi}\left[\cos 0 - \cos(-n\pi) - \cos(n\pi) + \cos 0\right] \\
&= \frac{2}{n\pi}\Big(1 - \cos(n\pi)\Big).
\end{aligned}$$

Since $\cos(n\pi) = (-1)^n$, this is 0 if n is even, and $\frac{4}{n\pi}$ if n is odd. Thus the nth Fourier polynomial (where, say, n is odd) is simply

$$F_n(x) = \frac{4}{\pi}\sin x + \frac{4}{3\pi}\sin 3x + \cdots + \frac{4}{n\pi}\sin(nx).$$

As $n \to \infty$, the nth Fourier polynomial must approach $f(x)$ on the interval $(-\pi, \pi)$. In particular, if $x = \frac{\pi}{2}$,

$$\begin{aligned}
F_n(1) &= \frac{4}{\pi}\sin\frac{\pi}{2} + \frac{4}{3\pi}\sin\frac{3\pi}{2} + \frac{4}{5\pi}\sin\frac{5\pi}{2} + \frac{4}{7\pi}\sin\frac{7\pi}{2} + \cdots + \frac{4}{n\pi}\sin\frac{n\pi}{2} \\
&= \frac{4}{\pi}\left(1 - \frac{1}{3} + \frac{1}{5} - \frac{1}{7} + \cdots + (-1)^{2n+1}\frac{1}{2n+1}\right).
\end{aligned}$$

But $F_n(1)$ approaches $f(\frac{\pi}{2}) = 1$ as $n \to \infty$, so

$$\frac{\pi}{4}F_n(1) = 1 - \frac{1}{3} + \frac{1}{5} - \frac{1}{7}\cdots(-1)^{2n+1}\frac{1}{2n+1} \to \frac{\pi}{4}\cdot 1 = \frac{\pi}{4}.$$